Geographical Issues in Western Europe

Andrew Hull
Stephen Kenny
Trevor Jones

Senior Lecturers in Human Geography,
Liverpool Polytechnic

BLOOMFIELD COLLEGIATE SCHOOL

This Book is the Property of the School

1. Keep it in good condition and do not write notes on it.
2. A charge will be made at the end of the year if it has been lost or has been badly treated.
3. Write your name, form and year of issue in the top available space.
4. Do not delete previous entries.

Name	Form	Year of Issue
CLAIRE MCKNIGHT	UVI	'90
Deborah Mooney	LVI	'91
Emily Maguire	LVI	93-94

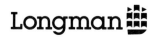

LONGMAN GROUP UK LIMITED,
*Longman House, Burnt Mill, Harlow, Essex CM20 2JE, England
and Associated Companies throughout the world.*

© Longman Group UK Limited 1988
*All rights reserved; no part of this publication
may be reproduced, stored in a retrieval system,
or transmitted in any form or by any means, electronic,
mechanical, photocopying, recording, or otherwise,
without the prior written permission of the Publishers or
a licence permitting restricted copying issued by the
Copyright Licensing Agency Ltd, 33–34 Alfred Place,
London, WC1E 7DP.*

First published 1988
ISBN 0 582 35487 0

Set in 10/12 point Times Roman, Linotron 202

*Produced by Longman Group (F.E.) Limited
Printed in Hong Kong*

Acknowledgements

We are grateful to the following for permission to reproduce photographs and other copyright material;

Alfa Fotoarkiv, Copenhagen, Fig. 5.10 (photo Soren Rud); Associated Press, Fig. 8.1; Bavaria-Verlag, Figs. 5.13 (photo Ernst Harstrick), 6.11 (photo Rudolph Dix) and 7.11 (photo Hans Schmied); Camera Press, Fig. 1.3 (photo Benoit Gysembergh); Deutsche Presse-Agentur, Figs. 2.9 (photo Leuschner) and 7.2 (photo Koll); Documentation Française, Figs. 2.11 (Photopress-Grenoble), 4.3 (photo EPAD), 5.5 above (photo Ministère de l'Agriculture) and 5.5 below (photo Alain Perceval); *The Economist*, Fig. 4.4; *The Financial Times*, Fig. 1.8; Information and Documentation Centre for the Geography of the Netherlands, Figs. 1.4, 4.1 (photo ANP) and 7.15 (photo Bart Hofmeester); Knudsens Fotosenter, Oslo, Figs. 5.8, 8.6 and 9.3; Grazia Neri, Milan, Figs. 3.10 (photo Uliano Lucas) and 7.10; The Nordic Council, Stockholm, Fig. 3.2; Rapho, Paris, Figs. 3.6 (photo Francois le Diascorn) and 8.11 (photo SAS); SEAT, Fig. 2.13; SNCF, Paris, Fig. 7.6; Swiss Association for Atomic Energy, Bern, Fig. 9.8; Swiss National Tourist Office, Fig. 9.12; Times Books Ltd, London, *Times Atlas of the Oceans*, Figs. 7.1 and 9.5; Times Newspapers, Figs. 7.17, 8.3 and 8.12.

Cover photograph by Camera Press.

We are grateful to the following for permission to reproduce copyright material:

The author, Geoff Andrews for his article 'France speeds up rail success' in *The Guardian* 23/4/85; The Economist Newspaper Ltd for the article 'Greening the Ruhr' in *The Economist* 6/10/84; Times Newspapers Ltd for the articles 'The miracle workers that Germany no longer wants' and 'Lupins set to flower as feed' in *The Sunday Times* 1/7/84 & 14/4/85.

Contents

Introduction	5
Chapter 1 Material Prosperity	**9**
1 Material prosperity	9
2 Distribution and access	11
3 The quality of life	34
4 Conclusion	43
Chapter 2 Industry and Regional Development	**44**
1 Regional inequalities and the production of wealth	44
2 Core and periphery	44
3 Geographical variations in the core-periphery relationship	59
4 Conclusion	69
Chapter 3 Planning and State Intervention	**71**
1 Intervention by supra-national agencies	71
2 Regional planning within Western European states	85
3 Other forms of state intervention	95
4 Conclusion: state intervention and regional planning in retrospect	97
Chapter 4 Problems of Urban Development	**98**
1 Characteristics of the modern Western European city	99
2 Urban growth and its consequences	104
3 Urban planning	117
4 Urban change in the southern periphery	126
5 Conclusion	130
Chapter 5 Rural Western Europe	**131**
1 Agricultural development	131
2 The transformation of rural life	146
3 The Common Agricultural Policy	154
Chapter 6 Population as an Issue	**162**
1 Demographic change as an issue	162
2 The transformation of Western Europe's population	168
3 Migration	176
4 Historical stages in migration	178
5 Conclusion	191
Chapter 7 Transport in Western Europe	**192**
1 The historical legacy	192
2 Transport and the future development of Western Europe	219
3 Transport integration – the way ahead?	220
Chapter 8 Environmental Concern	**224**
1 Environmental concern	224
2 Origins of environmental degradation in Western Europe	224
3 Environmental pollution	229
4 The impact of environmental pollution	233
5 The costs of pollution	235
6 A response to environmental degradation	247
7 The growth of environmental awareness	254
8 Western Europe – an environmental future?	256
Chapter 9 The Resource Issue	**258**
1 The depletion of global resources: optimistic versus pessimistic scenarios	258
2 Resources and resource depletion	261
3 Resource issues in perspective	284
Conclusions	**285**
Index	**287**

Introduction

Issues in geography

Among the many motives for writing this book one of the foremost has been the authors' desire to escape from a purely factual approach to geography and instead to present the human geography of Western Europe as part of a living debate. The typical concerns of geographers – the location and movement of people and activity, human use of the land, distribution of natural resources, regional variety – are often treated as purely intellectual concerns to be discussed in the abstract. This is perfectly valid and we do not wish to demean it. Geographical analysis in its own right is as mind-expanding as any other kind of academic training. Hence, this book pays due attention to the spatial concepts which are essential to the student and includes full discussion of those which are relevant to the Western European issues covered in its pages.

Even so, spatial concepts can be more than mental exercise. Our view is that they are relevant to the everyday lives and work of all Europeans, and everyone else for that matter. An individual's life chances (physical and mental well-being, enjoying a social life and leisure and so on) depend to a considerable extent on where he or she lives. Geographical location is one of the many factors (others are skills, qualifications, gender, race, nationality and sometimes inherited wealth or property) which affect each individual's material living standards. Location helps to determine whether or not we enjoy full access to the benefits or suffer the disadvantages of living in a modern affluent society.

Thus, geographical space has an impact on peoples' lives and minds and as such it is an *issue*, i.e. an item on the agenda of public and political debate, something which helps to shape current events. Among the similies provided by *Roget's Thesaurus* for the word 'issues', the phrase 'bone of contention' is closest to what we are discussing here. An issue arises when two or more parties are in contention (disagreement, conflict, dispute). At root, some of the great disputes of modern Western Europe, conflicts which from time to time erupt into the news headlines, are actually disputes over geographical space and the resources it contains. This being the case, the geographer has as much to say about them as any economist or political scientist.

One very clearcut example of the geographer's interest in current issues is the environmental debate – the continent-wide concern about the effects of pollution, especially (but not exclusively) the risk of nuclear contamination. Anxiety here has been so intense (heightened by the Chernobyl disaster of 1986) as to give rise to a new environmentalist party, the 'Greens', who now occupy several parliamentary seats in West Germany and the Netherlands. Since this environmental debate is largely a disagreement between the various political parties and interest groups about the use and abuse of natural resources, it falls very much within the geographer's province – though it is by no means exclusively a geographical issue. Some might even argue that it is the geographer's public duty to bring his/her specialised knowledge to bear on so critical a problem.

Other issues are less obviously 'geographical'. They are hardly ever connected in the public mind with factors such as location, land and resources. One example of this is unemployment, the issue which worries most Europeans at the present time and about which there is bitter political strife in almost every nation. What can geographers say about this, a crisis apparently created by technological change and government policy? The short answer is – plenty. In the first place, unemployment has a marked geographical character, affecting some regions (notably the older industrial areas of north-west Europe and the agricultural regions of the Mediterranean) more acutely than others. In the second place, the plight of these regions results from drastic changes in the economic geography of Western Europe. Economic and technological change has made many regions uncompetitive as their traditional products have become unwanted, but in other regions these same changes have created new life.

Today different parts of the world, countries, regions, places are increasingly interconnected so that individual regions no longer are exclusively in control of their own destinies. Traditional industries decline due to foreign competition, the impact of national government policy, the effect of multinational corporations and so on. Yet, every place is unique and the individual characteristics of a place or region can still have a significant effect on its present and future well-being. It is in trying to explain the interplay between these two levels, which gives rise to change, problems and issues, that the geographical perspective becomes important.

The content and layout of the book

Quite clearly, there is a host of everyday issues and problems whose understanding can be aided by a geographical approach. In this book we have tried to group them together in themes. The first three chapters all relate to the common theme of *social inequality*, the continuing existence of poverty and deprivation in the midst of general prosperity. For geographers it is the occurrence of regional variation in prosperity and opportunity – usually refered to as *regional disparity* – which explains our involvement in this question. Chapter 1 is concerned with the *symptoms* of the problem, the way affluence is unequally shared out between the regions of Western Europe and the conflicts to which this gives rise. The purpose of Chapter 2 is *diagnosis*, it examines the processes which have brought these regional variations into being. Chapter 3 completes the trio, being concerned with *remedies* – the action taken by European national governments and the European Community (EC) to combat the problems of poor and underdeveloped regions.

Not only have the changes of the modern world – industrialisation, growing wealth, ceaseless technological progress – opened up great gaps between regions, they have also revolutionised the relationship between town and country. Now the living environment for the majority is urban rather than rural, as had been the case from the beginning of human life until the Industrial Revolution. This *switch from country to town*, a veritable revolution in its own right, took place because the new jobs of the modern age were almost all urban-based and people left their farms and villages for factories and offices. In many respects, they have become better off by doing so. The living standards of the new urban Europe are many times greater than those of the old, rural, in many places feudal, Europe. But there have been social costs as well: strains of moving to new homes and adjusting to new circumstances, pressures on space as population crowded into confined cities are two of the more obvious social costs. Chapter 4 deals with the rise of towns and cities, in particular the social and environmental conflicts this has created. Chapter 5 focuses on the areas left behind, the rural sector of Western Europe, depopulated by the exodus of the city-bound migrants and left to survive on the basis of an agricultural economy which in some regions had hardly improved since the Middle Ages.

A further matter that has long interested geographers is *population change*, a subject which for well over a hundred years has given rise to public debate, much of it extremely anxious. In the last century, the chief anxiety for many nations was that numbers were rising so rapidly that poverty and hardship were bound to result. At present several governments are anxious about the very opposite problem – falling birth rates, ageing and the threat of population decline. Chapter 6 explains these trends and also focuses on the problems linked to *migration*, which in modern society has reached an unprecedented pitch. Included in this chapter is a section on international migrants, the foreign workers who have become the focus of heated emotions in West Germany, France and Switzerland, in particular. Just as permanent, long-term migration has increased, so too has the short-term movement resulting from industrialisation and urbanisation. Here, we are talking about the great flows of traffic betwen nations, between regions and within urban areas as goods and people move from place to place. Chapter 7 examines the issues that have arisen both in trying to improve the efficiency of *transport networks* and over the impact of such movements on society and the environment.

The final major theme to be tackled is that of the relationship between *society and the environment*. This can be viewed from two positions; the increasing damage and potential for *destruction* that modern industrial society poses for the continuance of the natural environment and, on the other hand, the *potential* that the environment has for society in terms of natural resources to fuel our homes and industries, feed us, power our cars and public transport and so on. Chapter 8 examines the first side of this equation and looks at the major environmental issues that have arisen due to the impact of pollution and the growth of urban areas. This is complemented by Chapter 9 which discusses the critical issue of whether modern society faces wholesale disruption on account of resource shortages as well as examining the geographical consequences of changing patterns of resource use.

Geographical diversity and the case study approach

Western Europe is a large territory of tremendous geographical variety. Firstly, it embraces a great range of climatic, geological and topographical types and a correspondingly large range of natural resources and hazards. Secondly, it is densely settled by a collection of separate nations and peoples, each of whom is historically and culturally distinctive. Finally, as previously emphasised, these peoples have been subject to profound, sometimes traumatic change in recent years which has affected places at different times and speeds, thus creating another type of regional contrast – the modern versus traditional region. Undoubtedly the various parts of Western Europe are becoming more alike, with the growth of foreign travel, international migration, mass consumption and the quick spread of information. Yet there is a very long way to go before it becomes culturally homogeneous.

This regional variety presents something of a problem for a book of this type. How can we discuss the whole of Western Europe and at the same time achieve deep and serious coverage of a wide range of topics? Our solution has been to use *regional case studies* as a means of illustrating general points by reference to a single place, thereby avoiding repetition and achieving economy of effort. For example, in Chapter 1 we consider the economic position of *minority nations*, people who (like the Welsh or Scots in Britain) have been swallowed up by a larger political unit and have virtually no separate legal identity. Western Europe abounds with similar cases – the Bretons and Corsicans of France; the Basques, Catalans and Galicians of Spain; the French speakers of the Italian Val d'Aosta.

Though each one of these is a unique case in its own right, they are also all examples of a common plight – the stateless nation. To write about all of them in detail would require a book on its own. Therefore we have taken Brittany as a representative case study to illustrate the common theme of sub-nations.

A similar approach has been followed throughout, with each Chapter containing a variety of regional case studies. Usually these have been selected because they *typify* a particular category of region, for example, the Ruhr to stand for a declining coalfield, the Randstad as a major conurbation. Occasionally they have been chosen because they illustrate a radically *different* approach to a problem. Thus, in Chapter 4 there is a study of the unorthodox form of town planning practised in the Italian city of Bologna. Whatever the motive for selecting them, we hope that these case studies combine the best of modern geography with the best of the regional approach to the subject.

Illustrations and boxes

As is customary in texts of this type we have included numerous maps and other illustrative material, often with a commentary giving additional information and explanation. On occasion we have also made use of boxes inserted into the text to convey supplementary information and interpretation for those not familiar with the particular topic.

Delimitation of the study area

One further problem is the need to define which 'Western Europe' we intend to study. In the broadest sense, Western Europe consists of all those European nation states outside the Soviet bloc. We have followed this definition but with certain important exceptions. The most vital exception is Britain itself, which is excluded except for occasional comparative purposes. To do this is in a sense to commit geographical heresy. Historically, much of the character of modern Europe has been formed by trends started in Britain two centuries ago. In the present, Britain's EC membership demonstrates her links with Europe are stronger than ever. Despite this, there are two pragmatic reasons for excluding Britain from this study; firstly, to cover Britain and continental Western Europe at this depth and breadth would require a book twice the size. Secondly, many of the regional syllabuses at A-level and above treat the British Isles separately – perhaps a hangover from Britain's insular past.

Other omissions are Iceland, Turkey and Yugoslavia whose identity as part of Western Europe is somewhat ambiguous even though some classifications include them. Thus our spotlight falls on the fourteen nation states of continental Western Europe. Within this, there is heavy emphasis on the countries of the EC. This is in recognition of the singular importance of this relatively new political-economic grouping, whose members include West Germany and France, two of the world's largest industrial powers; and whose heartland – the densely populated industrial concentration which spills over the borders of Germany, France, the Netherlands and Belgium – exerts an influence throughout the length and breadth of the sub-continent.

1 Material Prosperity

1 Material prosperity

a) Wealth as an issue

In a very real sense the threat of nuclear holocaust is far and away the gravest single issue which faces the citizens of Western Europe today. In another sense, however, the leading questions of the age are about economics. Whereas for most people 'the Bomb' is a vague, distant and barely credible nightmare, matters of money and work impinge directly on everyday life. Nuclear Armageddon is seen as hypothetical, but unemployment (with its attendant poverty, hardship and loss of social standing) is seen by many Europeans as a direct personal threat, one which could strike without warning at any time. With over one in ten of the West European labour force without work in the mid 1980s, there are visible grounds for such fears. As geographers we should be immediately aware that jobs and unemployment, wealth and poverty are shared out unequally from place to place, nation to nation, region to region. In this sense the great economic questions of our time are also geographical ones.

From an historical point of view, the economic pre-occupations of modern Western Europe are easy to understand. Throughout recorded history and until approximately the last 100 years, the European economy provided most of the population with little more than the bare means of survival. Yet, to all intents and purposes, the present century has finally banished mass poverty. As Table 1.1 shows, the last two decades in particular have been a time of dramatic increase in material abundance. Even now, however, there are many among the over 50 age groups

Table 1.1 Rising living standards for selected countries in Western Europe

a) Increases in per capita income

	Average annual % growth in per capita income, 1960–81	Per capita income 1981 (US dollars)
Austria	4.0	10,210
Belgium	3.8	11,920
Denmark	2.6	13,120
France	3.8	12,190
Italy	3.6	6,960
Netherlands	3.1	11,790
Norway	3.5	14,060
Spain	4.2	5,640
Sweden	2.6	14,870
West Germany	3.2	13,450

b) Use of consumer durables (cars, televisions, telephones per 1,000 population)

	Cars			Televisions			Telephones		
	1960	1970	1980	1960	1970	1981	1960	1970	1981
Belgium	86	213	312	67	216	301	124	211	417
Denmark	88	218	314	118	266	370	174	345	570
France	121	251	346	41	216	301	95	173	541
Italy	40	190	300	42	181	240	77	175	404
Netherlands	47	191	312	69	237	303	140	262	575
West Germany	78	222	367	83	272	362	107	228	571

with vivid memories of the hunger and want which prevailed in the 1930's depression and during and immediately after the Second World War. For these and others, the unprecedently high living standards of the present (despite distinct geographical variations) are fondly cherished and jealously guarded, and looked upon as form of compensation for past endurance.

b) Industrialisation and living standards

Clearly the present age of abundance represents a clean break with centuries of scarcity. Not only is this cause for celebration, it is also basic food for thought for the geographer. In particular, it highlights the continually changing relationship between human society and its physical environment. One hundred and fifty years ago, the land area covered by Western Europe supported little more than 137 million people and, although few reliable statistics exist for the living standards of the period, we can be certain that material consumption for most people was limited to little more than the bare necessities of survival. Now, the natural resources of the same land area support a population of around 350 million, almost a three-fold increase. Moreover, the average member of West European society today enjoys a high consumption lifestyle, placing incomparably greater demands upon the agricultural and mineral resources of the land. It is also true that increasing wealth is won at the cost of a *decreasing* amount of physical effort on the part of workers. Indeed, a keynote of the present age is the continuous reduction in the amount of work performed, for example the average number of hours per employee in manufacturing fell from 45.7 to 40.3 in France and 45.3 to 40.8 in the Netherlands between 1967 and 1981. The short explanation for this prolonged rise in productivity is contained in the one word *industralisation*. (See Table 1.2) In reality, this word refers to a far from simple concept which describes a complex series of interconnected changes and a complete transformation in the economic and social organisation of every Western European nation. For the present discussion, we need to note two major transformations brought about by industralisation:

I) METHOD INNOVATION

Beginning with the nineteenth-century Industrial Revolution, a continuous series of technological innovations has raised the productivity of both land and labour. For example, the harnessing of steam, electricity and other forms of inanimate energy has both increased the natural wealth extracted from the land and improved the manner in which it is transformed for human benefit.

II) PRODUCT INNOVATION

Existing forms of economic activity are incessantly replaced by new industries producing ever more efficient products. The earliest expression of this was the replacement of agriculture by manufacturing indutry as the principal economic activity (see Fig. 1.1) but the process of change has continued with the decline of the earliest manufacturing industries – textiles, metals, shipbuilding among others – in areas such as the Ruhr industrial district of north west Germany, the Saar-Lorraine iron and coalfield region of the Franco-German border and the Sambre-Meuse region in Belgium. Elsewhere, there has been a rise in newer industries – aircraft and motor manufacture, electrical engineering and chemicals – in south-western France, southern Germany and northern Italy for example. At present, Western Europe stands on the brink of yet another technological era which, it is generally accepted, will be dominated by newcomers such as electronics, biotechnology, robotics and information technology. The impact of these and other developments will, undoubtedly, have a significant impact upon the changing fortunes of many regions throughout Western Europe.

Thus one of the critical differences between pre-industrial and present day Europe is technological in nature. Due to its lack of sophis-

Table 1.2 Stages in the industrialisation process

	Pre-industrial society ⟶	Industrial society ⟶	Post-industrial society
Characteristics	Agriculture dominant. Labour intensive. Small-scale organisations serving local markets	Manufacturing becomes leading sector. Capital intensive with large production units serving regional and national markets	Most employment in the tertiary sector but labour inputs fall dramatically due to high technology (robots and information technology). Multi-national corporations dominate the economy
Impact on society	Local elites and the aristocracy control society. Many small peasant producers, the majority engaged in subsistence production with hard labour and long hours for little reward	Private capitalists control the economy. A waged workforce in factories. Mechanisation replaces labour with machines. Progressive unionisation, shorter hours, higher rewards	Increased state intervention, the public sector becomes a major employer. Further replacement of unskilled labour by machines (robots, computers). Short hours and high pay for highly qualified workforce. Large problem of finding sufficient work for unemployed
Quality of life	*Benefits* Small, integrated communities. Little damage to environment *Costs* Poor diet and housing. No health care. Little educational provision	Improved diet and housing. Gradual improvement in health care and educational facilities. Civic facilities developed Widespread pollution of urban industrial environment. Public health/sanitation conditions in cities initially appalling	Greater time and facilities available for recreation, universal education and health care. Social security systems/pensions. Consumer durables widely available, e.g. cars, TVs, washers, fridges Pollution more widespread and potentially harmful (e.g. acid rain, nuclear waste). Longer journey-to-work due to movement of wealthy out of cities. Industrialised agriculture destroys traditional rural landscape

ticated tools, machines and motive power, pre-industrial society was obliged to devote most of its time and energy to the drudgery of producing the rudimentary elements of survival. One hundred years ago, for example, no less than 46 per cent of German workers were engaged in agricultural employment – and this percentage was fairly typical of the entire continent at that period. Now, as a result of technological advance, it requires only 5 per cent of the German labour force to produce almost all the nation's food needs. The remaining 95 per cent are available to produce goods and services satisfying far less elementary needs.

2 Distribution and access

a) Material inequality

As is now abundantly clear, Western Europe

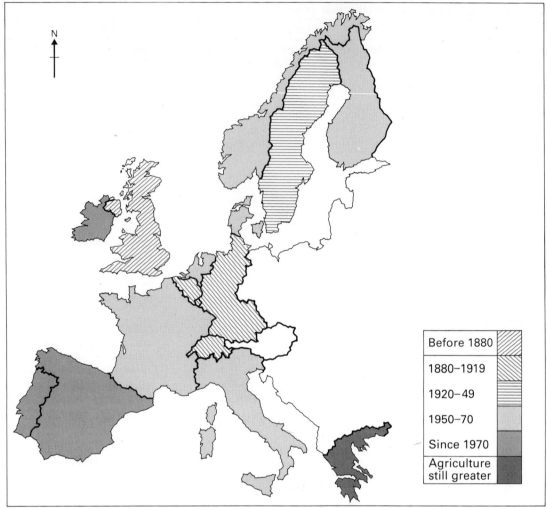

Figure 1.1 Date when manufacturing employment overtook agriculture

COMMENTARY

Industrialisation in Western Europe has been an extremely long process, initiated in Britain almost 200 years ago but only coming to fruition in much of the rest of the continent during the post-war era. The switch from agriculture to industry represents a major change in the structure of the West European economy. With the exception of Greece, all West European nations now derive considerably more jobs and income from manufacturing industry than from agriculture.

produces and consumes truly vast quantities of material wealth. Its affluence is massive in comparison both to past standards of living in Europe itself and to those which, even now, pertain over much of the rest of the world. Western Europe is included among that fortunate minority of world regions where mass poverty and hunger have been abolished and where the majority of citizens are concerned with acquiring more and better consumer durables than with a desperate struggle to obtain enough to eat. Yet this is not to say that all its citizens benefit equally from collective prosperity. On the contrary, widespread inequalities exist in all the countries of Western Europe, with a considerable minority of citizens failing to obtain what they regard as their 'fair slice of the loaf'. A recent publication by the European Commission admitted that:

> Although the standard of living has gone up considerably and social protection is wider spread, poverty (shanty towns, poor food, lack of medical care, backwardness in education, lack of information) still exists in the community.

Such social inequality has given rise to much public concern, political debate and government action in post-war Western Europe. In the modern nation state, most victims of poverty are not utterly lacking in the physical essentials of food, clothing and shelter. Their poverty usually consists of a lack of quality and choice, a level of living inferior to that which the majority of their fellow citizens would deem to be acceptable (see Box 1.1). To experience poverty in modern Holland or Italy is not to starve, nor in most cases to lack a home of sorts – it is to experience a sense of exclusion. It is no exaggeration to say that a sense of belonging depends on enjoying full access to the living standards taken for granted by one's fellow citizens. To be denied this is to be excluded from full membership of the community. In a sub-continent where all the constituent states are committed in some way to the principles and practice of democracy, social inequality – the persistence of poverty amid plenty – is widely regarded as morally unacceptable and widely feared as a threat to political stability.

Of special interest to the geographer is the manner in which access to prosperity is influenced by place. Fig. 1.2 highlights the sharp spatial variations in real income which are now a characteristic of Western Europe. It is evident that the individual citizen's standard of living is at least partly determined by where he or she lives. In general it is true to state that material life chances vary according to both one's nation of residence and one's region of residence within that nation.

Box 1.1. The nature of poverty

As with many other apparently straightforward concepts, there are many alternative ways of defining poverty, especially in advanced societies where the vast majority of citizens do not lack the basic essentials of life but where there is, nevertheless, considerable dissatisfaction with the way wealth is shared out.

- *Destitution* This occurs when the satisfaction of needs is so low that it affects the health and ultimately the survival of those affected by it. This kind of poverty – lack of food, clean water and basic shelter – is precisely the kind which is now rare in Western Europe. Note however, that it has not disappeared altogether as any one of the thousands sleeping rough on the streets of Paris or London will testify.
- *Absolute poverty* This affects people whose standards fall below those consistent with human dignity. In other words there are certain broadly agreed minimum standards of life which everyone has a right to expect, and in modern societies it is now generally agreed that all members of the society should have access to certain minimum standards of housing, health care and education in addition to food and shelter. In this sense, then, poverty is defined by cultural standards rather than bodily needs and some Europeans – such as the 'street urchins' of Naples – fall well below such standards.
- *Relative poverty* This is undoubtedly the sense in which poverty has most meaning in a modern advanced society. Relative poverty is defined as the inferior income of the poor person in relation to other members of the population. In Western Europe today poverty is more often than not a state of exclusion from the full fruits of the consumer society. This does not make it any less real as the squatters in Amsterdam or the long-term unemployed in Germany will witness.

14 GEOGRAPHICAL ISSUES IN WESTERN EUROPE

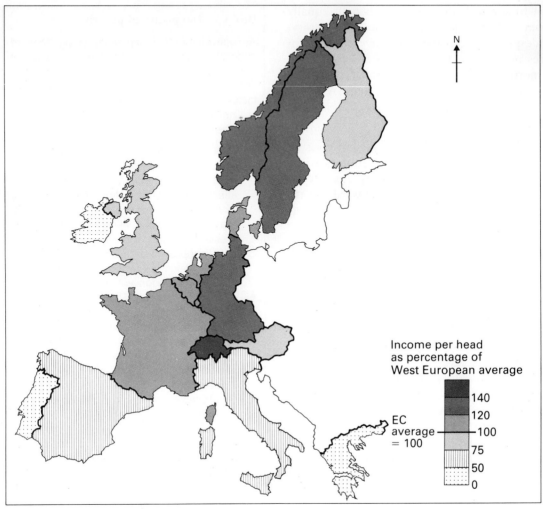

Figure 1.2 International income disparities, 1984

b) International inequality

While it is certainly the case that economic scarcity has now been largely banished from the face of the continent, nevertheless Fig. 1.2 warns us against portraying Western Europe as a place of uniform abundance. At the international level acute inequalities exist between countries. At the extremes of contrast, the average Swiss citizen enjoys a personal income more than eight times that of his Portuguese counterpart. Summarising the overall pattern (Fig. 1.2) we see that a contiguous group of nations extending from Scandinavia in the north through West Germany to Switzerland in the south form a class of their own with average incomes significantly above the continental norm. As a general rule, average income declines with increasing distance to the west and south, culminating in a group of nations – Italy, Portugal, Spain, the Irish Republic and Greece – where living standards are visibly lower and where a far larger proportion of people are excluded from the full fruits of the affluent society.

c) Regional inequality

Poverty and affluence are also geographically

MATERIAL PROSPERITY 15

Figure 1.3 Poverty in Naples

Figure 1.4 Affluence in the Netherlands

concentrated *within* each of the nation states. Regional pockets of relative poverty exist even in such affluent states as Swedish Lapland or the Ticino canton in Switzerland (a situation made worse by the high cost of living in both these countries) while regional pockets of affluence are to be found in Greece, in the region of Athens or in the northern Spanish province of Alava or in the regions of Lisbon and Setubal in Portugal. Fig. 1.5 stresses the way in which national averages tend to disguise the very real spatial inequalities prevailing at the regional level. This is particularly relevant to countries like France, where the distribution of income is so uneven that Paris (Ile de France) is the only region with a per capita income above the national average, or Italy, where the contrast between northern affluence and southern deprivation is so great as to virtually divide the country into two separate nations. In Spain, too, average personal incomes in the northern Basque provinces are almost twice those of the poorest provinces of the southwest (Andalusia).

d) Employment: access to the consumer society

Why are living standards affected by place of residence? What is the mechanism which distributes wealth so unevenly between countries and regions? To answer this we must recognise that in the first instance an individual's living standards are generally determined by *what he or she does*. The economic system rewards its participants according to the role they play (or are allowed to play) in the creation of the total stock of wealth.

It is important to note here that most Europeans exert very little direct control over this process of wealth production. Unlike the pre-industrial era, when small farmers, peasants, craftsmen and traders accounted for the bulk of production, the self-employed now form only a minute fraction of the workforce (about 12 per cent in the EC). It is almost a definitive characteristic of the modern industrial society that production and trade are controlled by a comparatively small number of extremely large firms. All over Europe, small family businesses have been declining in number throughout the present century, although at different rates in different countries. In Italy and Ireland, for example, the traditional economy of small independent entrepreneurs has survived far more strongly than elsewhere, mainly due to their continuing dependence on agriculture. Be that as it may, there can be no doubt that the vast majority of present day West Europeans gain their livelihood not by working with their own tools and property but by selling their labour to a relatively small number of employers in exchange for a wage or salary (see Box 1.2).

Job prospects depend significantly on one's position in the industrial division of labour. Agricultural workers, miners and employees in the older manufactures are increasingly at risk as their industries continue to shed labour. Among them, the oldest and the youngest are the most seriously threatened. Workers in certain service industries and in the newest branches of manufacturing (notably the high-tech sector) have better prospects, though even here rapid changes in skill requirements can eject those who are unable to adapt.

The geographical relevance of all this is tied in to the concept of the *regional division of labour*, i.e. the spatial concentration of economic activity has created many highly specialised regions, localised communities of workers geared to producing a relatively small range of goods and services for international and national markets. Hence, the social impact of contraction in agricultural, steel or cotton employment is magnified in those regions whose very livelihood rests on these activities and whose people have few other jobs to turn to.

It is also true that, irrespective of industrial specialisation, any individual's job prospects are conditioned by his or her *occupational position*. To put it bluntly, the industrial economy values its workers according to the

MATERIAL PROSPERITY 17

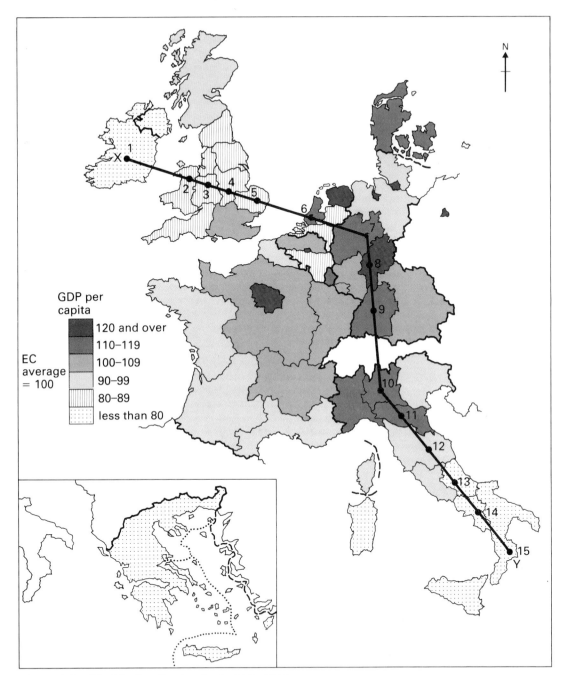

Figure 1.5 Wealth distribution in the EC, 1981

skills and expertise they can bring to the creation of wealth. Table 1.3, based on the official French occupational classification outlines the recent changes in the occupational composition of the French population. Since the French classification is similar to those adopted by other European Governments, it gives rise to certain general

18 GEOGRAPHICAL ISSUES IN WESTERN EUROPE

> **Box 1.2 Social stratification versus class polarisation**
>
> According to conventional sociological wisdom, the nations of Western Europe should be seen as *class-stratified* societies, whose populations are graded into various levels of wealth, privilege and power. Social class composition in Western Europe (as in all advanced capitalist societies) resembles a kind of pyramid, with a small number of the richest and most powerful at the apex and increasing numbers in each descending layer. Thus the share of the national income earned by the top 10 per cent of the population varies from over 30 per cent in France to 21 per cent in Finland and Sweden: the share of the bottom 20 per cent varies from 5 per cent in France to 8 per cent in the Netherlands. For the geographer, the vital consideration is that this socio-economic inequality expresses itself regionally, because the rich and powerful tend to be concentrated in certain regions, notably the great metropolitan centres, while the poor and dispossessed are also unevenly distributed.
>
> It is frequently argued by sociologists, economists and politicians that inequality of this type is acceptable or even desirable. It is an expression of the need to pay high rewards to those in responsible positions of leadership, especially when these are businessmen who create employment for other people. We might also note that Western European inequality has lessened in the present century and is much less harsh than in the rest of the world. Highly organised trade unions exist to protect the living standards of workers and, at the bottom line of poverty, the state intervenes to protect the retired, the unemployed, low earners and other potential victims. It also protects poor regions who, in the absence of state benefits for their unemployed and financial aid to their firms, would be far worse off than they actually are.
>
> Against these arguments, there is a strong (dominantly Marxist) European political tradition which argues that capitalist society is not stratified but polarised into two opposed classes. The key division is between the bourgeoisie, those who own large businesses and other productive property; and the proletariat, those who own little of significance apart from their own labour power which they are forced to sell to the bourgeoisie in return for a wage. The relationship is an essentially exploitative one, with the bourgeoisie accumulating profits by selling goods produced by sweat of the proletariat. Such a viewpoint holds that despite rising mass living standards, the power of the bourgeoisie to exploit others is as absolute as ever. The rise of the welfare state, with its subsidies for poor people and problem regions, is seen primarily as a means of appeasing social discontent and averting the proletarian revolution foreseen by Marx in the last century.
>
> With respect to regional inequality there is a similar theoretical debate. Is regional inequality a matter of a few privileged regions exploiting the rest? Or is Western Europe best seen as a hierarchy of regions, grading down from the richest and most dominant through a series of intermediate levels to the poorest most remote and backward agricultural regions?

principles which have a bearing on the present argument.

- *Status (i.e. social standing or prestige) and rewards tend to vary with skill and specialism.* Non-manual workers are generally ranked above manual, since their work requires considerable education and training. In particular, workers with intellectual, managerial and administrative skills tend to be in relatively short supply and so can command high rewards (salaries, working conditions, fringe benefits and so on) from their employers. Within the manual sector, skilled craftsmen and technicians are ranked above operatives and labourers.
- Parallel to the skill hierarchy there is a *hierarchy of authority*, with employers and business owners at the top descending via various ranks of management through supervisory workers to the rank and file. The closer to the top, the greater the material rewards and the greater the

Table 1.3 France: changing occupational structure

Per cent in each category	1968	1975	1982	Comment
Employers and self-employed in industry and commerce	9.6	7.9	8.2	Centralisation of production – i.e. the continuing amalgamation of farms, manufacturing firms and other businesses into fewer larger units, thereby reducing opportunities for individual ownership
Farmers	12.1	7.6	6.0	
Top professionals and managers	4.9	6.7	12.0	Ever-growing emphasis within the large corporation and government organisation on administration, marketing, research etc. Growth of educational, medical and social services
Intermediate professionals and managers	9.8	12.7	15.0	
Clerical	14.7	17.5	12.0	Post war expansion of office activities but routine tasks now beginning to be replaced by automation
Supervisors and skilled manual	14.7	15.7	23.3	Increasing premium on technical skills though some tasks susceptible to automation
Semi-skilled manual	15.3	14.5	15.2	Heavy and routine tasks now increasingly performed by machines and automated systems. Great post-war boom in demand for semi-skilled factory labour up to the 1960s but now a strong reverse trend. Farm labour continuing to be replaced by mechanisation
Unskilled manual	7.8	7.4		
Farm workers	2.9	1.7	1.5	
Service workers	5.7	5.7	3.8	Continuing disappearance of the servant class
Others (e.g. armed services, clergy, police)	2.6	2.5	3.0	
Total	100.0	100.0	100.0	

power to determine not only one's own life but the lives of others.

- In France, as in the rest of Western Europe, *the occupational structure of the workforce is undergoing gradual but profound alteration*. As the modern economy advances, so it is able to replace many of its more routine, menial and unpleasant jobs with machines: at the same time it places an ever-growing premium on skills and qualifications to carry out the increasingly complex tasks that are necessary to its functioning. Manual work makes less and less of a contribution to wealth creation, while scientific, technological and creative skills become ever more imperative.
- This last point is particularly crucial because it emphasises the manner in which large sections of the workforce are becoming *marginalised* i.e. totally or partially unnecessary rejects with no real part to play in economic life. The unqualified school leaver and the middle-aged redundant factory worker are both outstanding examples of the new marginality.

Once again it is possible to translate the occupational division of labour into geographical terms, to trace the manner in which it operates spatially and the uneven impact which it makes on the regions of Western Europe. Just as industrialisation has created a pattern of industrial specialisation by region, so it has given rise to specialist occupational roles for each region. Just as the typical region tends to concentrate on a distinctive range of products, so also it occupies a distinctive role in the organisation of economic life. Power itself is geographically concentrated. The great decision-making organs of the modern world – the apparatus of state government, the policy-making headquarters of banks and giant corporations – are localised within a handful of dots on the map of Europe (the reasons for this are fully discussed in Chapter 2; for the present we are concerned with effects rather than causes).

The social consequence of regionally concentrated power is illustrated by the occupational structure of the Parisian workforce which is quite distinct from the rest of the nation, being distinguished by above average proportions in the upper ranks of the status hierarchy. As a government, financial, industrial, leisure and cultural capital, Paris offers a unique array of openings for the professionally qualified, the skilled, the highly educated, the gifted and the ambitious. This is reflected in its marked over-representation of professional and managerial workers; and in the great gulf in personal income which separates it from the rest of the country. A similar phenomenon is seen throughout Western Europe in Madrid, Vienna, Stockholm and a host of other cities which act as the foci of power and opportunity.

e) Unemployment and exclusion

Employment is the principal passport to an income for the citizen of the industrial society. Yet according to 1984 unemployment figures there were over 12 million in the 14 countries of continental Western Europe who were effectively disqualified from holding this passport. Apart from the tiny minority who have voluntarily opted for the relative poverty which accompanies joblessness, most of the unemployed fall into two categories: *1)* those made redundant or compelled to retire early, as in the case of the workers laid off from the EC coal and steel industries since 1970; and, *2)* school leavers entering the labour market but not finding a job. In 1984 the proportion of unemployed youths aged between 15 and 24 in the EC varied from 13 per cent in West Germany to 35 per cent in Italy.

A significant proportion of this mountainous unemployment is *cyclical* in nature and, it is hoped, temporary. It results from the current world recession, a short-term slow-down in trade and business activity, creating a shortage of demand, which has driven countless firms to extinction or to reduce output with fewer workers. More ominously, however, we must class much of Western Europe's unemployment as *structural* (i.e. caused by the structural changes in the economy itself). As we have seen the industrial economy is propelled by a compulsive drive for modernisation and efficiency, replacing established products and methods by new ones. This results in impressive production growth but at the cost of serious human casualties. By rendering a product obsolete industrialisation also renders obsolete the men and women who make it. Thus, workers in many of the traditional industries of Western Europe face a crisis either because their product has been superceded and/or because it can be produced cheaper elsewhere (often in the newly developing countries such as Brazil or South Korea). By way of example, the shipbuilders of Sweden and West Germany faced crippling competition from the Far East, and the Swiss watch industry was brought to near extinction when digital technology introduced an alternative product.

It is true that *in the long term* and at the national level, the composition of the labour force changes in tune with industrial change. Figure 1.6 demonstrates for France the manner in which the composition of the national workforce has changed in response to the decline of one set of employment opportunities and the birth of a new set of alternatives. As almost every commentator on modern France tells us, a nation of peasants, farm labourers and small town merchants has been reborn as a nation of industrial workers and functionaries of streamlined, large international corporations. It is also true that other traditional rural nations of Western Europe – Spain, the Irish Republic, Greece – are straining to emulate the French feat of modernisation. But to see this transformation as a smooth and painless process of adjustment would be to treat people as machines and to ignore the human cost. Industrial change has always been a supremely painful procedure, whichever part of Western Europe it has affected.

There are a host of ways in which material progress collides head on with human values. In Britain we are accustomed to the view that many people are loathe to change, have ingrained attitudes, habits and working practices; are reluctant or unable to learn new skills; are afraid of the unknown; value security and continuity above exciting but risky prospects of material gain; reject the material for the spiritual; act emotionally rather than rationally. Although other Western European nations, notably the French and Germans, have accepted the consequences of modernisation much more readily, aside from this cultural and psychological inertia, there is another more mundane element which ensures that progress is painful – that element is geography itself. Factors of distance, location and accessibility are among the critical obstacles to painless economic progress.

The question of space as a brake on modernisation is taken up again in Chapter 2 and 6. Essentially, the problem is that new technology tends to seek out new locations and to demand new geographical distributions. Early industrialisation, with its shift from an agricultural to a manufacturing emphasis, created a geographically centralised pattern of production, with new and growing economic activity heavily concentrated into a small number of strategically placed regions, to the inevitable neglect and exclusion of other regions. Our own era has seen further shifts, with the decay of some of the earliest industrial centres and the rise of newcomers.

f) Excluded regions

This geographically uneven distribution of economic activity and unemployment is vividly captured by Fig. 1.7 which plots the pattern of unemployment in the EC. The map identifies two major areas of high unemployment: underdeveloped agricultural regions and decaying industrial areas. The former comprise a group of regions on the southern and western fringes of Western Europe where high unemployment coincides with extreme dependence upon agriculture as a source of livelihood. These areas – southern and western France, Spain and Portugal, Greece, and the Irish Republic – are among the worst hit by the concentration and localisation of the modern economy. In effect they have

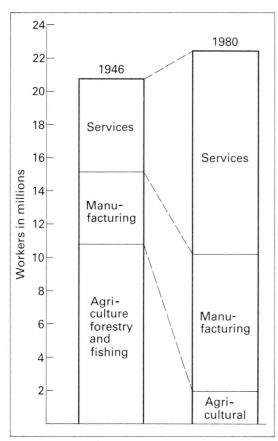

Figure 1.6 France: changing employment structure

22 GEOGRAPHICAL ISSUES IN WESTERN EUROPE

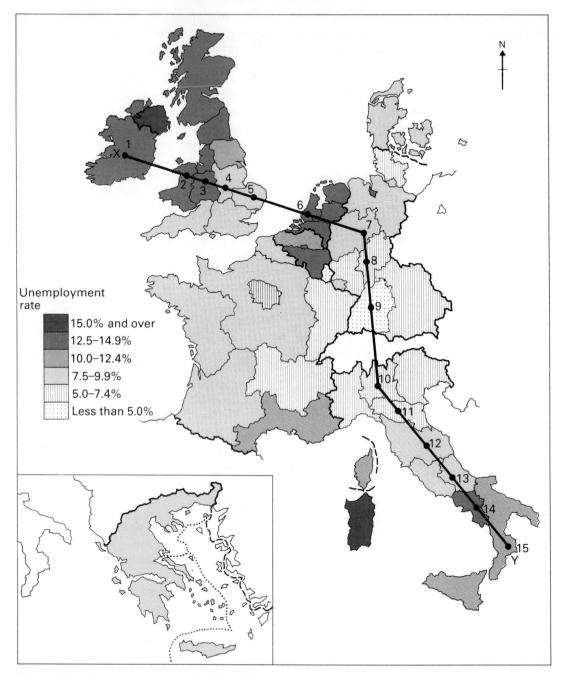

Figure 1.7 Unemployment in the EC, 1983

been left out in the cold, unable to participate fully in national industrial growth. As a result, they have been obliged to continue their reliance upon agriculture which results in a chronic wastage of labour.

Self-assessment exercise

Figures 1.5 and 1.7 show geographical variations in wealth and employment but as they use average values for each region, this can be misleading in a number of ways – for example, they assume an even distribution throughout a region whereas there may be considerable variations within it, also they assume abrupt changes at boundaries which are not found in reality.

Another way of depicting these data which still allows the trend of variations across boundaries to be illustrated is a *transect* (you could also use an isopleth map). The line X-Y on the two maps has been chosen as a north-west–south-east transect through the regions of the EC. Construct a graph to show variations in wealth and unemployment along this transect following the instructions below:

- The horizontal scale on your graph paper will be distance; you should measure this between the transect points along the line.
- The vertical scale, in the first instance, will be GDP per capita (make sure that you leave enough room by locating the maximum and minimum values first).
- The data to be plotted are given in the table below. Plot the GDP value for the first transect point at the left-hand margin of your transect graph with a cross and label it.
- Repeat the process for each transect point in turn (scaled along the horizontal axis) then connect the resultant crosses to show the overall trend.
- Replace GDP per capita with the unemployment rate on the vertical scale and repeat the exercise using the unemployment data given below.

You should now have not only an indication of the geographical trend in the two variables, but also a visual comparison between the wealth of a region and its level of unemployment.

What are the main features of the individual transect lines?

What is the relationship between wealth and unemployment?

Are there any anomalies? If so, can you give reasons for these?

The figures for Switzerland and Austria are also given – do they change the trends shown in the transect?

	GDP per capita (1982)	Unemployment rate (1984)
1 Ireland	68	16.9
2 Wales	88	12.8
3 West Midlands	81	13.0
4 East Midlands	88	9.6
5 East Anglia	89	7.9
6 Western Netherlands	109	10.8
7 North Rhine – Westphalia	110	8.7
8 Hesse	123	5.8
9 Baden-Württemburg	118	4.4
10 Lombardy	108	8.4
11 Emilia-Romagna	110	8.7
12 Centro	94	8.6
13 Abruzzi-Molise	71	8.3
14 Campania	62	14.3
15 Sud	58	12.5
Switzerland	160	1.1
Austria	113	3.9

Case study: Underdeveloped agricultural regions – Sicily

Lying immediately off the south-western extremity of Italy's toe, the island of Sicily exhibits strong locational, historic and economic similarities with other offshore islands and remote peninsular regions of Western Europe. In many senses it is a member of a 'family' of ultra-remote and underdeveloped economies, which includes Western Ireland, Brittany, Corsica, Sardinia and Galicia. There are, of course, vast cultural and physical disparities between the members of this family but their major common factor is their sheer *marginality*. Their contribution to the total production of European wealth is a slender one and throughout the entire modern era their economies have been too weak to sustain an adequate level of living for their populations. Consequently, ever since the mid-nineteenth century many of their people have been obliged to migrate in search of a livelihood. Arguably their greatest contribution to modern economic development has been the great quantity of able-bodied labour which they have 'exported' to the industrial economies of Northern Europe and North America.

Table 1.4 offers an outline profile of the present state of the Sicilian economy. The first pointer to the marginality of this economy is the extremely low income per head in comparison with national and EC norms – Sicily is ranked 19th among the 20 Italian regions in income terms and compares even worse in the EC at large. The root cause of these low incomes is the chronic labour surplus which has bedevilled Sicily for more than a century – in Sicily people are abundant but jobs are in short supply. People are abundant as a result of persistently high birth-rates which, though diminishing considerably since the 1950s, are still well in excess of a death rate which is also falling. In the short period from 1977 to 1980, Sicily added an extra 105,000 to its population through natural increase (excess of births over deaths), equivalent to a 2 per cent population increase.

Job scarcity stems principally from two main factors:

- *The inability of agriculture to absorb the extra population into employment.* For the peasant (small land-owning) class, extra population has been accommodated by subdividing small farms into even smaller units. This has created inadequate dwarf units capable of yielding only a meagre livelihood. No amount of industry and sweat on the part of the Sicilian peasant can disguise the fact that there is simply not enough land to go round. For the landless farm labouring class, population increase has meant too many workers

Table 1.4 Sicily: socio-economic profile

	Sicily	*Italy*	*EC*
Population 1982 (millions)	5.0	57.2	272.1
Population density (persons per km²)	196.0	190.0	163.0
Population growth 1975–80 (per cent annually)	0.7	0.4	0.3
Birth rate 1981 (births per thousand population)	13.9	10.9	12.3
Infant mortality 1981 (deaths per thousand births)	16.9	14.0	11.4
Income per head 1980 (per cent EC average)	60.0	88.0	100.0
Activity rate 1981 (labour force as per cent of total population)	28.3	35.2	41.6
Female activity rate 1981 (as above, women only)	11.9	20.9	29.3
Unemployment rate 1981 (workless as per cent of registered labour force)	7.6	5.6	7.8
Per cent in agriculture 1981	18.1	11.0	6.7
Per cent in manufacturing 1981	26.5	38.6	40.2

chasing too few jobs on large farms, driving down the wage rate in their desperation to work even for a pittance.

The agricultural sector, then, is incapable either of supporting extra workers or of offering a satisfactory livelihood to its existing participants. Indeed, it is crystal clear that Sicilian agriculture needs fewer rather than more workers, a slimmed down labour force using productive modern mechanised methods of production in place of its present huge army of cultivators struggling with antiquated methods. In recent years, Sicily has begun to move in line with agriculture elsewhere in Europe, but by modern standards much of its agricultural workforce is still surplus to requirements, marginal in the true sense of the word (Table 1.5).

- *The inadequacy of non-agricultural alternatives.* Outside agriculture, Sicilians have traditionally looked to small-scale industry and services for employment. Yet here once again the story is often one of expendable labour: small textiles, leather and woodworking firms, with insufficient capital to invest in modern equipment and surviving by courtesy of cheap labour. This traditional manufacturing economy has for long been in retreat in the face of competition from mass-produced goods imported from the mainland. The largest employment sector on the island is service industries (e.g. hotels), employing over half the total workforce. Apart from government employment, which offers a number of well-paid non-manual jobs, this sector consists largely of small retailing, catering and trading businesses, often surviving only by virtue of unpaid family labour.

Like so many other remote underdeveloped regions Sicily's essential problem is that its long established economic base is gradually withering away. Those opportunities which it does offer are extremely unrewarding in comparison with those offered by large

Table 1.5 Surplus agricultural labour in Sicily

1 Numbers engaged in agriculture (1981)	246,000
2 Number of Sicilian agricultural workers required to produce the average output of 100 EC workers	131
3 Number of Sicilian agricultural workers required to produce the average output of 100 Lombardy workers	154
4 Number of Sicilian agricultural workers required to produce the average output of 100 Danish workers	218
5 Surplus Sicilian labour at EC standards	58,000
6 Surplus Sicilian labour at Lombardy standards	86,000
7 Surplus Sicilian labour at Danish standards	133,000

COMMENTARY

In Sicily, due to smaller farm size, lower levels of mechanisation and other relatively inefficient practices, it takes more workers to achieve a given output (items 2–4). On this basis, items 5–7 are estimates of the number of workers who would be surplus to requirements if Sicilian agricultural standards were raised to EC, Lombardy or Danish standards. It is obvious from these observations that Sicilian agriculture is harbouring very large numbers of economically unnecessary people. More than anything else it is this which accounts for the island's low incomes – too many people queueing for a share of a small cake.

There is, of course, a major dilemma here. The *commercial* problems of agriculture could be cured by a modernisation programme but this would generate massive *social* problems – an army of unemployed agricultural workers and small farmers displaced from the land. This is an issue faced by all under-developed agricultural regions throughout Western Europe.

modern organisations using efficient methods to produce goods which are in high demand in European and world markets. For Sicilian workers displaced from agriculture or other traditional livelihoods there are very few

ITALIAN REGIONS III

Italy's distorting mirror

Sicily
JAMES BUXTON

THERE IS an English ice cream van outside the magnificent temple of Concord at Agrigento on the south coast of Sicily. It still bears the well-known English name Tonibell and sounds the chimes that are part of everyday life on the housing estates of England. Its Sicilian owner, who left Agrigento at the age of 16 and spent 22 years as an ice cream man in England, drove it back to Sicily two summers ago. He is an unhappy man.

"Coming back here was the biggest mistake I ever made," he says. "I had a Tonibell franchise in Buckinghamshire, a fleet of 12 vans and a nice house of my own. Here I've just got one van and I make my own ice cream. It was difficult enough to arrange even that.

"The licence just to put my van here costs L750,000 (£312). Everything requires a permit and it's not like England where the official just says or no—here you have to find the right man, succeed in getting in to see him and then he'll probably say 'maybe.' There are health controls here like there are in England, but there you knew they were being applied fairly. You need a different mentality here and I'm afraid I've lost it after 22 years. My son said to me the other day: 'What the hell did you bring us to a horrible place like this for?'"

Returned emigrants are not always the best guide but this man's story tells much about Agrigento and western Sicily. Superficially Agrigento appears to be booming, with clusters of new high rise flats on the hill where once stood the ancient Greek city. Construction is spreading haphazardly and destructively down through what should be an archaeological park to the sea. Traffic roars on multi-lane highways.

Yet rampant building speculation, which is what it is, only means prosperity for some. For the small man who is not a client of a rich patron, life in Agrigento may be miserable. Every summer, part inefficiency and corruption combine to deprive the city of a reliable water supply—instead water is sold at an exhorbitant price. Other services are poorly run, if they exist at all.

The ice cream man did not say it but what he and other small businessmen in Agrigento are up against is the old southern Italian practice of 'clientelismo' which in its more ruthless form means the Mafia.

Police in Palermo set up a road block after a Mafia killing. Some progress has been made in using a new law against suspects.

It is often said that Sicily is a distorting mirror of Italy, that whatever Italy has is to be found in exaggerated form in Sicily. Unfortunately that now seems to be particularly true as far as negative things go. If Italy suffers from weak and inefficient governments, then Sicily, despite have been a region with special powers since 1946, has an even feebler administration.

Regular crises

The regional government lurches in and out of crisis with appalling regularity and the consequence is that many of the funds which Rome pours into the island remain unused in the bank since the Government in Palermo, when there is one, cannot agree on how to spend them and the islands bureaucrats are very bad at implementing whatever decisions are made.

In this vacuum real power tends to be wielded by the big men of business. In Western Sicily, including the capital Palermo, they are frequently the leaders of the Mafia, and the organisation is now also active in Catania, the once relatively "clean" city at the eastern end of the island. Where its writ runs, the Mafia decides what factory, dam, irrigation scheme and housing development does or does not get built. It also, as has been increasingly obvious in the past few years, has the power of life and death over most Sicilians.

In the past the Mafia was a form of resistance to foreign rule. By the first part of the 20th century it had become an organisation for the protection of the often absentee landlords of western Sicily. It retarded economic development but maintained order of a kind. But the landlords' power was reduced in the post-war land reforms and the new Mafia drew on the ingrained traditions of its feudal past to become a mainly urban affair.

Even in the late 1960s books were still being produced that painted the Mafia as a more honourable and romantic organisation than it ever deserved. At that time it was predominantly involved in building speculation and other rackets, such as receiving stolen property and illegal gambling, and its tentacles were spreading up Italy. But now the processing and trading of drugs is thought to be the Mafia's most important activity, accounting for half its income which was estimated a year or two ago at about L16,000bn.

The amounts of money involved are now so great that they have triggered off a series of other developments: a far greater ruthlessness by the Mafia in dealing with those politicians, policemen and magistrates who stand in its way, almost permanent warfare between the different Mafia gangs producing more than 70 murders so far this year and 99 last year, an accelerated pace to the building speculation which has virtually destroyed the beauty of Palermo and other cities, as the Mafiosi rush to invest their gains and faltering efforts of the authorities to do something about it.

For years governments in Rome dared do little effective about the Mafia, lacking the powers that Mussolini had used against it with some success in the 1930s and conditioned by the organisations's political influence. Last year, the Government sent General Carlo Alberto dalla Chiesa, a Carabinieri officer who had played a major part in defeating left-wing terrorism, to Palermo as prefect with a special brief to make inroads on the Mafia.

He was gunned down with his young wife in a horrifying assassination after only four months. Parliament then swiftly passed an anti-Mafia law that had been prepared a decade before, giving the authorities special powers to track down Mafia suspects, including examining their bank accounts, and making membership of the Mafia an offence.

Then last July Sig Rocco Chinnici, the state prosecutor of Palermo who was apparently on the point of naming those responsible for the death of General dalla Chiesa, was blown up in a massive bomb explosion in the centre of Palermo. Sig Emanuele de Francesco, the prefect who had taken over from dalla Chiesa with even greater powers, has become the centre of a political row for his alleged lack of progress against the Mafia.

Progress

Some progress has been made in using the new law against suspects, but the legal bureaucracy that must process the evidence is slow and often (with justification) fearful. Nevertheless the Mafia must be worried as shown not just by the cold-bloodedness of its response to threats to its position but also by the fact that bank deposits in Sicily have dropped by L8,000bn in the past few months, apparently as the suspects get their money out.

The Mafia, well represented in much of the rest of Italy, especially the north, is a national issue for Italy. In Sicily it has brought an ill-gained prosperity to some of the towns, at the cost, in the western part of the island, of stunting that spontaneous growth of small industry and commerce which has been perhaps the major development of the rest of the Italian economy in the past 20 years. Happily that is not the case in the east of the island where the crowded coastal strip from Messina to Catania is a hive of activity, while Catania itself has a sound economy based on its rich citrus growing hinterland and some industry. Syracuse is also doing well.

But for the most part the productive part of the Sicilian economy presents a depressing picture.

Partly because of the imperative in Rome and Milan to be seen to be "doing something" about the south, Sicily has a fine collection of "cathedrals in the desert"—the vast, capital intensive, heavy industrial plants which do not produce much indirect employment and whose sites are usually chosen on political rather than economic criteria. From the late 1950s onwards when oil was cheap the Sicilian coast became dotted with oil refineries and chemical plants. The plants, largely built for Montedison and ENI, were mainly concentrated on the east coast near Augusta and on the south coast at Gela. In both cases, they caused immense pollution which governments for long ignored.

The Italian oil refining industry had its heyday in the 1960s but the subsequent drop in European oil demand, the low technical sophistication of many of the Sicilian plants and their distance from the main European markets told badly against Sicily in the later 1970s.

In consequence much of the Sicilian population, already having a lower income than most of the rest of the country, is drifting deeper into "assistenzialismo"—living off the welfare state, often on unjustified disability pensions. Already at least 20 per cent of the island's income comes from the provision of services by the public administration, against an average for Italy as a whole of 13 per cent.

There are some bright spots, though. Heavy spending on roads has given the island a fine infrastructure. Tourism supplies a reasonable income, but is threatened by the remoteness of Sicily and the bad but justified publicity about the Mafia. The trans-Mediterranean pipeline, bringing gas from Algeria, passes through the island on its way to northern Italy, and Sicily is entitled to 3.6bn cubic metres of gas a year. But here again the potential is not being lived up to: the idea is that a network of gas mains in 46 towns will give Sicily a great boost. But despite the delays in the pipeline coming on-stream the national and local authorities have failed to get more than a skeleton network installed.

But there is better news on another energy front. Oil-fields off the south coast of Sicily have long been small producers, helping the refining industry and the Cicilian economy. Now Montedison is developing what promises to be the biggest oilfield in the Mediterranean, the Vega field, with a flow rate estimated between 60,000 and 80,000 barrels of oil per day.

Although the oil is heavy it is still valuable to Italy which imports almost all its crude, and will be refined in Sicily. The Vega field should come on-stream in 1985, by which time two or three much smaller new oil and gas fields will also be producing. But crude oil alone is not going to change the bad old ways of Sicily.

Figure 1.8 *The Financial Times*, 24 September 1983

realistic alternatives. Most of the available alternatives are in some way unpalatable for the workers concerned or socially undesirable for the community as a whole.

ALTERNATIVE 1 – INACTIVITY

Perhaps the most telling Sicilian statistic is its activity rate (Table 1.4), which at just over 28 per cent is the lowest in the entire EC. The activity rate is defined as the number of registered workers expressed as a percentage of the total population. A low figure is indicative of an underdeveloped economy with little potential for employment. In the Sicilian case, fewer than one in three residents (one in six females) are active members of the labour force. Of these, a further 7 per cent are currently unemployed. Many of this army of non-participants are dependent on state benefits and welfare provision, one of the many ways in which Sicily is obliged to rely on external life support.

ALTERNATIVE 2 – MIGRATION

The classic method of avoiding passive dependency is to search for work elsewhere, a somewhat drastic response which usually involves permanent exile in a distant region or foreign country. In Sicily, emigration on an epic scale has been a continuing fact of life for generations, so that foreign cities such as Boston in the United States of America contain more people of Sicilian descent than most of the towns on the island itself. The drain continues mainly now to the more prosperous countries of the European Community and in the brief period from 1977–1980 over 10,000 migrants left the island. It is often argued that migration is a valuable economic safety valve but in Sicily's case migration is only running at one-tenth the rate of natural increase and is thus failing to reduce that quantity of superfluous labour on the island. Perhaps its most valuable contribution is the cash remittances sent home by migrant workers to their families on the island. This is another example of Sicily's dependence upon lifelines from the outside world.

ALTERNATIVE 3 – 'THE BLACK ECONOMY'

Despite this external dependency, Sicilians have also shown themselves capable of self-help, though in a rather desperate form. The so-called 'black economy' (also known as the 'hidden' or 'submerged' economy) is by no means confined to Sicily. All over Western Europe there are citizens unregistered for work but who, nevertheless, are prepared to undertake 'jobs on the side' unknown to the State which continues to pay them welfare benefits. By definition, the black economy is impossible to measure but it is thought that southern Italy contains overlarge numbers of such invisible (and technically illegal) workers, one reason why Sicily's *official* activity rate is so low. Often the black economy shades into organised crime and no discussion of Sicily can pass without reference to its historic role as nerve centre for the Mafia with its global networks of smuggling, drug dealing and racketeering. Vile and repulsive though this may seem to some readers, we nonetheless must regard organised crime as a significant Sicilian economic activity, a perverted form of capitalism which creates jobs and redistributes wealth but which exacts barbaric penalties from those it preys upon. In truth, of course, the very existence of this flourishing gangster industry is a sad commentary on the paucity of *legitimate* opportunities on this isolated and backward island.

The second major group of high unemployment regions are the *decaying industrial regions*. Here the problems are acutely familiar to British readers. Figure 1.9 identifies a belt of regions running across and adjacent to Western Europe's main coalfield zone – northern France, southern Belgium and the West German Ruhr – where heavy unemployment is linked to a high dependence on indutries which have lost their former glory. Mining, textiles and heavy metal manufacture, the heralds of the first industrial revolution and the staple industries of these regions, are now in decline. As a source of job opportunities they have diminished spectacularly since the early 1950s, their decline accelerating in the 1970s and 1980s, hit by a combination of falling demand for their products and technical and organisational change.

Case study: Decaying industrial regions – the south Belgian coalfield

If the underdevelopment of agricultural areas is akin to a slow wasting disease to which the patient has grown accustomed, then the symptoms of industrial decline have flared up much more abruptly and the pain is obvious for all to see. In March 1982, a protest march through Brussels by 10,000 striking steel workers led to pitched street battles in which 180 policemen were injured. Most of the protesters came from the districts around Liège and Charleroi, two of the main centres of the south Belgian coalfield (also known as the Walloon coalfield and the Sambre-Meuse coalfield), an industrial region contained mainly within the two provinces of Hainault and Liège (but overlapping into the adjacent regions of Namur and Limbourg Fig. 1.9). Like many other old industrial centres of Western Europe, this is a region whose working population feels increasingly threat-

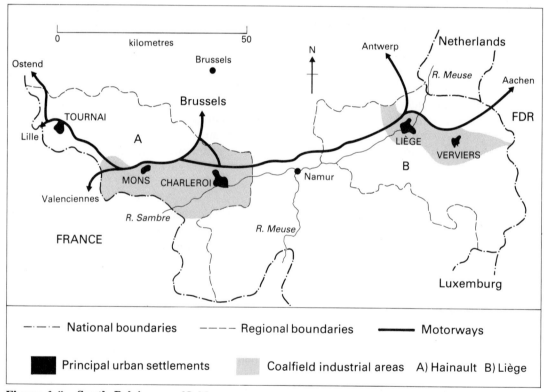

Figure 1.9 South Belgium coalfield

COMMENTARY

The map emphasises the alignment of the old coalfield along the Sambre-Meuse axis, and its separation into two wings located in the regions of Hainault and Liège respectively. It also stresses that, although this is an industrial region in decline, its potential for future development is enhanced by its great accessibility – a location on a major international east-west motorway route at the centre of Western Europe's leading concentration of population and activity.

ened. The growing industrial unrest of the past decade is the outward symptom of the fears of a labour force facing mass redundancy and falling living standards.

The present century and particularly the recent post-war period has seen a drastic reduction in the economic stature of regions of this type. In sporting parlance, they have been relegated from the upper to the lower divisions of the West European league. From acting as one of the spearheads of the first Industrial Revolution, south Belgium is now a somewhat marginal contributor to the Western European and world economies. Its products are no longer in universal demand and in some cases the industries which produce them are close to the point of extinction.

It is generally recognised that early nineteenth-century Belgium was the first child of the British Industrial Revolution, the earliest continental nation to wholeheartedly embrace the then ultra-modern steam and metal-working technology. The first Belgian coke-fired iron furnace was in action in 1823, almost a quarter of a century before the adoption of this technique in the Ruhr. In establishing this lead, Belgian capitalists borrowed freely of British expertise and enterprise. Indeed, several of the new iron and coal enterprises were wholly or partly British-owned; one of these was the ancestor of today's Cockerill-Sambre corporation, Belgium's leading state-owned steel making concern. For the technology of the nineteenth century, which required all coal and steam-using activities to consume truly colossal quantities of coal, the Sambre and Meuse valleys provided the ideal physical foundation for industrialisation – coking coal and iron ore in mutual proximity, together with potentially navigable waterways.

Not surprisingly the region (already a famous centre of armament manufacture and metal working) emerged as a formidable and expanding complex of coal mining, iron and steel making and heavy engineering. From 1831 to 1900, its population doubled to just over 2 million, as the rapid rate of new job creation attracted migrants from neighbouring areas in Belgium, France and Germany. Yet it is now clear that the past few decades of the present century have thrown its development into reverse and that this is now an exceedingly unhealthy regional economy (Table 1.6). Unemployment is the most striking and tragic feature. At over 12 per cent it compares with the worst regional levels anywhere in the EC and in certain localised pockets such as Liège it exceeds one in five of the registered workforce.

South Belgian decay arises from certain global trends which are largely beyond the power of local people – workers, managers and administrators alike – to influence.

- *Falling demand for its coal.* This is a

Table 1.6 South-Belgian coalfield: additional commentary

Decline in population 1971–80	−16,000 (0.7 per cent)
Loss of population by migration (1978–81)	−7,300 (0.3)
Decline in total number employed (1971–80)	−29,000 (3.4)
Decline in iron and steel employment (1976–9)	−9,000 (15.8)
Decline in other manufacturing (1976–9)	−26,000 (15.5)
Loss of underground mining personnel (1978–9)	−600 (46.1)
Growth in total unemployment (1976–79)	24,000 (23.5)
Growth in male unemployment (1976–79)	10,500 (40.4)
Rate of unemployment (1981)	126,000 (14.8)

typical example of product obsolescence, as a formerly vital energy source it has been superceded by alternative forms. In south Belgium, the problem is aggravated by geological difficulties which have raised the cost of extraction to the point where it is cheaper to import American coal for use in the furnaces of Liège. Working in combination, geological and demand factors have produced a truly disastrous industrial collapse: production has diminished from 30 million tonnes at its 1955 post-war peak to a mere 5 million tonnes in 1982. The resultant job loss has been crippling, with over 8 out of 10 jobs disappearing since 1960. An industry employing over 100,000 in 1960 can now support only 17,000 miners.

- *Uncompetitive steel production.* Among the factors which have brought about a parallel (though less total) contraction in the steel industry is the lack of competitiveness of south Belgian steel in world markets. The world's second oldest iron and steel industry is now paying the price for past glories, as it struggles to rid itself of obsolete plant and equipment so as to match the technical efficiency of newer competitors such as Japan. Compared to Japan, the Belgian industry needs twice the manhours to produce each tonne of steel, a fact which makes Belgian steel extremely costly. Costs are driven even higher by the need to import both coal and iron to an inland location: Belgian steel is *locationally* as well as technologically uneconomic. Consequently, lost markets have led to a contracting output and heavy job losses (Table 1.6). The Cockerill-Sambre group alone has lost 4,000 out of 19,000 workers between 1974 and 1984.

Many business economists would applaud the efforts made by Belgian steel to increase its productivity. Thanks to new methods and equipment, steel productivity (i.e tonnes per man-hour) has almost doubled in the past decade. For local workers, however, it is hard to appreciate the virtues of such modernisation. Belgian trade unions are caught on the horns of a dilemma: if they do not co-operate with modernisation programmes they risk the collapse of the entire industry; if they do co-operate, they voluntarily sacrifice the jobs which the new machines are designed to replace. Despite reasonably generous state welfare and unemployment payments, unemployment represents a drastic fall in living standards and in any case few workers are willing to opt for a life of economic uselessness. In towns like Liège where as many as half the population are dependent on steel-related industries, the current situation is an inevitable cause of personal insecurity and political tension.

By now it will have become clear that regional inequality is an important *political* issue in Western Europe. There is now a widespread public awareness that material life chances are unevenly distributed between the regions of the continent. This inequality is resented as being socially unjust: inhabitants of less favoured regions are conscious of themselves as 'have-nots' and tend increasingly to demand what they see as their fair share of national wealth. Regional inequality is a source of political conflict, with poor regions ranged against the rich, whom they see as gaining at their expense.

Nowhere is this conflict more sharply defined than in the case of regions whose people, often with a different language and culture, see themselves as a separate nation subject to the political control of an alien nation state. Ever since the break up of feudalism onwards, the political map of Europe has been increasingly shaped by *nationalism* as 'nation-states' have been built up from the amalgamation of smaller independent units. For example, in the course of the nineteenth century both Germany and Italy were created out of a number of smaller states and a national identity was imposed upon the hitherto independent separate states. In this process, many minorities (often distinguished by cultural and linguistic differences) have been incorporated into the territory and subjected to the jurisdiction of larger nations. The map of Europe is peppered with regions which are homelands of such minorities – for example, the Basque and Catalan provinces of Spain, the French-speaking Val d'Aosta in Italy, and Corsica and Brittany in France.

Many, though not all, of these regions are to be found on the economically marginal periphery of Western Europe. Just as in other peripheral areas, there are good grounds for economic dissatisfaction, but in regions such as Brittany or Corsica a sense of economic loss is linked to one of political loss also. The stateless nation has no power to determine its own affairs. Economic and political grievances are seen as part of the same question: in the eye of minority nationalists, political subjection is the root cause of economic underdevelopment, since the large nation is free to exploit the small nation within its borders. They argue that development and prosperity can never be achieved unless the minority people are autonomous, free to govern themselves in their own interests. Freedom is thus seen as both a virtue in its own right and a necessary condition of material prosperity.

Case study: Sub-national minorities – Brittany

Throughout the post-war period, Brittany has been in the forefront of the resurgence in minority nationalism. Within modern France, Brittany's official status is equivalent to that of the other 21 regions, with no special powers and no independent organs of government (although limited powers have been decentralised to all French regions in 1982–3, see Chapter 3). This is a source of tension between Brittany and France, because in many important cultural and historical respects Bretons are simply not French. Their ancestors were Celtic refugees fleeing the Anglo-Saxon conquest of Britain in the sixth century and even now Bretons tend to emphasise Celtic heritage, Celtic identity and links with other Celtic nations (see Fig. 1.10). Despite many centuries of rule from Paris, the Breton language – a variant of the ancient Brythonic tongue whose other variants are Welsh and Cornish – is still spoken by about one in five of the 3.5 million Bretons.

Despite Brittany's physical remoteness, the ease of modern communication has drawn the region into an ever closer relationship with mainstream France. This relationship has often been discordant. Its post-1945 phase has been marked by increasingly vocal discontent and by the rise of political movements dedicated to Breton independence. Breton farmers, too, have become notorious for the forcefulness of their demonstrations against French and EC farm policy. As seen by Bretons, French domination has damaged their homeland in two major respects.

- *Cultural oppression*. Many Bretons resent the manner in which Parisian governments have for generations pursued a policy of 'Frenchification', for example, trying to eliminate their language by punishing any child heard to speak Breton on school premises (an experience not unknown to Welsh-speakers two generations ago). In recent years such crude harassment has ceased and Breton political pressure has won various concessions from Paris: an increase in Breton language broadcasting and the establishment of bilingual primary schools in certain areas.
- *Economic underdevelopment*. Table 1.7 depicts Brittany as exhibiting all the classic symptoms of the rural periphery – low incomes, high unemployment, agricultural labour surplus and few viable alternatives except migration. Breton nationalists argue that much of this results from the 'colonial' status of the region within France. As a colony, its functions were to supply cheap primary products (vegetables and fish) together with cheap migrant

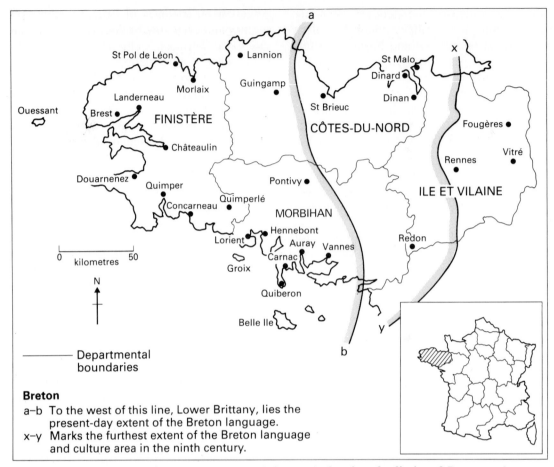

Figure 1.10 Brittany (départements and chief towns) showing the limits of Breton culture

labour, to the benefit of firms and workers in metropolitian France but to the detriment of Brittany. Migration is a special source of resentment. The region lost no less than one quarter of a million migrants in the first post-war decade, a drain of mainly youthful workers which has continued (though more slowly) up to the present mainly to Paris and especially the Montparnasse area in the south-west of the city.

Friction between France and Brittany persists even in the face of the rapid economic growth and modernisation of the past two decades. We should note that this has been brought about by *more rather than less contact* with France and the outside world. Yet this contact has been double-edged, conferring short-term prosperity but imposing hidden costs.

INDUSTRIALISATION

Since the 1950s Brittany has been a major beneficiary of state regional aid, whereby firms are given government grants and loans to expand in underdeveloped areas (see Chapter 3). By 1970 over 23,000 new manufacturing jobs had been created by this means, many of them in large plants such as Citröen at Rennes. Yet at no time has job creation been sufficient to cope with growing numbers of workers displaced from traditional activities. In any case, nationalists tend to regard industrialisation as a form of intensified col-

Table 1.7 Brittany: Socio-economic profile

Brittany value as a percentage of	France	EC
Population growth 1970–79	96	146
Birth rate 1980	101	117
Population density	99	57
Infant mortality 1981	99	85
Income per head 1980	82	90
Activity rate 1980	99	104
Female activity rate 1980	104	121
Agricultural proportion of employment	244	325
Manufacturing employment	75	68
Employment growth 1971–80	116	200

COMMENTARY

Quite clearly Brittany is neither as socially deprived nor as economically retarded as Sicily in relation to the rest of its nation and to the EC. Demographically, Brittany's profile is remarkably similar to that of the rest of France: no longer is it a region with a rapidly growing surplus labour force swollen by high birth-rates. Definite progress has been made with the creation of new jobs and a new economy, as is shown by the figure for employment growth – twice the rate for the EC as a whole. On the debit side, the tell-tale item is agricultural employment, whose proportion of the workforce is two and a half times that of France and over three times that of the EC. Despite amalgamation and modernisation agriculture remains a relatively unproductive sector containing far too many workers. In 1979 it occupied over one-fifth of the region's labour force to produce just over one-tenth of its total output.

onialism, using Breton labour for the benefit of outside capital and management and increasing Breton dependence on decisions taken outside the region. Particular resentment developed over a proposal to build a nuclear power station in Western Finistère in the early 1980s (see Chapter 9, pp. 280).

AGRICULTURAL DEVELOPMENT

Government aid has helped to accelerate agricultural modernisation (see Chapter 5), redistributing land to create larger farms and providing increased mechanisation. An inefficient retarded agricultural economy has been substantially transformed, with impressive results in yields and output (Table 1.7). It has, however, displaced large numbers of peasant families and created severe depopulation in many village communities.

TOURISM

A great increase in the number of holiday visitors (boosted once again by central government investment in roads and resort facilities but also by local initiatives, for example, Brittany Ferries is owned by a large agricultural co-operative) has greatly increased the incomes of those selling tourist services. For Breton nationalists, however, tourism is another form of cultural colonialism, opening up the Breton homeland to alien influences ('driving the language off the streets'). Seaside and rural communities are also under threat from the large numbers of newcomers buying holiday and retirement homes and driving property prices beyond the reach of local families.

g) Other deprived minorities

Not all disadvantaged sections of the West European population can be neatly placed into specific regions or types of region. There are certain economic and social disqualifications which apply universally throughout the continent, though their impact may vary from place to place.

One such disqualification is *foreign origin*. The advanced industrial countries of the EC, Switzerland and Scandinavia now contain large numbers of foreign immigrants originating mainly from Iberia, Turkey, Greece, the Balkans and North Africa. There are also important groups of long distance migrants, such as West Africans in France and Indonesians in the Netherlands. Immigrants are to be regarded as a generally disadvantaged group in that: a) they are for the most part assigned to the least desirable and lowest paying jobs and to inferior housing accommodation; and b) as aliens they are excluded from the franchise and various other citizenship rights; and c) they are frequently the target of racist hostility and discrimination on the part of the native population (see Chapters 4 and 6 for fuller discussion).

A second disqualification is *gender*. Traditionally women have been excluded from full participation in the wage-earning economy, being for the most part confined to roles within the family and dependent upon male wage earners. All this is gradually changing but despite a substantial post-war upsurge in the female labour force there is still a great gulf between male and female participation in the labour market. Currently the female activity rate (see above p. 24) in the EC stands at only 55 per cent that of males. Geographically speaking, this sexual discrepancy is most marked in the southern rural periphery of Western Europe, as in the Mezzogiorno (Italy), for example, where men out-number women in employment by almost three to one but where female family workers (i.e. helpers in small family businesses, often unpaid) out-number male family workers by two to one. It is also the case throughout Western Europe that women are under-represented in the 'plum jobs' – management, the professions, skilled work.

A third category whose access to the fruits of the affluent society tends to be restricted is the *aged*. The elderly are a vulnerable group firstly because many of them are totally or partly reliant for their economic support on state pensions; and secondly because of the very nature of the ageing process they require greater health care and social provision than any other section of the community. As the elderly proportion of the population steadily increases throughout Western Europe, so the income transferred from wage earners to the retired must also increase if the latter are to retain even their present (often unsatisfactory) living standards.

3 The quality of life

a) Human well-being in its broadest sense

During the 1970s Western Europeans became increasingly conscious of the broader meaning of phrases such as 'social well-being' and 'standard of living'. As the fear of poverty and want faded into a distant memory for most citizens, so it came to be realised that wealth in the monetary sense is not necessarily equated with well-being. It is not just that the Dutch market-gardener earns more than the Portuguese small-holder or that the Austrian ski-instructor's income is greater than the Greek or Spanish waiter's but also the quality of education and standard of health care enjoyed by the Dutch or Austrian families far exceeds that experienced by their Greek, Spanish or Portuguese equivalents. Indeed, there are a host of attributes that can be considered as contributing to the quality of life ranging from life expectancy to housing conditions and to whether one lives in a polluted environment.

Ultimately, of course, each human being's happiness is an entirely subjective matter, which can only be pronounced upon by the individual him or herself. Human emotions are intangible and notoriously difficult to measure statistically. In consequence, no study of the 'good life' can ever be entirely scientific. However, an important distinction can be made between those positive factors the presence of which helps to improve social well-being and negative factors which may pose a threat to the quality of life.

I) POSITIVE FACTORS
In modern society there are several elements which are difficult to value in monetary terms but which are generally considered vital to leading a full life. Among the items are:

- *Health*, a factor not necessarily determined by private affluence but more dependent upon state provision of doctors, hospitals and other care and medical research facilities;

- *Education*, the key to literacy and to higher forms of intellectual and creative self-fulfilment;
- *Social provision*, welfare and social work institutions caring for the more vulnerable members of the community;
- *Leisure and cultural opportunities*.

We should note that the above four items depend heavily and increasingly upon public (i.e. state government) finance. State provision of these benefits varies significantly from one West European nation to another according to the size of government revenues and proportion of these revenues devoted to various purposes. Countries like Spain, Portugal and Greece clearly suffer in comparison to their neighbours, chiefly as a result of their lower national income and consequently lower tax revenues, whereas certain nations, notably Sweden, Norway and the Netherlands, have built up comprehensive networks of state provided services of a very high quality. At the regional level, too, there are significant contrasts, with the large urban regions of the core enjoying higher standards of public provision.

II) NEGATIVE FACTORS

Increasing material wealth may result in non-material losses. While the urban-industrial areas of Western Europe enjoy the highest levels of private affluence they also suffer severely from 'stress factors', reflected in high rates of crime, suicide, alcoholism, divorce and road accidents; and *a poor physical environment*, reflected in polluted rivers and high levels of air pollution and land dereliction.

All this is to say that material progress is achieved at the cost of numerous less tangible and measureable benefits – the point being that the ceaseless quest for increased production places great strains upon people and upon the physical environment. For every material gain there is a hidden cost, the flip side of economic progress.

Much industrial development, especially the basic heavy activities such as mining, metal smelting, oil refining and heavy chemicals, are despoilers of landscape, water and atmosphere. Traditionally, the people of industrial regions such as the Ruhr, the French Nord and Lorraine or the south Belgian coalfield have been forced to live among noxious emissions from factories, the scarred derelict landscapes of mining, the ugliness of industrial buildings and the pollution of their rivers. To say that this detracts from their quality of life is an understatement. In a very real sense the industrial firms have enjoyed the profits while the public has been obliged to bear the costs of industrialisation. Ironically, the development of a conservation ethic has coincided with the decline of many of these industries and some now seek to preserve relics of this industrial heritage. For example, the Bochum Mining Museum in the Ruhr is the sole remnant of the industry in a city which at its peak in 1956 had 65,000 mineworkers.

Industrialisation also requires individuals to make painful adjustments in their personal and family lives. In the post-war period many hundreds of thousands of men, women and children migrated permanently from countryside to town. This rural-urban migration was inevitable in view of the run-down in agricultural employment and the expansion of large urban-based industries and services. Most of the rural-urban migrants have undoubtedly benefited from better wages and better opportunities but they have paid a high price in personal upheaval. Life in the secure tranquil village community has been exchanged for the stresses of the crowded, anonymous city. It is no coincidence that the highest stress factors are recorded in the most urbanised areas of Western Europe.

b) **Geographical variations in the quality of life**

A recent assessment of social well-being throughout all of Europe has been carried out by Ilbery, who used 27 variables as indicators of the quality of life. These were reduced to just six key variables by complex statistical manipulation which were then used to calculate an overall index of social well-being for each country. (The technicalities of the exercise need not concern us here as you will be asked to carry out a similar exercise shortly.) The resultant pattern depicted in Fig. 1.11 illustrates the sharp differences in the quality of life that exist on a national basis with Sweden emerging as having the highest (best) levels of social well-being (closely followed by West Germany).

The geographical variations shown in Fig. 1.11 are usually described by geographers as a *core-periphery* pattern. Core regions are the prosperous 'heart' of Western Europe (both geographically and economically). They exhibit all the characteristics of a healthy, dynamic economy and their citizens enjoy all the benefits that accrue in terms of job opportunities, higher incomes and better social facilities as well as some of the costs such as a congested and polluted environment. The periphery, on the other hand, is much less prosperous, lacks modern industrial and agricultural development and is deprived of many social services and amenities. Examination of Fig. 1.11 reveals that the pattern of inequali-

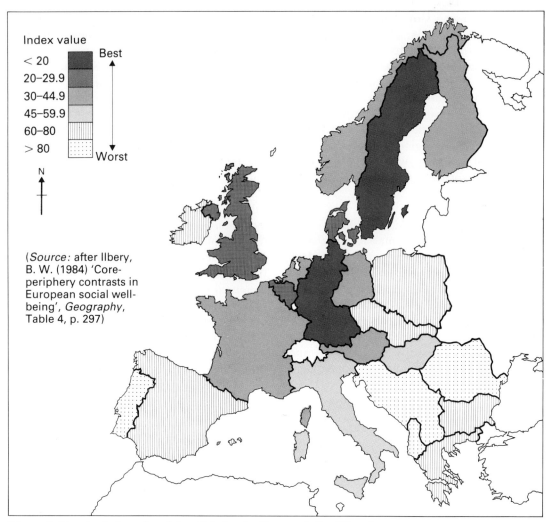

Figure 1.11   An index of social well-being in Europe (after Ilbery)

ties in Western Europe closely fits this description with successive rings of less prosperous peripheral countries radiating outwards from the core, especially to the south and west. Taking Sweden and West Germany as the core, an inner ring of countries with intermediate levels of social well-being can be discerned consisting of Switzerland, Belgium, the United Kingdom and Denmark at a slightly higher level than Austria, France, Luxemburg, the Netherlands, Norway and Finland. The peripheral countries with the lowest quality of life are clearly shown on the southern and western margins of the sub-continent with Italy faring slightly better than Greece, Spain and Ireland, and Portugal having the poorest standard of social well-being in Western Europe.

The causes of these gross inequalities will be examined in detail in Chapter 2 but a number of qualifications to this form of analysis need to be made here. Firstly, the choice of indicators of social well-being is crucial and very subjective. If a different variable had been chosen, a different pattern may have resulted. Secondly, the overall index conceals important variations in a country's performance, for example, West Germany was ranked second in terms of housing conditions but only thirteenth in terms of infant mortality. Thus, to have a high standard of social well-being overall does not imply a rosy picture in all aspects of life. Finally, by using national data important variations *within* countries are masked. Within West Germany, for example, there are significant differences in the quality of life between the decaying, polluted industrial regions such as the Ruhr and the rapidly expanding, environmentally attractive areas in the south of the country. Whatever the country, the citizen's quality of life will vary according to place of residence: urban or rural, core or periphery, industrial or agricultural, as the following case study will show.

Figure 1.12 Thirty regions of West Germany

Case study: Regional variations in West Germany

Just as in the case of Western Europe, we would expect inequalities in social well-being within individual countries – an uneven access to the ingredients of the good life. Fig. 1.13 portrays geographical variations in the quality

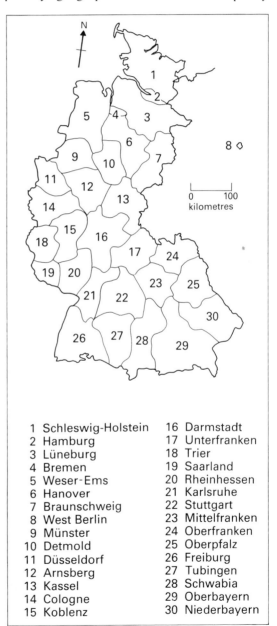

1 Schleswig-Holstein
2 Hamburg
3 Lüneburg
4 Bremen
5 Weser-Ems
6 Hanover
7 Braunschweig
8 West Berlin
9 Münster
10 Detmold
11 Düsseldorf
12 Arnsberg
13 Kassel
14 Cologne
15 Koblenz
16 Darmstadt
17 Unterfranken
18 Trier
19 Saarland
20 Rheinhessen
21 Karlsruhe
22 Stuttgart
23 Mittelfranken
24 Oberfranken
25 Oberpfalz
26 Freiburg
27 Tubingen
28 Schwabia
29 Oberbayern
30 Niederbayern

38 GEOGRAPHICAL ISSUES IN WESTERN EUROPE

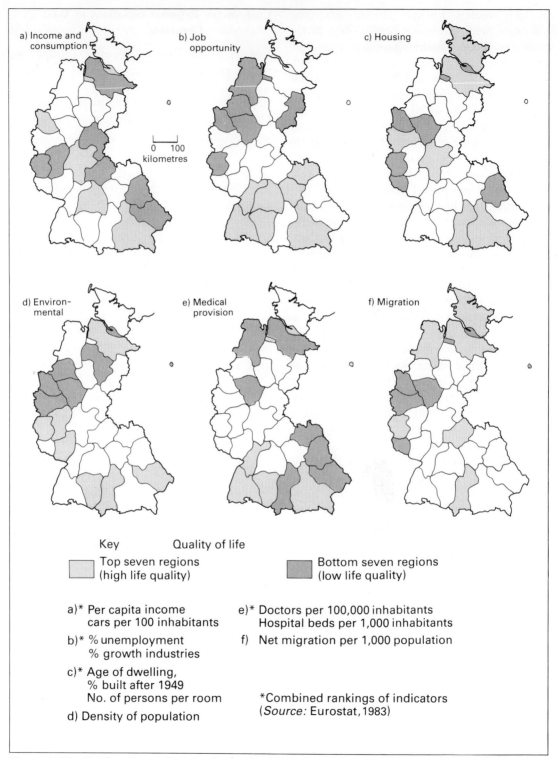

Figure 1.13 Regional variations in six quality of life indicators in West Germany

of life in West Germany. The 30 regions in West Germany are ranked according to six quality of life indicators: affluence, economic opportunity, housing, medical provision, physical environment and migration. Among the very detailed information given by the maps, two important general points stand out.

Firstly, life in advanced society is essentially contradictory. Monetary and career success are usually only gained by the sacrifice of some other desirable object, such as environmental or housing quality. Thus, for example, we see that several of the old-established urban and industrial regions of northern Germany – Düsseldorf, Hamburg, Bremen and West Berlin – enjoy some of the highest levels of income and medical provision (a typical urban amenity) but suffer some of the worst housing and environmental conditions in West Germany. At the other end of the scale, the very best physical environment – spaciousness, absence of pollution and dereliction – are to be found in the predominantly rural regions like Schleswig-Holstein, Lüneburg or Schwaben. Usually, however, the price to be paid for living in such idyllic but economically peripheral areas is lower wages, fewer job opportunities and poorer public amenities. Lüneburg, for example, is ranked second in the national environment ratings (Fig. 1.13d) but also has the lowest per capita income (Fig. 1.13a). Despite this, however, it is a highly attractive proposition for new migrants (Fig. 1.13f), many of whom are moving out to escape the congested living conditions of Hamburg and Bremen.

The *second* general point which can be made from the maps in Fig. 1.13 is that ideally the most desirable residential location is one which somehow combines the virtues of city and country, work and leisure, old and new. In practice, of course, no such earthly paradise exists but some places will come closer to it than others. However, it is possible to carry out an exercise which combines all six indicators into a composite quality of life index.

Self-assessment exercise

Essentially the problem is how to combine the wide range of criteria used to judge the quality of life into just one measure of social well-being. Traditionally, geographers have used the tried and trusted 'eyeball' method. Visual inspection and interpretation of a series of maps such as Fig. 1.13 yielded an individualistic assessment of which regions were 'best' and 'worst'. However, a more objective method is available and we would like you to try it. It involves a series of statistical manipulations, some of which we have done for you to simplify the task.

i) Raw data are gathered for a set of social well-being indicators considered most appropriate. We will use those shown in Fig. 1.13 which relate to the regional planning divisions of West Germany.

ii) These data are then ranked according to whether they indicate a high or low quality of life. Great care should be exercised at this stage to ensure correct results, for example, the region with the *highest* average income or number of doctors per person would be ranked first as would the one with the *lowest* unemployment rate or level of air pollution.

We have carried out these first two steps for you and the rankings are shown in Table 1.8.

iii) You now need to consider the six indicators. Some may be considered more important than others, for example, you may think that income is more important than housing quality and much more significant than migration. Accordingly, you need to decide *weighting factors*, so income may have a factor of six, housing quality two and migration one. Do this for your assessment of the relative importance (weighting) of all six indicators. (Hint: choose whole numbers between 1 and 10.)

40 GEOGRAPHICAL ISSUES IN WESTERN EUROPE

Table 1.8 Composite ranking of quality of life indicators for West German regions, 1979

Overall Ranking	Region	Quality of life indicators (ranking)						Total rank score	
		A	B	C	D	E	F		
1	Tübingen	13	3	5	8	7	10	46	Top 7 regions according to the composite ranking
2	Oberbayern	7	4	22	12	3	2	50	
3	Karlsrühe	8	7	15	20	4	8	62	
4	Freiburg	16	5	14	12	5	16	68	
5=	Darmstadt	5	9	21	22	10	4	71	
5=	Stuttgart	5	1	12	13	19	21	71	
7	Mittelfranken	9	5	18	14	18	9	73	
8=	Schwabia	19	11	7	6	25	6	74	
8=	Rheinhessen	10	13	6	16	9	20	74	
8=	Lüneburg	30	10	1	2	30	1	74	
11	Schleswig-Holstein	21	17	17	7	14	3	79	
12	Niederbayern	27	8	2	3	27	13	80	
13	Cologne	11	17	27	25	8	5	83	
14	Oberfranken	20	2	8	8	28	18	84	
15	Kassel	26	17	8	11	13	11	86	
16	Koblenz	24	11	4	15	19	15	88	
17	West Berlin	3	20	24	30	1	17	95	
18	Detmold	15	15	19	20	20	7	96	
19	Hanover	12	13	29	18	12	14	98	
20	Unterfranken	25	23	10	4	17	22	101	
21	Hamburg	1	16	27	28	2	28	102	
22	Trier	29	22	2	1	23	26	103	
23	Bremen	2	23	20	25	6	29	105	
24	Braunschweig	13	25	16	16	16	21	107	Bottom 7 regions according to the composite ranking
25	Weser-Ems	22	26	10	5	26	19	108	
26	Saarland	16	30	10	24	15	30	125	
27	Düsseldorf	4	28	30	28	11	27	128	
28	Oberpfalz	29	21	18	10	29	24	131	
29	Münster	18	26	25	23	22	22	136	
30	Arnsberg	17	29	28	27	24	25	150	

(*Source: Eurostat*, 1983)

Key
Quality of life indicators, 1979
A = Income & consumption (per capita income and cars per 100 inhab)
B = Job opportunities (% unemployment, % in growth industries)
C = Housing (% of dwellings built after 1949, no. of persons per room)
D = Environmental (population density)
E = Medical provision (doctors/100,000, hosp. beds/1,000)
F = Migration (net migration/1,000)

iv) Compile another table of rankings for the planning regions by multiplying the rankings shown in Table 1.8 by the weighting factors decided in step iii). For example, if you choose a weighting factor of two for housing, all the rankings in the third column should be multiplied by that figure and all the rankings in column one will be multiplied by six if this was selected as the weighting for income, to give *weighted rankings*.

v) You now need to add up the six new rankings calculated in step iv) across the rows so that each region ends up with a *total score*.

vi) The total scores calculated in step v) are

now ranked so that the lowest score is ranked 1 through to the highest score which will be ranked 30. It could be that you have two or more regions with identical total scores tying for the same rank, for example, if two regions tie for rank 3 you take the ranks they should cover (i.e. 3 and 4) add them together and divide by the number of regions (i.e. 7 ÷ 2), this gives them both a rank of 3.5 and you then proceed with the next region as rank 5. This gives *composite rankings*.

vii) These overall composite rankings weighted according to the perceived importance of the indicators can now be mapped so that geographical variations can be examined – use the identification map on Fig. 1.12 to plot your results.

viii) Compare your map with Fig. 1.14 which we have compiled by using the same data but all the indicators were weighted equally. Do the two maps show the same geographical pattern overall? What are the main similarities and differences? What reasons can you suggest for this, did your chosen weightings have any effect?

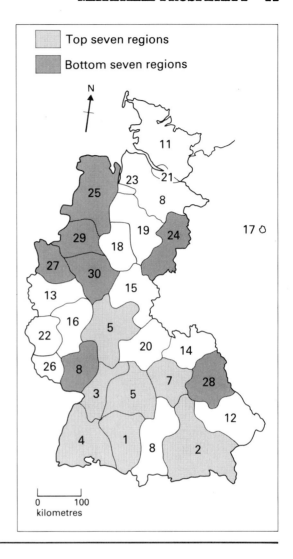

Figure 1.14 Composite quality of life index, West Germany, 1979

Our interpretation of Fig. 1.13 is that the seven regions which come closest to the ideal are all located in the southern half of the country. They form a cluster shaped roughly like an inverted 'Y', whose stem begins in the central region of Darmstadt (a region which includes Frankfurt and the Rhine-Main industrial complex); whose western branch runs southward along the upper Rhine axis and includes all four regions of Baden-Würtemburg; and whose eastern axis runs along the Neckar valley and thence to Mittelfranken (Nüremburg) and Oberbayern (Munich). The attractiveness of this group of regions emphasises one of the two great post-war changes in the geography of West Germany – the shift of economic activity, prosperity and people towards the south of the country. Conversely, the seven regions with the poorest quality of life can be categorised as old, declining industrial regions (Arnsberg, Düsseldorf and Münster which centre on the Ruhr in the north-west, and the Saarland in the south-west), frontier isolation (Brunswick and Oberpfalz in the east) and rural underdevelopment in the north-west (Weser-Ems).

According to the life chances which they offer inhabitants, the West German regions

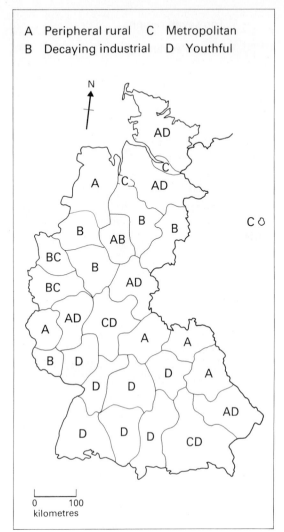

Figure 1.15 Regional typology of West Germany

fall mostly into four broad types (though there are several intermediate cases which do not fit easily into a single type – see Fig. 1.15).

PERIPHERAL RURAL

Geographically, this category includes the eastern border zone, much of the north German plain and the westerly upland province of Trier. Relative remoteness has isolated these areas from the mainstream of national economic life, so that incomes and opportunities are very restricted in comparison with the rest of the nation. Their assets are an unspoilt (and often scenic) environment and good housing conditions but these are insufficient to compensate for low wages and lack of urban amenities. No region of this group is ranked higher than fourteenth on the composite index.

OLD INDUSTRIAL

These regions are the German equivalent of south Belgium or the French Nord stretching in a broad belt across northern Germany with an outlier in the south-west (the Saarland). Until recently, long-established heavy industry continued to supply well-paid employment and these regions are still among the most prosperous. By contrast, housing and environmental conditions are among the worst in West Germany. Most seriously of all, these are 'sunset' regions with a very heavy stake in dying and degenerating industries (for example, the Ruhr and Saarland's decline as coal and steel producers). Between them, the six regions contain over one-third of the nation's unemployed and none of them is ranked higher than nineteenth on the composite index. These factors combine to produce an insecure working environment, whose unpopularity is confirmed by a rapid exodus of people leaving to take up residence elsewhere.

METROPOLITAN

The three metropolitan regions (Bremen, Hamburg and West Berlin) emerge in their own right mainly because the boundaries have been drawn tightly around them, whereas other major cities such as Frankfurt and Munich are part of much larger regions to the south (see below) and Düsseldorf and Cologne remain prosperous areas within older declining industrial regions. Metropolitan centres have extremely dense populations, they are also top level trading, banking and office centres, offering many very highly paid career opportunities and an unparalleled standard of urban amenities. All this is somewhat offset, however, by intensely congested housing and living conditions. Consequently, the metropolitan regions rank only moderately on the quality of life index and, in the case of Hamburg and Bremen, are being deserted by large numbers of their residents searching for more spacious living elsewhere.

YOUTHFUL

These regions (all of which are located within the southern half of the country) are 'youthful' in the sense that much of their economic development is recent in origin, modern in character and relevant for the future, with much hi-tech manufacturing, services and tourism. In addition, the group also contains two of the Federal Republic's greatest banking centres – Munich and Frankfurt. This is not to say that there is a lack of industrial tradition. Any regional geography text on West Germany will testify to the ancient craft skills of the Black Forest and the Upper Rhinelands. Even so, the critical feature of their economic geography is the absence of the heavy, dirty and now declining manufactures which were so vital in the early phase of German industrialisation. This is an attractively mixed living environment, combining easy access to the opportunities provided by medium/large free standing towns with plentiful open space and an absence of vast, ugly, sprawling conurbations. These youthful regions are magnets for new migrants and their population has grown on average by three times the German national rate since 1970. All this would seem to defy the adage that 'you cannot have your cake and eat it', for the youthful regions of southern Germany have succeeded remarkably in achieving the benefits of modern industrialisation without paying a heavy environmental price.

4 Conclusion

In this chapter we have demonstrated the wide geographical disparities in living conditions (both monetary and non-monetary) which persist even in this most prosperous of ages. We have also identified the kinds of places and people upon whom the worst burdens of inequality tend to fall. In concluding, we must admit that future prospects for the lessening of social and regional inequality are not bright. Indeed there are good grounds for believing that the gulf between the haves and the have-nots may well widen still further.

The greatest single threat on the horizon is unemployment, stemming from the capacity of the advanced modern economy to create more and more output with fewer and fewer workers. In West Germany, for example, economic researchers have estimated that production of goods and services must now increase at 3 per cent a year simply to prevent any job losses. Throughout Western Europe there is a clear danger of a widening discrepancy between those workers whose skills are necessary to the new economy and are thereby destined to share in its material benefits, and those who will be forced to join the swelling ranks of the redundant. The largest gains of all, of course, will be made by the owners and controllers of the large corporations in the forefront of technological progress. There is no built-in economic safeguard which can compel them to share their profits with those whose labour is no longer of use in production. Any sharing out which occurs takes place through political means – the fiscal and welfare organs of the state itself.

On the regional dimension, the probability is that rewards will be increasingly concentrated in the metropolitan nerve centres, where almost all critical political, finanical and top management decisions are taken: and in regions where the crucial hi-tech industries and research activities are becoming concentrated. Such key regions have the power to determine developments elsewhere in Western Europe. It is now quite literally the case that the 'sunrise' industries which have developed in regions like Bavaria can spell death to jobs in decaying areas such as the Ruhr – one region's gain is another's loss. Our task in the next chapter is to explain the processes which have created (and continue to create) regional inequalities and disharmonies.

2 Industry and Regional Development

1 Regional inequalities and the production of wealth

The geographical variations in prosperity described in the previous chapter can only be explained by reference to the geography of production, since to enjoy wealth one must first create it. Broadly, regional disparities in consumption and living standards are caused by regional disparities in production. An inordinately large share of West European business activity – notably its most modern, productive and highly rewarded branches – is concentrated within a relatively small number of high intensity zones of industry and commerce. Because of the important economic role which they perform, workers in these regions usually enjoy a higher level of prosperity than those in areas of little industrial development or where industry is outdated, run-down or inefficient.

The economies of West European nations tend to be biased in favour of a single dominant region. The degree of dominance varies from one case to another. For example, French economic life is overwhelmingly centred upon the Ile de France (Paris region), which accounts for over a quarter of the nation's output of goods and services, although it comprises only a little over 2 per cent of the national territory. Other countries such as Italy, West Germany and the Netherlands show a less striking concentration but even so, every country contains one outstandingly active region which makes a disproportionately large contribution to its total output. Throughout Western Europe this is a common feature whether it be Scandinavia (where, for example, the Stockholm and Oslo regions dominate their respective national economies), the Alpine countries (Vienna's central role in Austria) or southern European nations (Athen's dominance of Greece).

2 Core and periphery

In order to describe and explain this contrast between first and second class regions, the concept of core-periphery is now frequently employed by geographers. The English word 'core' derives from the same root as the French 'coeur' meaning 'heart', and so core regions are those at the heart of national affairs. They are central in a *functional* sense; the hubs around which the entire economy revolves. They are also central in a *geographical* sense since the core region is generally located at the most accessible point within the national space. It is this very accessibility which is the key to their economic importance: they enjoy an economic advantage over other regions in that they are located in closest proximity to the vital resources which business and industry need in order to thrive profitably (see Box 2.1).

Case study: The anatomy of a core region* – the Paris region

In many respects the dominance which this region exerts over the economic life of its nation is quite exceptional in Western Europe – it is a core region par excellence. It also

---

* Similar treatment will not be given to a peripheral region in this chapter, but the reader is referred to Chapter 1 and the Sicilian case study, and also the Brittany case study in Chapter 6.

INDUSTRY AND REGIONAL DEVELOPMENT

> **Box 2.1 The profit motive and industrial location**
>
> In the modern West European economy the great bulk of production is carried out by privately owned firms whose object is to realise profit. This has long been the governing principle of economic geography and you may have come across theories such as Weber's which seek to explain this. Firms will tend to locate in areas which offer the best opportunities to reduce costs and maximise revenues. They will gravitate towards regions which offer cheap access to necessary resources – materials, energy, manpower and markets.
>
> It follows from this that the firm's location decision must be rational and unsentimental if it is to survive competition from its rivals. Furthermore, no firm can afford to remain in regions which do not permit it to adapt to the continually changing circumstances of the modern economy. In the matter of their own survival private firms are not generally moved by consideration of local loyalty or the welfare of local workforces.
>
> This is, of course, one of the basic tensions within a capitalist society such as Western Europe. Regions *are* people and the avoidance or abandonment of a region by industry makes itself felt in unemployment, poverty and deprivation. It is important to note here that during the post-war period there has been an increasing level of government intervention in an attempt to soften these blows – public intervention is the subject of the next chapter.

ranks highly in international terms, notably as a centre for decision-making within the EC. Thirty years ago the French planner, Gravier, coined the phrase 'Paris et le désert Français' to denote the extreme concentration of economic, social and political life in the capital city and the consequent inferior status of provincial France. For decades Paris has attracted many of the most ambitious, best qualified people from elsewhere in France, leaving the outlying regions denuded of much of their talent and vitality. Even after more than 30 years of government policy aimed at decentralising economic and cultural activities, the Paris region retains a formidable grip on the nation (Table 2.1). Its economic position is deeply entrenched and not easy to overturn.

Geographers would naturally stress the location of Paris as the key to its economic success. Positioned at the heart of the nation's most densely populated region, with natural routeways radiating in every direction, the city and its region offer clear advantages to government, industry, commerce and indeed every form of activity which depends upon communicating and interacting with others. Yet all this must be seen in its historical context. Much of the present dynamism of the region results from an accumulation of economic power over a very long period of time. For centuries before the Industrial Revolution it was an international capital of great eminence, the social hub of the European elite in a Europe where French was the language of the educated and cultured. As a result Paris derived immense power, prestige and economic stimulus. Subsequently, every stage of industrialisation has allowed it to enlarge and strengthen its economic role. Consider the following key economic functions which it performs:

- As a populous and wealthy centre it has always been a natural location for the manufacture of consumer products. Until the mid-nineteenth century this took the form of craft industries such as high fashion, jewellery and perfumery. Since then there has been a huge expansion in mass consumer products. Today the region accounts for over a third of national employment in the key motor vehicle and electrical engineering industries.
- As a centre of higher learning and research it has attracted the lion's share of 'hi-tech' industries, outranking even new 'sunrise' regions such as Grenoble in the manufacture of office and data processing equipment and precision instruments.

Table 2.1 The Paris region in its national setting

Economic Indicator	Definition	Paris (Ile de France)	Rest of France	Second-ranked region	Bottom-ranked region
Population density	Inhabitants per km²	836.0	81.0	316.0 (Nord)	26.0 (Corsica)
Activity rate	Workforce as % of total population	49.1	41.8	45.4 (Centre)	38.0 (Languedoc)
Unemployment	% of registered workforce	7.3	11.3	8.1 (Alsace)	14.8 (Languedoc)
Income	Annual wage/salary per head of population ('000 francs)	83.0	75.0	64.0 (Provence)	54.0 (Corsica)
Savings	per head ('000 francs)	36.0	21.0	26.0 (Provence)	19.0 (Franche-Comté)
Tertiary employment	Service industry workers as % of total workforce	65.9	52.7	see note	42.7 (Basse Normandie)
Commercial birthrate	New firms 1975–82 as % of total firms 1975	4.2	2.8	n.a.	n.a.

Note: In the case of tertiary employment, Paris is second ranked, coming below Provence (66.8). The tertiary sector is of vital importance, since it is the only sector of the economy which is actually creating a growth in the number of jobs. Most branches of manufacturing are replacing human labour with mechanised and automated production methods and it seems that the modern economy will rely increasingly on service activity to provide job opportunities. Thus, one of the chief assets of metropolitan regions like Paris, Brussels and Amsterdam is their ability to attract high order service activities. Regions like Provence, with their highly developed tourist and leisure industries are also strongly placed to cope with future developments.

According to a recent survey, it outranks all other regions in respect of the proportion of its firms which are engaged in electronics and information technology – industries at the forefront of economic progress.

- With the expansion of government activity in the present century, so the number engaged in civil service and other official posts has mushroomed. In Paris, government service alone accounts for half a million jobs since 1950, many of a managerial or technical nature. Additionally, Paris is also the headquarters of several major international organisations – such as UNESCO, ILO and UNICEF.
- The presence of government and financial institutions, together with Parisian social prestige has acted as a magnet for major industrial companies. The resultant office expansion since the 1950s – much of it located in new purpose-built complexes such as La Défense to the west of the Arc de Triomphe – has created yet another layer of high grade occupations.

Paris, then, is a capital city in every possible respect – political, financial, commercial, cultural and artistic. For the French job-seekers, especially those in search of a satisfying career and financial rewards, it is without parallel. Despite this, however, its economic growth appears recently to have gone into reverse. Since 1970 the region has lost over one million manufacturing jobs and over 150,000 residents. This outward migration of people and firms is inevitable in view of the intense overcrowding and congestion which has built up as the result of rapid growth. It is vital to note, however, that this job loss is mainly an exodus of routine industrial work: as such it hardly impairs the viability of the regional economy. Even as it has been losing industrial jobs, the metropolis has gained from the rapid growth in the tertiary sector – the expansion of company and government offices, research institutions, media, communications and financial institutions. Ancilliary activities such as advertising agencies, marketing consultants and legal services have also multiplied.

I) CENTRALITY

The first lesson to emerge from this Parisian study concerns the sense in which core regions are geographically central and the importance of this centrality to their economic development. Paris of course is by no means central in a geometric sense and indeed there are numerous other national core regions – Lombardy in Italy, Stockholm in Sweden and Vienna in Austria – which are even more eccentric in their physical location. Yet all these regions are central in relation to the population distribution of their country. Large concentrations of people represent large resources of potential customers and workers, the principal attractions for modern industry.

By contrast, peripheral regions like Calabria in Italy, the northlands of Sweden or the eastern frontier regions of Austria, offer no such basis for industrial development. Not only do they contain relatively sparse populations within themselves, but they are also surrounded by adjacent areas of sparse population (or empty sea) or by inpenetrable frontiers and are extremely remote from major population centres. It is this low *economic potential* which may be held to explain their low levels of economic activity (see Box 2.2 before proceeding to the next section).

II) SELF-SUSTAINING GROWTH

It is now clear that the commercial strength of the West European core regions derives from their centrality. We should also recog-

> **Box 2.2 Economic potential**
>
> Regional economic potential may be defined as a given region's centrality in relation to the market for its products. It has been suggested that the market represents the most *vital* of all resources for the modern business, thus for manufacturing and service industries the most obvious advantage of centrality is accessibility to markets for products. Clearly, a region which contains a large proportion of rich consumers (firms as well as households) is a more attractive location for producers than one which is sparsely populated and impoverished. Moreover, if the populous, rich region is close to other similar regions, then its market situation is further reinforced.
>
> These factors are incorporated in the calculation of regional economic potentials (see Fig. 2.4). By using a complex formula the potential for every region in the twelve nations of the EC can be calculated. In the case of the Düsseldorf region, for example, the economic potential value was 8,082 million European Units of Account (EUA) per square kilometre. This means that each 1 kilometre zone surrounding the central city of Düsseldorf contains, on average, a total income of 8,082 EAU. This very high figure reflects both the high total income of the Düsseldorf region itself and its proximity to other high income regions within the EC. Other things being equal, any large business interested in exploiting the EC market is likely to find Düsseldorf a highly advantageous location. By contrast the figure for Sicily is only 1,305 (well under a fifth of that of Düsseldorf), reflecting a sparse, low income, local population and the great distances between Sicily and the other major EC centres (see Fig. 2.4).

nise that as these regions have attracted additional workers and industry, so over time they have become even more central. Historically, the core zones have benefited from a process of self-sustaining growth (also known as the regional multiplier process – see Fig. 2.1), in which an initial advantage becomes subsequently enlarged. In the case of Paris, a heavy concentration of population

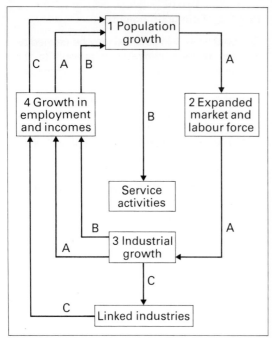

Figure 2.1 The regional multiplier

EXPLANATORY NOTES

The diagram is designed to show, in a general abstract sense, how economic activity and population are linked together in a circular fashion, with growth in one stimulating growth in the other, and vice versa.
Circuit A illustrates how population growth (1) expands market opportunities and labour force (2) thereby creating an attraction for industrial expansion (3) which provides additional jobs and income (4) to attract still more population (1). The process is in a sense self-perpetuating, with people attracting industry attracting people: once set in motion it tends to continue of its own momentum, unless disturbed by some external force.
Circuit B shows the further tendency for population growth to attract service activities – shops, transport, education, public administration – which set up an additional chain reaction by creating extra jobs and income to attract yet more population to the region.
Circuit C incorporates the principle of *industrial* linkage, by which the presence of a leading industry attracts other activities which need to buy and sell from it: e.g. textiles tends to be located in close proximity to clothing manufacture (a market outlet) and engineering (the supplier of special machinery). These linked activities also set up a chain reaction.

The entire process has been labelled the *regional multiplier* to describe the manner in which growth at one point in time creates the conditions for further growth in the future.

and wealth acted as an attraction for many of the new industries of the nineteenth century – in their turn these industries attracted migrant workers from outside the region, with the resultant population growth providing a basis for yet more industrial development. A similar cumulative growth process has also been evident in the other major cores, particularly the Ruhr in West Germany, the Wallonian and Nord coalfields in Belgium and France and also in Scandinavia where the capital cities drained the economic vitality from the regions to the north.

As Fig. 2.1 also shows, self-sustaining growth is further reinforced by linkages between different branches of economic activity. The expansion of one branch of industry in a particular place will tend to attract other activities which need to use its products or to sell their own products to it. The geography of Western Europe abounds with examples of these *agglomerations*, tight-knit regional concentrations of closely related activities – the coalfield areas where smelting of metals and metal-using industries are clustered together for mutual benefit; the capital cities, where government, finance and private company head offices are all closely tied together (see Box 2.3).

III) SPATIAL INTERACTION

A further lesson to be learnt from the Paris case is that core regions win their economic success largely at the expense of the periphery. In France, as elsewhere, the

INDUSTRY AND REGIONAL DEVELOPMENT   49

Box 2.3   Linked agglomerations – the case of the French aerospace industry

Recent years have witnessed greater co-operation with aerospace manufacturers in other countries of Western Europe, exemplified by Airbus Industrie, a consortium of aircraft firms from six nations, set up in 1970 to produce the A.300 series Airbus, the wide-bodied, sub-sonic airliner built as a competitor to the major manufacturers in the United States (see Fig. 2.2).

Today, the French aerospace industry employs just over a 100,000 people throughout the nation although there are four major centres of concentration – Paris, Bordeaux, Toulouse and Marseilles – with secondary agglomerations in the extreme south west and in the vicinity of Bourges (Fig. 2.3).

The French aerospace industry remains the most successful in Western Europe, second only to the United States in the western world. Since 1945 the industry has combined individualism and flair in the production of a wide range of aero equipment and, apart from one major commercial failure – Concorde – in the 1970s, has gone from

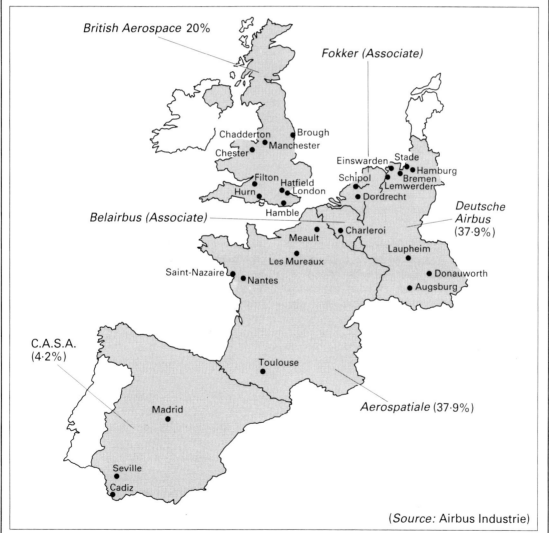

Figure 2.2   Airbus industrie consortium. Location of production facilities

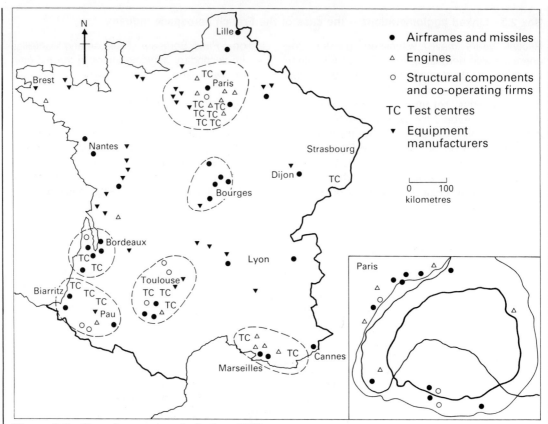

Figure 2.3 French aerospace industry, 1985

strength to strength and today supplies most of France's military aerospace requirements and a significant part of her national airline fleets. In addition to the huge domestic market, the quality of French aerospace products has opened the industry to markets worldwide.

One of the industry's strong-points is its diversity: the production of a wide range of military aircraft (such as the Mirage, Super Etendard), civil airliners, business jets and light aircraft, helicopters, guided weapons (such as Exocet), aero-engines for both the civil and military markets, ground and airborne systems installations and, finally, space systems and launchers such as Ariane.

progressive concentration of economic activity in and around the metropolis has meant a declining share for the rest of France. The very success of Paris has been based on taking away markets and workers from other regions. According to many writers this process of regional exchange (or interaction) is fundamentally unequal, one which benefits the already strong core regions and further weakens the position of the periphery. In a very real sense the regions of the modern nation state are in conflict with one another, competing for a share of the national market.

It should be emphasised that this system of regional competition is a product of relatively recent history. Indeed it is part of one of the great geographical revolutions of the period during and after the Industrial Revolution (see Table 2.2 on the spatial impact of the Industrial Revolution). Prior to this the

Table 2.2 The spatial impact of the Industrial Revolution

	Pre-industrial economy		Modern industrialisation	
	Structure	*Spatial form*	*Structural change*	*Spatial outcome*
Type of Production (What?)	Agriculture dominant Minor role for manufacturing and services	Dispersed production → Industrial location dependent upon agricultural location	Manufacturing the leading sector	→ Spatial concentration
Technology (How?)	Labour Intensive • Animate energy (human and animal power) • Limited range of materials • Limited transport output	Localised energy → and material resources exerting little pull → Production for local markets based on local materials	Capital Intensive • Inanimate energy (fossil fuels, HEP etc.) • Widening range of materials, many imported • Transport revolution and the conquest of distance	Steam power, → industry attracted to coalfields → Attraction of ports, iron fields etc. → National and global markets. Regional specialism of production
Organisation (By and for whom?)	Feudal Aristocracy dominant. Small peasant producers and domestic craftsmen	→ Dispersal. Subsistence production for local needs	Capitalist Large production units. Private firms controlled by boards of directors. Majority of workers employees of above	→ 'Rational' location in lowest cost regions. Regional competition

economy of pre-industrial Europe was characterised by a *very low degree of spatial interaction* – i.e. contact between regions and nations. It was largely made up of very small-scale firms, whose goods and services were destined for sale in local markets. Because of the expense of transport, only a very few high-value products were sold in national or foreign markets. Consequently, each locality needed to be highly self-sufficient, producing most if not all of its own necessities.

This geographical isolation was swept away by the series of advances in transport and communications inaugurated by the canal and railway age; and by the consolidation of nation-states, such as Germany and Italy, and the resultant abolition of customs barriers between regions. The modern era thus became characterised by *integrated systems of exchange*, through which goods could be cheaply distributed both nationally and internationally. The regional consequences of this were profound indeed. Previously self-contained regions were now exposed to open competition with the outside world: in such a situation, the least efficient producers in the least advantageous regions were unable to survive competition from lower-cost firms in more advantageous regions. One of the classic instances of this was the emergence of a national market in Italy, following the political unification of the nation in 1860 and the spread of railways in the mid-nineteenth century. In the south of the country, the small handicraft producers, who had traditionally enjoyed the dual protection of customs barriers and distance cost, were eliminated by competition from the modern mass-manufacturers of the north. Present-day Italy continues to bear the legacy of these events.

Regional trends initiated in the nineteenth century have continued well into the present century, with national economic growth persistently biased towards the north of the country. Currently, the three leading regions of Piedmont, Lombardy and Liguria (sometimes labelled the 'Turin-Milan-Genoa Triangle') produce 35 per cent of the national output on 19 per cent of the area. In contrast the three southernmost provinces of Basilicata, Calabria and Puglia produce only 7 per cent of the national output on 15 per cent of the land area. Such bias reflects the degree to which the Mezzogiorno has been denuded of its traditional small industries and forced to rely upon backward agriculture for its livelihood. It is cheaper for the population of southern Italy to buy manufactures from distant Turin or Venice, than to buy high-cost products manufactured locally.

Though perhaps an extreme version, the Italian experience is broadly typical of a universal trend in which the outlying regions of Western Europe have lost many of their traditional industries and people to the emerging core zones. Other examples of regions affected by this early wave of industrialisation include much of western and south-western France in a broad arc from Brittany to Languedoc-Roussillon, the northern Netherlands, and the northern lowlands of West Germany including Schleswig-Holstein. The Industrial Revolution has had the effect of enlarging the economies of nation states, whilst simultaneously diminishing (in relative or absolute terms) the economies of many regions within them.

IV) THE SUPRANATIONAL LEVEL

During the post-war period with the formation and expansion of the EC and the European Free Trade Association (EFTA), it is no longer sufficient to consider the effects of regional competition *within* the nation-state. Historically, of course, nations have never been entirely isolated. For the past two centuries or more international trade, both within Europe and outside, has grown continually, and has been interrupted only by wars. But the more recent abolition of tariffs and other border restrictions has given an unprecedented boost to the flow of goods, services and people between EC member states and between EFTA trading partners. This means that producers in any given region are now competing in a large tariff-free international market.

One effect of this is that producers in peripheral regions are even more geographically disadvantaged than previously. By way of illustration, Breton producers are now operating in an integrated system of over 100 regions in 12 EC nations instead of in a domestic system of 22 regions. Brittany's largest city, Rennes, which has always suffered from a location 320 kilometres west of Paris, now has to contend with a distance of 640 kilometres to the heart of the EC in the lower Rhineland. Clearly, regions like Brittany – and even more so in Spain, Portugal, southern Italy and Greece – are increasingly disadvantaged by widening integration.

By the same token, the established core regions have consolidated on their dominant position. Producers in Cologne, for example, have unhindered access to the huge volume of suppliers and consumers just over the border in Benelux. Whereas, immediately after the last war, they were part of a German core zone of about 30 million people, with the formation of the EC they became part of a continuous belt of regions containing 50 million. In 1985 there were nearly 30 million people residing within 150 kilometres of Cologne – for the manufacturer of consumer goods or the seller of services, this is an irresistible attraction.

Thus the meaning of Western European space has been transformed and Fig. 2.4 shows the new pattern of core-periphery. The map depicts a regional system in which the core areas of the most centrally placed nations – West Germany, Belgium, northern France and the Netherlands – have coalesced to form a new international supercore. Also included are certain detached regions – Paris, Hamburg,

INDUSTRY AND REGIONAL DEVELOPMENT

Self-assessment exercise

On a map of north-western Europe draw (in pencil) a circle whose radius represents a distance of 200 kilometres from Cologne. On a sheet of paper list all towns and cities with a population of 100,000 or more. For this part of the exercise you will need to go to your library and use an international gazeteer (such as *The Statesmans Year-Book World Gazeteer*) which will give you the population of all towns and cities required. When you have done this, draw a map showing the location of Cologne and all the major towns and cities you have identified.

Repeat the exercise for the city of Rennes in France and compare the two maps.

Can you find a more 'central' place than Cologne with a greater number of towns and cities over 100,000 population?

West Berlin – which enjoy a comparably high economic potential. For brevity's sake we shall refer to this regional amalgamation as the *Eurocore*.

The Eurocore forms the economic heart of the new Europe, accounting for 46 per cent of EC production on 14 per cent of its space. This truly overwhelming concentration stems from the outstanding locational advantage the Eurocore offers to producers. For any major manufacturer or service firm catering for the EC market, location within the Eurocore guarantees a high degree of access to that market, together with incomparable opportunities for close contact with customers and suppliers.

This notion of a supranational core area is by no means new. Throughout the post-war era as international barriers have gradually been broken down, the image of a 'Golden Triangle' has often been used by commentators to describe the Eurocore. Sometimes the points of this triangle have been placed on London-Paris-Ruhr; on other occasions Birmingham-Milan-Hamburg. Whatever its bounds, the triangle is conceived as a giant hub of the West European economy, a pole of attraction for many of the largest, most important, modern and go-ahead firms. In addition to its vast internal market, the triangle offers further commercial attractions – major international airports, major seaports and some of the most important nodes of the international postal and tele-communications services throughout the world. The Eurocore model retains most of these concepts but is less vague than the triangle model – it is defined fairly precisely by economic potential values.

We also need to note that the opening up of domestic markets to international competition does not only apply to the formation of the large supranational trading blocks such as EFTA and the EC. In practice, every economy in Western Europe has become increasingly exposed to competition from outside the sub-continent, particularly from the so-called 'newly industrialising countries' (such as Japan, South Korea, Taiwan and Brazil). The Swedish shipbuilding industry, for example, faces extinction because of this intense competition. Industries such as coal, steel and car manufacturing are particularly vulnerable to competition from outside Europe. International competition can also adversely affect previously strong industries in prosperous countries. Watch-making is probably Switzerland's best known industry, famous for everything from ornate cuckoo-clocks to precision timepieces made from precious metals. For many years it was the country's major export earner with as much as 97 per cent of production leaving Switzerland. However, in recent times it has faced a dual international threat. On the one hand economic recession in its traditional markets (the luxury nature of much of the output makes it particularly vulnerable) has led to a reduction in sales. On the other hand, their market share has been reduced by competition from new innovations in the same market sector manufactured largely outside Europe – in other words the digital watch produced in vast numbers at low cost in the

54 GEOGRAPHICAL ISSUES IN WESTERN EUROPE

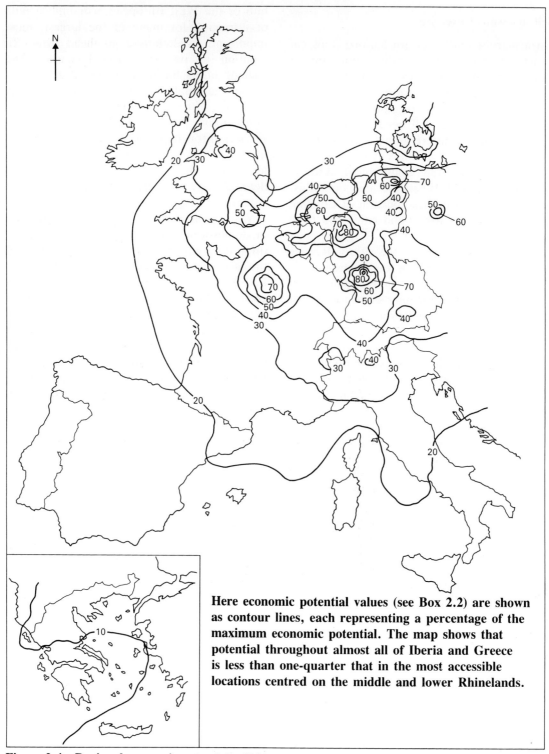

Here economic potential values (see Box 2.2) are shown as contour lines, each representing a percentage of the maximum economic potential. The map shows that potential throughout almost all of Iberia and Greece is less than one-quarter that in the most accessible locations centred on the middle and lower Rhinelands.

Figure 2.4 Regional economic potentials, 1982

Far East. This twin threat led to a major crisis in the Swiss watch industry which was particularly severe in the Jura region – the major watch-making area. By 1970 the industry had been relegated to third position in terms of export earnings, between 1966 and 1977 the number of factories fell by 30 per cent and employment decreased from 72,000 to 48,000. Today the industry survives, but on a much smaller scale and its future has only been secured by adopting digital technology and by vigorous marketing methods.

Clearly, the fate of Western European regions is not solely determined by the opening up of national markets into supra-national trading blocks. It is also increasingly affected by changing world markets and the international pattern of production. As previously unindustrialised nations begin to develop their economies the parts of Western Europe which have traditionally specialised in these products will come under increasingly severe competition in both their domestic and export markets with potentially disastrous consequences.

V) POWER AND DECISION MAKING

During the course of this chapter we have introduced – either directly, or by implication – several alternative definitions of core-periphery: accessible-remote, productive-unproductive, rich-poor, strong-weak, active-passive. In reality all these definitions complement one another, since it is from their geographical accessibility that core regions are able to derive their industrial strength and prosperity. But there is yet another dimension to core-periphery relations – that of *power*. As we have seen, Paris is a core region not simply in the sense that it has captured for itself a vast share of French and European economic activity, but it is also a *command* region, one which contains the decision makers who have won the power to determine the economic fortunes of other regions. Hence the core-periphery relationship is one of dominance and subordination and it is this aspect which is of the utmost importance in the present-day world. The core region is not only in a position to benefit from economic growth; it also enjoys the power to dictate the course of that economic growth. This applies to Madrid, Athens, Vienna, Copenhagen, Stockholm and many other core regions throughout Western Europe.

Among the various reasons which might be

Box 2.4 CIBA-CEIGY – transnational corporation

Many of the typical characteristics of the transnational corporation may be demonstrated by reference to the Swiss-based firm, CIBA-GEIGY, whose headquarters are located at Basle. The scale and extent of its operations can be gauged from Fig. 2.5 and the following items taken from its 1982 company report.
- Annual sales valued at 13,808 million Swiss francs (£3,945 million) and a profit of 622 million Swiss francs (£178 million).
- A huge range and diversity of products including chemicals, pharmaceuticals, plastics, photographic and electronic equipment. The giant corporation is rarely restricted to a narrow range of products.
- Worldwide employment of 79,000 workers of all grades. Note that only 21,000 of these are Swiss and that Swiss workers are outnumbered by those elsewhere in Europe. Furthermore, the dominant geographical trend in company employment over the past decade has been a shift away from Western Europe towards the United States and the Third World, with the number of West European employees falling by over 4,000 (1973–1982), while the employment of workers elsewhere in the world rose by over 9,000.
- A multitude of branch plants and subsidiary companies located in numerous countries and regions (see Fig. 2.5 for the West European distribution). In effect the corporation embodies a core-periphery system within itself, with Basle the controlling centre ultimately benefiting from the labours of workers in all the many outlying locations. It is also, of course, determining their welfare.

Figure 2.5 Ciba-Geigy. Location of principal branch establishments, group companies and associated companies in Western Europe, 1982

Table 2.3 The largest twenty West European industrial firms, 1984

Rank	Company	Headquarters	Number of Employees	Activity
1	Royal Dutch-Shell	The Hague	163,000	Oil and petroleum
2	British Petroleum	London	143,350	Oil and petroleum
3	Unilever	London	283,000	Food products, detergents etc.
4	Veba	Düsseldorf	80,474	Electricity, lignite, petroleum
5	Cie. Français des Petroles	Paris	42,290	Oil and petroleum
6	BAT Industries	London	178,000	Tobacco, retailing, paper and packaging
7	Ste. Nationale Elf Aqitaine	Paris	58,800	Petrol, oil, natural gas and sulphur
8	Siemens	Munich	324,000	Electrical, general engineering
9	Philips	Eindhoven	336,200	Electrical and electronic products
10	Daimler-Benz	Stuttgart	187,824	Motor vehicle and engine manufacturing
11	Volkswagen	Wolfsburg	243,417	Motor manufacturers
12	Renault	Paris	214,000	Motor manufacturers
13	The Electricity Council	London	141,385	Electricity production and distribution
14	Hoechst	Frankfurt	182,154	Chemicals, dyes and plastics
15	Bayer	Leverkusen	179,463	Chemicals
16	Nestlé	Vevey (Switzerland)	152,653	Chocolate, milk and food products
17	Agip Petroli	Rome	5,950	Petroleum and petroleum products
18	Electricité de France	Paris	108,150	Electricity production and distribution
19	Thyssen	Duisburg	144,715	Steel, capital and manufactured goods
20	ICI	London	123,800	Chemicals, plastics, textiles, petroleum

cited to explain the rise of this localised economic power, one of the most important is the rise of the giant corporation. The West European economy is no longer dominated by countless small and medium-sized locally based firms. The key unit of production is now the large corporate enterprise with a stake in dozens of separate factories and other establishments, usually spread over several regions. Frequently it is *multinational* (or *transnational*) in scope, owning branches in many countries outside its home base (see Box. 2.4). Table 2.3 shows that the top 20 industrial firms in Western Europe alone control over 3 million jobs throughout the sub-continent and beyond. Furthermore, a majority of these have located their headquarters in the Eurocore. Company headquarters show a preference for large metropolitan cities, the seats of finance, trade and, in some cases, government. Such a pattern is confirmed in Fig. 2.6 which plots the location of the 100 largest industrial corporations in Western Europe. The picture is one of massive concentration of economic power, one in which the metropolitan regions – both inside and outside the Eurocore – are outstanding.

The giant corporation is hierarchical in structure. All truly critical policy decisions are taken at head office and percolate downwards and outwards through a network of national, regional and branch offices. The obvious regional significance of this is that policy decisions affecting the livelihood of workers in Palermo or Valencia are likely to be taken in remote centres like Zürich or Amsterdam;

58 GEOGRAPHICAL ISSUES IN WESTERN EUROPE

Figure 2.6 Location of the headquarters of the 100 largest industrial companies in Western Europe, 1984

INDUSTRY AND REGIONAL DEVELOPMENT 59

Self-assessment exercise

Headquarters of the 50 largest banking and insurance companies in Western Europe, 1986

Banks		Insurance companies	
Location	*No.*	*Location*	*No.*
Amsterdam	3	Amsterdam	2
Basle	1	Cologne	3
Bilbao	2	Edinburgh	1
Brussels	5	Hamburg	2
Düsseldorf	1	Hook of Holland	2
Edinburgh	1	Koblenz	1
Frankfurt	6	Le Mans	1
London	4	London	5
Madrid	3	Milan	1
Mainz	1	Munich	3
Munich	3	Norwich	1
Paris	11	Oslo	1
Santander	1	Paris	9
Stuttgart	1	Perth	1
Stockholm	3	Rome	2
Wiesbaden	1	Rotterdam	1
Vienna	1	Rouen	1
Zürich	2	Stockholm	2
		Stuttgart	1
		Turin	1
		Utrecht	1
		Vienna	1
		Wiesbaden	2
		Winterthur	2
		Zurich	3

From the information provided above, plot the location of the headquarters of the 50 largest banking and insurance companies in Western Europe. Use a different symbol to identify the two different types of company Compare your map with Fig. 2.6 and note carefully any differences or similarities in the two patterns.

or even New York or Nagasaki. It is noticeable from Fig. 2.6 that very few of Western Europe's largest industrial corporations direct their operations from any point within the outer peripheral arc, although major banking and insurance institutions tend to be rather more localised in their nature as they serve national markets.

To a great extent, it is now true to say that regional development patterns in Western Europe reflect a *regional division of labour*

(see Chapter 1) in which decision-making, management and high-skill functions are carried out by the metropolis while the routine tasks are performed by the periphery. Figuratively, the metropolitan core zones act as the heart and brains of the economic system, the periphery acts as its limbs. This unequal partnership of dominance and subordination clearly works to the disadvantage of the latter in several respects.

- *Profit transfer*: where ownership of production is concentrated in the core, then much of the profit earned by production will be transferred out from the periphery.
- *Quality of work*: because it performs mainly manual and lowgrade clerical tasks, the periphery will be able to create few rewards for the skilled and qualified. The prestigious and well rewarded tasks are disproportionately concentrated in the core. In France, for example, over a third of all professional and scientific occupations are located in Paris.
- *Dependence*: peripheral workers are extremely vulnerable to major policy decisions taken at head office.

3 Geographical variations in the core-periphery relationship

a) Problem regions of the core

Up to this point we have written as if the Eurocore was a monolithic block of regions each occupying a position of impregnable power, controlling and manipulating the West European economy and exploiting the periphery to its own advantage. In truth this is far from the case. Within the Eurocore there are now widening contrasts between those regions which continue to thrive on their centrality and those which have reached a fairly advanced stage of economic decay. Indeed there exists a belt of ten regions running across the very heart of the Eurocore from the French Nord to the West German Ruhr where by 1985 unemployment was running in excess of 10 per cent of the total labour force (see Fig. 2.7). These regions share the following characteristics which go a long way towards explaining their high unemployment.

- *Age*: they represent the earliest industrialised areas of continental Europe, the entry points through which the British innovations of the Industrial Revolution were first disseminated.
- *Coal*: they are situated on, or adjacent to, coalfields and based on industries typical of such areas.
- *Deindustrialisation*: industrial decay first became evident here as early as the 1920s. Although disguised by a lengthy period of post-war growth, decline has reasserted itself during and since the 1960s.

In practical terms these regions are kept in a vice. They are afflicted with the decline of their ageing industries – coal mining, iron and steel, textiles and (in the case of the Nord) shipbuilding. At the same time they are unable to attract sufficient new-generation replacements. Hence their unacceptably high levels of unemployment. When products become obsolete, those who make them also become redundant. During the past 30 years, countless European miners, cotton spinners and steel workers have been forced to swallow this bitter truth. The following case study demonstrates the localised effects of structural change in a region which, until very recently, would have been considered the leading industrial centre of Western Europe.

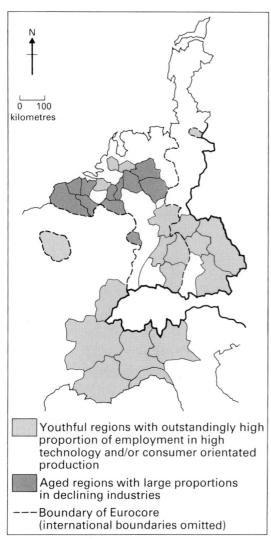

Figure 2.7 Structural change in and around the Eurocore

Case study: Structural change in the Ruhr

From an economy founded 150 years ago, based almost entirely upon coal mining and the production of iron and steel, recent years have shown a dramatic change in the industrial structure of the Ruhr region as these traditional industries have had their power and influence severely eroded. In the case of steel production, the growth of cheap foreign steel, over-production and falling sales are responsible, while the coal industry was seriously disrupted by competition from cheaper hydrocarbon fuels which began to flood the European energy market in the late 1950s. From 1957, comparatively inexpensive oil and natural gas started to replace coal as the traditional source of energy for Ruhr industry and from that time, production continued to fall dramatically from 125 million tonnes in 1956 to 70 million tonnes in 1981. Inevitably, the number of jobs associated with the industry has also declined from 393,831 to 140,536 in 1980, while the number of pits has been reduced from 140 to just 31 over the same period (Fig. 2.8). From producing 6 per cent of the world's coal in 1960, the Ruhr produced only 2 per cent by 1981. The great bulk of the closures have occurred south of the River Emscher, and nowhere is this better illustrated than at Bochum, once at the heart of the coal mining empire which employed some 65,000 at its peak; today mining employs just over 200 workers. Active coal mining is to be found in the vicinity of such towns as Gelsenkirchen and Recklinghausen, extending northwards to the semi-rural and rural districts of Dorsten, Haltern and Marl.

The iron and steel industry, traditionally the second most important industry in the Ruhr has also contracted, although the decline has been more recent (Fig. 2.8). Raw steel production started to fall in 1974 from a figure of 32.2 million tonnes to 22.3 million tonnes in 1981 and from producing seven per cent of the world's raw steel in 1960, the Ruhr's share of the world market fell to just three per cent by 1980. The 1970s witnessed a vigorous programme of rationalisation in the region which resulted in a concentration of production and investment in the areas of Duisburg and Dortmund, at opposite ends of the Ruhr, in

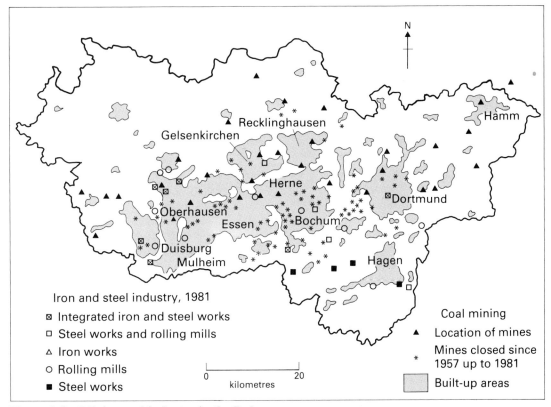

Figure 2.8 **Mining and industry in the Ruhr**

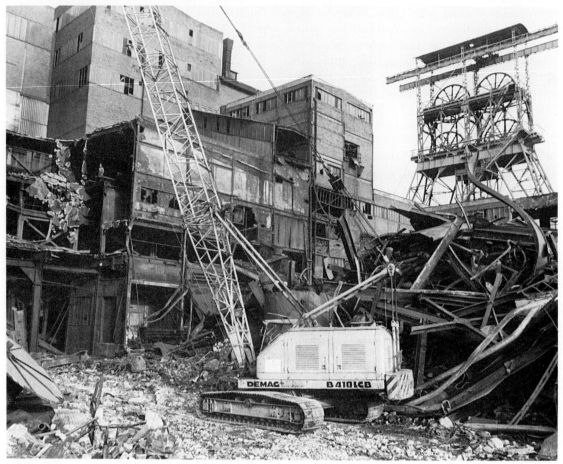

Figure 2.9 Declining industry in the Ruhr

contrast to the central areas of the region which experienced the major share of closures. Despite the slight increase in production in recent years, more than 13,000 jobs were lost in the industry between 1977 and 1980. During 1983 the problems of the steel industry in the Ruhr were further intensified with Krupp Stahl, one of Europe's largest steel makers, announcing 3,660 redundancies in the Ruhr, while at the same time more than half of the remaining workers started working short-time.

Although a wholesale decline has taken place in both the coal and iron and steel industries, employment in these sectors still accounts for 38.9 per cent of all employment in non-service jobs. Recent years have also witnessed a decline in the relative importance of other sectors of the Ruhr's economy so much so that of the ten leading industrial sectors, only three have recorded an increase in employment between 1977 and 1980, with a net loss of 35,000 jobs. Arguably, the wide range of technological advances which have occurred in areas such as the chemical, mechanical engineering and electronics industries have permitted a diffusion of industrial growth throughout the Republic and industries traditionally bound to the Ruhr for historical reasons have been able to locate elsewhere.

Although the physical environment has been improved (more often as a result of the changing structure of the economy rather than direct planning involvement), serious and, at

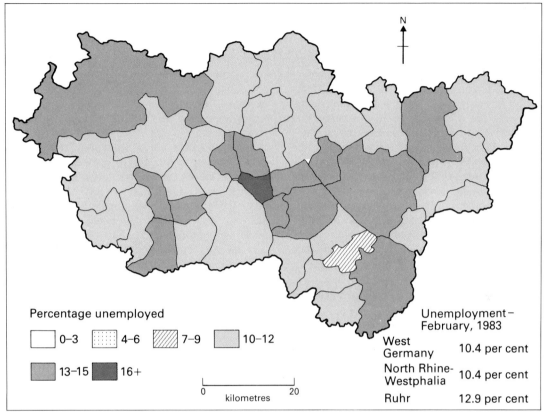

Figure 2.10 Unemployment in the Ruhr, 1983

present, insurmoutable, problems have begun to develop. Of these, unemployment is perhaps the most serious in this once prosperous industrial region, which in 1970 could boast fewer than 20,000 people out of work. By 1983, the situation was significantly more acute with over 250,000 jobless recorded. As would be expected, the spread of unemployment is uneven and important intra-regional variations (Fig. 2.10) exist, reflecting the success or failure of the towns and cities of the region to adapt to the changing economic climate. Areas along the Emscher valley – Bottrop, Gladbeck, Castrop-Rauxel and Lünen – and the towns of Duisburg, Bochum, Dortmund and Hagen exhibit unemployment rates in excess of 13 per cent whilst lower levels are recorded in the south of the region adjacent to the River Ruhr in Mülheim and Ennepe-Ruhr Kreis. Throughout the area there is further concern for the future as the number of job vacancies has continued to fall and long term unemployment has continued to increase. Furthermore, a disproportionate share of unemployment is being borne by those under the age of twenty and over forty-five, females of all ages and foreign workers (*Gastarbeiter*) drawn primarily from Turkey, Yugoslavia and other Mediterranean and European countries.

The general principle to emerge from the Ruhr case study is that modernisation – the replacement of one generation of products and technology by another (see Box 2.5) – is almost always accompanied by geographical change. In effect the pattern of regional advantage and disadvantage becomes altered. A location eminently suitable for the industry of one era may well be utterly unsuitable for succeeding generations of economic activity.

Box 2.5 Modernisation and structural change

In Table 2.2 the Industrial Revolution was presented as structural change as new technologies and products have ousted old ones. At this point it is vital to note that the Industrial Revolution was not a 'one-off' event but simply the beginning of an incessant process of changes which have continued and accelerated to the present day. Modernisation of production is a continuous process, the results of an unceasing search for greater financial returns on the part of the capitalist firm. Table 2.4 outlines the chief directions which modernisation has taken in the past and will probably take in the near future. One form of change which is of immediate concern to the regional geographer is *product innovation*, the manner in which established products are continually rendered obsolete by the development of new ones. Many now commonplace products were completely unknown in the last century, or even in the last decade (home computers, video games, video recorders, for example). At the same time, other products which were once at the very heart of the European economy – coal, steel, textiles and ships – have for half a century suffered from diminishing demand and overseas competition. Metaphorically, the process is akin to the natural cycle of life and death, in which successive generations replace one another. Expressions such as 'industrial birth and death rates', 'economic rejuvenation', 'ageing industries' and 'sunrise and sunset industries' are commonly used to describe structural change.

Table 2.4 The structural transformation of the West European economy

Dominant Emphasis	First Industrial Era (until interwar period approximately)	Second Industrial Era (to the present)	Spatial Outcomes
Energy Base	Coal →	Electricity, oil, gas and other alternatives →	Greater industrial location choice
Products	Narrow range →	Widening range – new innovations in engineering, vehicles, pharmaceuticals, consumer goods →	Need for regional adaptation
	Heavy processing →	Light assembly and fabricating →	Increasing attraction of large markets
	Capital equipment →	Consumer goods →	
	Relatively simple technology →	Increasingly complex →	Attraction of pools of skilled labour, education and research
Employment	Manual →	Professional, admin, technical, white collar and skilled →	Increased importance of metropolitan centres of office expansion
	Manufacturing →	Services	
Organisation	Family firm, small and medium sizes →	Multi-plant, multinational corporation	Enhanced power of metropolitan centres
Transport	Rail, canal →	Road. Reduced costs of overcoming distance	Greater locational flexibility

In particular the latest wave of 'sunrise' industries is extremely unlikely to be attracted to the old industrial areas created by the Industrial Revolution. The Ruhr and other coal-based regions of that older generation, with their scars of dereliction, poor housing, congested transport systems and out-dated labour skills are repellent rather than attractive to the streamlined industries of the future. As the custodians of the past, they have lost much of their former greatness as industrial leaders. Arguably they should no longer even be considered as core regions, since in an economic sense they have been marginalised, relegated to an increasingly peripheral position in the functioning of the West European economy despite their high degree of geographical centrality.

b) New industrial growth areas

The post-war period has seen a marked shift in the distribution of economic activity within and around the Eurocore, as old industries have contracted in established areas and newcomers have arisen elsewhere. In West Germany there has been a general shift southwards with the decline of the Ruhr and the rise of Bavaria and Baden-Württemberg. In Belgium, the northern Flemish-speaking section of the country has usurped the economic leadership of the southern coalfield based zone. These and other industrial leaders are depicted in Fig. 2.7 as 'youthful regions' and are generally located on, or a little outside, the fringes of the Eurocore. The following case study analyses the rapid rise of one such youthful region which, until the 1950s was virtually virgin territory in an industrial sense.

Case study: Grenoble – a youthful region

An outstanding example of spontaneous (as against state-planned) industrial expansion in post-war Europe has been the transformation of the French city of Grenoble in the region of Rhône-Alpes, from a small provincial town – noted for glove-making – into one of Western Europe's leading centres of high technology. The scale and speed of its growth has identified Grenoble as a pioneering example of the potent and powerful link between investment in research and technological progress on the one hand, and the growth and development of new industry on the other.

At first sight it appears strange that a city isolated in the Isère Valley, high in the Alps and far from the main lines of communication,

Figure 2.11 Grenoble in its alpine setting

markets and research institutions should become one of the leading industrial and scientific centres in Western Europe. Indeed, one might have expected the regional capital, Lyons, and France's second largest city, to have attracted this development, lying as it does on the main north-south axis 100 kilometres north-west of Grenoble. Yet paradoxically, Grenoble's geographical situation was a crucial factor in originating and sustaining its phenomenal growth. The reason for this can be explained by the harnessing of hydroelectric power from the surrounding mountains during the nineteenth century. This led to the establishment of a number of key engineering firms specialising in electro-metallurgy, civil engineering and the production of turbines and tubing – all of which were destined to become market leaders.

These developments established the technological basis of Grenoble's post-war development which was fostered by the growing importance of the city as a research centre in both the private and public sectors. Not only were a large number of industrial research laboratories set up, but the Science Faculty of the University was rapidly expanded during the 1960s and a number of important state research institutions were located there, notably France's principal nuclear research centre. Thus the expanding pool of highly skilled labour, the unique role in promoting liaison between university and industry and the abundance of research and development activities constituted a major attraction to advanced industries. Today, manufacturers of electrical goods, electronics and computers are of major importance having flocked not only from other parts of France (notably Paris) but also from abroad where a number of foreign multi-nationals, of which Hewlett-Packard and Caterpillar are just two examples, have successfully located branch factories. The rapidity of Grenoble's industrial expansion can be judged from its rapid population growth from 98,000 in 1946 to 471,000 in 1975. Many of the migrants to the city, being highly skilled and educated technicians, were attracted not only by its lucrative employment opportunities, but also by its avant-garde reputation and attractive environment.

In many ways the growth of Grenoble can be seen as the classical example of the *multiplier effect*, but with one significant difference. Here the principal impetus for investment and expansion came not from manufacturing but from new dynamic, job-creating sectors producing skilled white-collar employment. Such a process is viewed by many commentators as a prototype for the 'post-industrial society' underlining the significance of research and development as a stimulus to modern industrial growth – an example that planners and politicians are attempting to copy elsewhere. Futhermore, Grenoble's experience suggests that other non-economic factors can influence industrial location in the modern world, in particular the ability to attract skilled labour to environmentally attractive areas which can offer a wide range of recreational opportunities.

Despite the success of Grenoble and its immediate vicinity, there are signs that it has not escaped the effects of the recession. The flow of new manufacturing and research centres to the city has dried up and the much praised co-operation between the university and industry has also dwindled. It is apparent that Grenoble has reached saturation point in both its physical and economic development but nevertheless experience has shown that the recession has had significantly smaller impact here in a centre of advanced industry that it has had elsewhere.

c) Industrialisation on the periphery

Economic decentralisation has not been confined solely to the fringes of the Eurocore and its adjacent regions. Since the 1960s it has drawn into its orbit many outlying areas, regions on the very bottom rung of the West European division of labour. Here we are dealing not simply with peripheral regions like the Mezzogiorno, but with entire nations

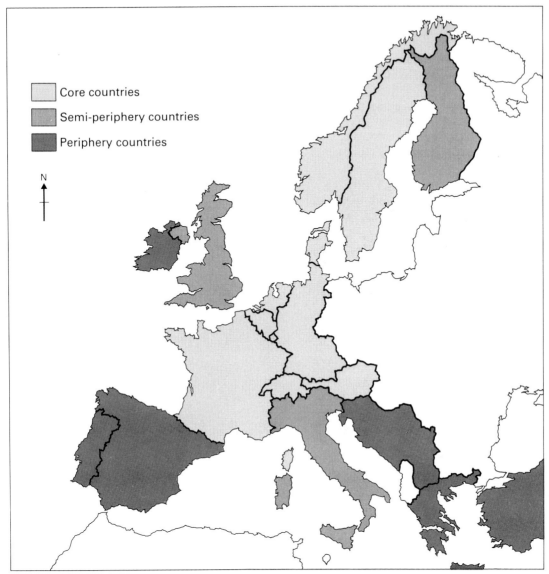

Figure 2.12 The core and periphery countries of Western Europe

whose status must be regarded as peripheral (Fig. 2.12). Included in this category are Ireland and the three most recent entrants into the EC – Greece, Spain and Portugal. These nations present something of a paradox. On the one hand they have made great economic strides in recent years, experiencing a belated industrial revolution with the rise of a wide range of new and modern industries. These have created new job opportunities for workers many of whom had little choice beyond agriculture. On the other hand, since much of this new activity is foreign owned – branch plants of companies based in the Eurocore or in the USA and Japan – the local populations are largely confined to the role of wage labour. Whether industrialisation is a net gain or loss to peripheral nations may be better assessed in the light of the following case study.

Case study: The Spanish 'economic miracle'

Until the 1960s Spain was regarded as virtually an underdeveloped country, with more in common with the Third World than with its European neighbours. Its economy was still heavily reliant, as it had been for time immemorial, upon agriculture (employing almost half the workforce in 1950) and on small scale traditional manufactures based on backward inefficient technology. Under the dictator General Franco, however, the national government launched a policy of boosting modern manufacturing by public investment in new industrial enterprises and by encouraging foreign firms to set up factories in Spain. A case in point was the establishment of the new national motor manufacturing company, SEAT in which the government held 35 per cent of the shares. Altogether the government had gained ownership of over 70 companies by the mid-1970s.

The results of this industrialisation drive were highly impressive. Between 1960 and 1973 manufacturing output more than doubled in value. The growth of many new industrial branches – chemicals, electrical engineering, vehicles and a range of consumer products – which had previously played only a minor role gave the economy a broader, more diversified and modern base. Consequently a range of new and better paid jobs became available to the Spanish worker and there was mass migration from the overpopulated rural areas to the new urban-industrial centres, notably Madrid, Catalonia and the Basque provinces. By the mid-1980s less than a fifth of the workforce remained on the land.

Closer examination, however, reveals this economic 'miracle' to have been somewhat less miraculous than official government publications would have us believe. In the first place it has been *insufficient* to close the development gap between Spain and the rest of Western Europe. For all the rapid growth in the Spanish economy, per capita income

Figure 2.13 Industrial development in Spain

remains well under half that of West Germany and France. Moreover, the creation of new jobs has failed to keep pace with the growth in the labour force and throughout the 1960s and 1970s hundreds of thousands of job seekers were obliged to migrate as temporary 'guest workers' to the countries of the Eurocore (see Chapters 4 and 6). Secondly, the rewards of the 'miracle' have been extremely unevenly shared out, both between classes and places. Salaried workers, who have been trained for the new managerial professional and clerical jobs in industry and government enjoy living standards comparable to those of their counterparts in richer Western European countries and some wage earners are also relatively well off. On the other hand the rural population live at much lower levels. Thus rural poverty and underemployment are still pressing problems, particularly in the remoter rural regions such as Andalusia in the south and Galicia in the north-west.

Since the mid-1970s the hidden weaknesses of the industrial economy have come increasingly into the open. The post-1973 recession in the French, Swiss, Belgian and West German economies meant that many Spanish migrant workers became redundant and were forced to return home. This highlighted the inability of Spanish industry to create employment for all job-seekers, a position which has steadily worsened from that point onwards. By late 1984, unemployment had risen to 2.5 millions, almost one-fifth of the workforce, a proportion well in excess of even Belgium and the Netherlands, the worst hit of the core countries of the EC. Since 1979 unemployment has more than doubled.

Entry into the EC (finally achieved in 1985) threatens still further problems. Up to that point industry had been able to rely on tariff barriers to shield it from foreign competition: indeed one of the major attractions for the foreign company investing in a new factory in Spain was that it would gain access to a large domestic market of almost 40 million people protected from outside competition. Now that producers must compete openly with other firms throughout the EC, it is feared that the disadvantages of a highly peripheral location will drive many out of business.

Our final assessment must take account of the criticism (voiced by many shades of opinion, not all of them on the political left) that the industrialisation of peripheral regions is the latest stage in a long-running battle between the trans-national companies and their labour forces. By cutting back established plant in the traditional core regions (Ruhr, south Belgium, west Netherlands) and relocating new plant in the periphery (Spain, Portugal, Greece, the Third World), the companies are seeking to boost their profits at the expense of their workers. Thus the future of Ford factories in Dagenham or Cologne is juxtaposed against that in Valencia, likewise General Motors can divide production between Luton, Frankfurt, or Zaragoza. Relocation enables them to avoid the high wage demands of powerful unions in the core regions, and at the same time to exploit new labour forces, whose material aspirations are low and whose political organisation is weak. Both sections of labour are losers in this contest, core workers through redundancy and unemployment, peripheral workers through low wages. Even though Spanish hourly wages are rising rapidly (at more than double the rate of most EC countries), the country still offers a huge pool of relatively cheap labour for trans-national companies.

4 Conclusion

For anyone concerned with the social well-being of Western Europe the message of the present chapter has not been a comfortable one. The dominant regional tendency has been shown to be the persistence of the core-periphery gap. Despite industrial and employment growth in areas such as Bavaria and the outlying regions of Mediterranean Europe, the Eurocore retains its pre-eminent position.

Such a position derives from: a) its centrality and high economic potential; b) its generally advanced economic structure; and c) its great metropolitan nerve centres with their influential decision-making institutions. Consequently, it is difficult to visualise how the remote peripheral regions are to overcome their threefold handicap of distance, structural backwardness and dependence upon distant firms and governments.

At the same time we should now be aware that economic weakness (and its attendant social problems) is not confined entirely to the periphery. The march of technology has created severe problems for the oldest manufacturing and mining regions, the traditional industrial heartlands of Western Europe. Even here, however, deindustrialisation does not stop at the coalfields. The displacement of labour by advanced technology and by relocation to the periphery is now widespread throughout the Eurocore and it is only those regions attractive to 'hi-tech' industry or high order service activities such as banking, finance, media and the professions which have succeeded in compensating for this job-shedding. Total unemployment in the Eurocore exceeded 3 million by 1984 and even Paris, a dynamic pole of attraction, was recording over 6 per cent of its labour force unemployed.

It must be acknowledged, however, that this chapter has told only part of the story – certainly a very important part, but not the whole. It has concentrated entirely on the activities of the private firm and has deliberately omitted to mention the *role of government* in regional development. Since the 1950s, most West European governments have followed active policies designed to bring new opportunities and prosperity to their peripheral regions. For many residents of regions such as Andalucia, Brittany, western Ireland and southern Italy, it is this public intervention which has held out the best hopes for a better future. It is the role of government that we turn to in the next chapter.

3 Planning and State Intervention

It is commonly accepted that one way of reducing and hopefully eliminating spatial inequalities (as outlined in Chapters 1 and 2) is the operation of an effective planning system through which state agencies influence the geography of employment, income, transport and so on. However, 'planning' is a much used and abused term which is often employed too loosely. In reality, planning is a rather narrow concept with limited powers. Yet, despite these limitations 'planning' and members of the 'planning profession' are all too often held responsible for the failure of public sector policies. Essentially planning, in its true perspective, is concerned with the control of land-uses at the urban scale and the promotion of economic growth at the regional scale. By exercising these basic principles at either central or local government level the freedom of individuals or institutions to develop where they want is controlled in the interests of society as a whole. However, despite these laudable aims, the true course of events is often frustrated by a variety of forces.

Planning is commonly regarded as a benevolent and positive activity shaping the practices of both private industry and government agencies and, in so doing, promoting social and economic well-being. However, there are occasions when state intervention can make things worse and reduce the quality of life. These often occur when one government agency adopts measures which work in the opposite direction to other policies: for example, efforts to revitalize the economy of a region heavily dependent upon shipbuilding may well be undermined by a central government decision to withdraw state subsidies to that industry in general. If we are to understand how the actions of governments and supra-national bodies (such as the EC) influence prospects and the quality of life in the various parts of Western Europe we must look beyond the narrow issues of urban and regional planning and examine the wider impact of state intervention in general. There are three different types of state intervention each operating at a different geographical scale (Table 3.1), and this chapter will focus upon the first two dimensions, whilst consideration of the urban scale will be discussed in Chapter 4 where urban problems and policies are examined together.

1 Intervention by supra-national agencies

As Table 3.1 shows, a variety of supra-national bodies have become increasingly involved in shaping the geography of Western Europe. Emphasis in this section will be given to the various actions of the EC as they are, arguably, some of the most potent forces for change in Western Europe today. However, firstly a brief overview of the activities of other bodies illustrates the scope and nature of supra-national intervention.

a) Cross-border initiatives in Scandinavia

The *Nordic Council* has introduced specific regional planning initiatives as well as taking direct action. The five countries of Denmark, Finland, Iceland, Norway and Sweden initially formed this advisory body on matters of common interest in 1953 but the role of the

72 GEOGRAPHICAL ISSUES IN WESTERN EUROPE

Table 3.1 A classification of state intervention and planning

Types of state intervention	Geographical scale of operations		
	Supra-national	*Regional*	*Urban*
1 *State planning* Attempts to coordinate the activities of both public and private sector to achieve the 'best possible' spatial arrangements of land-uses and economic activity	Rarely found, the major example in Western Europe is the Regional Development Fund of the EC. Also the Nordic Council have attempted to coordinate development	Virtually all European countries have regional planning measures to influence the level of development *between* regions, although there is great diversity in the nature and level of such activity	Almost all major urban areas are subject to regional planning policies *within* their region primarily to allocate land uses, e.g. Green Belts, New Towns, etc.
2 *Direct state investment and development* Governments provide facilities considered economically or socially necessary (e.g. road, schools, hospitals) *outside* the planning system	Relatively small compared with national schemes, e.g. aid for retraining and new industry from the European Coal and Steel Community	This is the major direct form of state intervention which has effects at both scales:	
		e.g. the construction of inter-regional motorway links	e.g. slum clearance schemes and the building of public sector housing estates
3 *Indirect state action* Taxes, subsidies, and regulations which influence the location of development but which have *no* spatial planning objectives	Increasingly important, EC activities are having far-ranging impact, e.g. the Common Agricultural Policy subsidises farming regions, also EFTA operates a large free trade area	The general activities of government can promote or retard development:	
		e.g. subsidies to industries geographically concentrated within a particular region, e.g. shipbuilding	e.g. subsidisation of public transport systems in urban areas

Self-assessment exercise

The framework given in Table 3.1 can be used to classify many types of state intervention. Into which categories do the following fall and what would be their likely impact in geographical terms?

i) EC grants to build the Channel Tunnel.
ii) A French government decision to subsidise the building of a 'stretched' version of Concorde.
iii) The imposition of much reduced fishing quotas and conservation measures on the Danish fishing fleet as part of the EC's Common Fisheries Policy.
iv) The construction of four new universities in the Ruhr region of West Germany where there had previously been none at all.

Council has been strengthened since 1971 with the introduction of a whole string of initiatives. Since that date there has been the development of a free labour market, the abolition of passport controls, a Transport Agreement, a Treaty on the protection of the environment, and the establishment of the Nordic Investment Bank, whilst more recently a Nordic regional planning policy has been inaugurated. After many years, during which time the emphasis has been on research and the exchange of information, the Nordic Council introduced a five-year programme in 1979 to enable cross-border co-operation and co-ordination of development backed by limited funding. Two types of regional action are being considered; firstly, support for public projects which straddle borders and, secondly, the direction of 10 per cent of the Nordic Investment Bank's funds for industrial

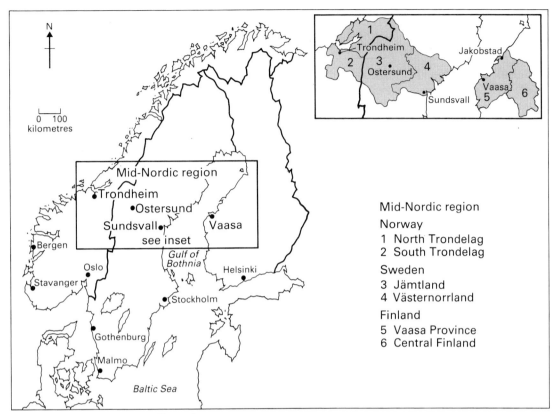

Figure 3.1 The mid-Nordic region

developments in Nordic 'development areas', which will cover the offshore islands of Greenland, Iceland and the Faroes as well as the northern parts of Scandinavia especially around the Finland–Sweden border.

An interesting small-scale development has occurred since the late 1970s to extend Nordic co-operation in more practical terms. Six regions in Norway, Sweden and Finland have come together to form the *mid-Nordic Region* (Fig. 3.1) with the aim of developing trade and industry which, for many years, has drained southwards to their respective capital cities. Furthermore, the barriers to east-west communication in Scandinavia had also produced a strange U-shaped pattern of trade whereby goods went south to Oslo, Gothenburg, Stockholm or Helsinki before coming north again. By way of illustration, fish deliveries originating in Trondheim were sent south via Oslo, Gothenburg and Stockholm before returning north to be marketed in Sundsvall – a process that did little for the quality of the product! Improvements in east-west road and rail links and better shipping connections in the Gulf of Bothnia as well as the discovery of the Haltenbanken oil field offshore from Trondheim awakened the mid Nordic Region to the potential of intra-regional trade and development. As a consequence, cross-border trade offices were set up in Trondheim, Sundsvall and Jakobstad, a daily air link was established between Trondheim, Ostersund, Sundsvall and Vaasa, regular regional trade fairs are organised, and regional banking networks are gradually being expanded from one country to another. In essence, this is small-scale action aimed at stimulating trade and industrial development across borders and it complements the

Figure 3.2 Debate in the Nordic Council

more wide-ranging initiatives of the Nordic Council.

b) The development of free-trade areas

Before moving on to the European Community, mention must also be made of the other major trading block in Western Europe – the *European Free Trade Association* (EFTA). Formed in 1959, EFTA comprised the 'outer' seven countries of Austria, Denmark, Norway, Portugal, Sweden, Switzerland and the UK (to be joined by Finland in the following year). Its aim was to remove artificial barriers on trade such as import tariffs and quotas between the member nations. This was achieved in 1967 but the organisation has since lost Denmark, the UK and Portugal upon their entry into the EC.

Commonly, one geographical consequence of economic integration such as this has been the widening of regional disparities. This has occurred where regions which are peripheral within nations become even more exposed to free market competition from growth areas located in countries elsewhere in the supranational free trade zone. Particular concern was expressed in Finland, for example, where research had shown that free trade affected some regions more than others due to variations in the composition of industry and employment. It was expected, therefore, that free trade would benefit Finland's four southernmost provinces (which already produced over 70 per cent of value added in manufacturing in 1974 – with only 59 per cent of the population) rather than the central, eastern and northern provinces – the so-called 'developing areas'. Contrary to expectations, however, growth has occurred in these developing areas where migration loss has been stemmed since the mid 1970s. Their share of total employment in manufacturing rose from 26.4 per cent to 31 per cent between 1971 and 1980, whereas industrial growth in the Greater Helsinki area declined, and an estimated 20,000 extra jobs have been created. Here then, is a case where the growth of a supra-national organisation has not resulted in the rich getting richer but has seen a reduction in regional inequalities in favour of the periphery.

c) The impact of EC policies

Table 3.1 implies that action to reduce regional inequalities by bodies at an international scale is rare. Yet Western Europe has an outstanding world example of a supranational institution (the European Community) attempting to revive the flagging economies of the poorer parts of its member nations by a wide range of regionally-based programmes. Outstanding, that is, in terms of application rather than effectiveness. The European Community has developed a series of policies which have an impact on regional affairs both directly and indirectly. These are summarised in Table 3.2 which gives examples and classifies the type of intervention according to the criteria set out in Table 3.1. All these complex and sometimes overlapping measures are designed to transfer resources either directly or indirectly from the rich core regions of the European Community to the least prosperous zones to correct imbalances in wealth, economic activity and social welfare.

Yet virtually *all* community decisions have an impact (either beneficial or damaging) for some or all of the disadvantaged regions. For example, the annual unrest shown by Community farmers prior to the fixing of food prices and support shows how critically important such intervention in the form of the Common Agricultural Policy has become – not only to the poorer, peripheral agricultural regions but also to the rich agri-businessmen of the Eurocore. Similarly, the entry of Spain into the Community brings with it potentially damaging competition to wine and citrus producers such as France, Greece and Italy. Other areas of activity such as external trade relations and industrial policy also have regional consequences. For example, coal and steel regions in the Community could be adversely affected by lower import tariffs on Polish or Australian coal or by the Commission imposing production quotas on each member state as part of an attempt to rationalise iron and steel output during the recession. Arguably, these general activities have a much more significant impact (both positive and negative) than those measures which are specifically designed to help the poorer regions as the following paragraphs and case study will show.

I) A COMMON REGIONAL POLICY?

The European Regional Development Fund (ERDF) is one of the more recent Community institutions. It was not founded until 1975 many years after the other major functions had been established. By this date it was realised that intervention on a Community-wide scale was necessary as the gap between rich and poor regions was widening dramatically, in part a response to the operation of the Common Market itself (as shown in Chapter 2). This belated reaction cannot be simply attributed to the idealistic objective of reducing socio-economic inequalities, indeed the creation of the ERDF owed much to political pressures resulting from changes which had taken place in the composition of the European Community. In 1973 the Community was enlarged to include Britain, Denmark and Ireland, and it was British membership in particular that posed a major problem in terms of how much the UK paid in budget contributions in relation to how much was received in return. In practice, Britain as a primarily industrial nation with a relatively small agricultural sector did not stand to benefit from payments from the Common Agricultural Policy sufficiently to cover its contributions to the Community's coffers. Thus, the development of the ERDF can be seen in large part as a means of channelling money back into Britain to try to 'balance the books', especially as Britain stood to gain much from a policy designed to assist decaying industrial regions of which it had many. Needless to say, the annual wrangles over 'Britain's budget rebate' has shown this to be a rather forlorn hope.

A common regional policy had long been advocated by the European Commission on the grounds that the competition between member states and individual regions to outbid each other so as to entice potential employers was extremely wasteful. Further, it was considered that the supra-national body could organise schemes that straddled the frontiers of two or more nations more effectively. It was also thought that disenchantment with the concept of European unity in the peripheral regions would quickly spiral if

Table 3.2 EC policies with a regional impact

Direct regional planning policy (specifically designed to aid less-favoured regions through planned development)	*European Regional Development Fund* (ERDF)	Established 1975, it gives financial support for regional development schemes within member states. Regions are selected as 'priority areas' and each country gets a fixed quota of aid each year to spend on schemes predominantly designed to create or maintain employment
Direct investment programmes (not exclusively regional, they have broader aims but have a significant impact on poorer regions)	*European Coal and Steel Community* (ECSC)	One of the founding institutions, it gives loans to modernise coal and steel industries or attract new jobs to coal and steel regions in decline
	European Investment Bank (EIB)	Another early EC development, it grants loans to industry and for public infrastructure (e.g. water supply schemes), many of which are part of regional development schemes
	European Social Fund (ESF)	Distributes grants for training and retraining workers unable to obtain jobs, mainly in poorer regions, often in conjunction with ECSC
	Guidance section of the European Agriculture Fund (EAGGF)	Direct measures (grants) to modernise farms, amalgamate small-holdings, improve marketing, and special assistance for hill-farmers and others in difficult areas
Indirect action (general taxes, subsidies, and regulations which have a particular impact in regions with concentrations of the activities affected)	The EC has a wide range of such measures but the following should be noted:	
	Common Agricultural Policy (CAP)	Subsidises the production of certain crops: generally regarded to benefit North European rather than Mediterranean farmers
	Common Fisheries Policy (CFP)	Regulates the amount of fish to be caught by member states by specifying the type of fish and where it can be caught
	ECSC (see above)	Sets production quotas for the iron and steel industries of member countries
	Transport Policy	Aims to improve the overall transport system by supplementing national networks with a fully integrated system

Table 3.3 Distribution of funds from the ERDF.

	Quota (% of ERDF total expenditure)	% of total EC population	Distribution of CAP receipts, 1982 (% total)
Belgium	1.11	3.6	4.3*
Denmark	1.06	1.9	4.5
France	13.64	19.9	23.0
Greece	13.00	3.6	5.6
Ireland	5.94	1.3	4.0
Italy	35.49	21.0	21.2
Luxemburg	0.07	0.1	n.a.
Netherlands	1.24	5.2	11.4
UK	23.80	20.7	10.4
West Germany	4.56	22.7	15.6

*inc. Luxemburg

Self-assessment exercise

Comparison of the quota with the percentage of total population will show deviations from an hypothetical even distribution. You can demonstrate this visually by plotting a graph with percentage of total population on the horizontal axis and the quota on the vertical axis. Draw a line at 45 degrees through the origin and this will represent an even distribution. Those countries above the line have a quota above their population share and vice versa. Italy and Greece in particular, and Ireland and the UK to a lesser extent, emerge as the principal beneficiaries. Why is this so? Draw up a list of reasons why such countries should receive a huge quota.

Unemployment is clearly a key factor in determining the amount of ERDF aid but not the sole criterion. The degree of dependency on agriculture and general level of wealth are also important as the statistics overleaf illustrate. In each country wealth and employment indicators are given for the region receiving most ERDF aid in 1982. It is evident that the situation of Greece is very different from that of the UK, that some regions with high unemployment (Belgium) or heavy dependency on farming (France) do not necessarily gain large amounts of ERDF assistance, and that wealth is perhaps the most sensitive indicator.

nothing was done to correct the ever-widening disparity between the prosperous core and declining rim of Europe. Unfortunately, the operation of the ERDF has so far failed to fulfil these commendable objectives. Funds are allocated to member states on a quota basis which takes account of the severity of their regional problems according to a formula calculated by the European Commission. Table 3.3 shows the present distribution of funds on a national scale.

This money can be spent in Regional Fund priority areas but these are the same as those regions which are designated for regional aid by national governments. In effect, the ERDF funds merely top up nationally-financed schemes as the Regional Fund only supports investments which are part of a national programme. As a result the regional distribution of ERDF expenditure closely parallels the aid given by individual countries, as shown in Figure 3.3. The major recipients are Greece, southern Italy, western and northern parts of the UK, and Ireland. In addition, substantial assistance has been given to Western France and, interestingly, to a number of the less prosperous regions in the Eurocore, notably the Saarland, northern Netherlands, and southern Belgium and Luxemburg. On the other hand, prosperous parts of the EC such as south-east England, Ile de France, coastal Netherlands, Hamburg, and northern Italy received no ERDF aid at all between 1980 and 1982. Figure 3.3 also

78 GEOGRAPHICAL ISSUES IN WESTERN EUROPE

Table 3.4 Region receiving most ERDF aid, 1982

	GDP per capita	Unemployment rate (%)	% in agriculture
Belgium (Wallonia)	80	12.8	2.9
Denmark (All)	111	10.1	7.4
France (Ouest)	90	6.4	17.2
Greece (Voreia Ellada)	51	2.9	42.1
Ireland (All)	63	10.5	17.2
Italy (Campania)	60	10.6	14.6
Luxemburg (All)	117	n.a.	5.0
Netherlands (Noord)	130	6.5	8.2
UK (Scotland)	88	10.3	4.1
West Germany (Bavaria)	110	2.3	9.4

Self-assessment exercise

Fig. 3.3 does not display the total amount of ERDF investment in each region (by proportional symbols, for example) as is conventionally done. This approach can be misleading if you wish to show the *intensity* of aid, as the amount given is not related to the population of the region. For example, between 1980 and 1982 the Flanders region of northern Belgium (population 5.64 million) received 19.4 million ECU's* from the ERDF whereas Wallonia (the southern half of Belgium with a population approaching 1 million) obtained 14.8 million ECU's of aid. Clearly the impact in Flanders with over five times as many people is much less – thus Fig. 3.3 shows the amount of grants per head of population. Moreover, the average for 3 years (2 years in the case of Greece) is calculated to help iron out any abnormal fluctuations.

The map might still be regarded imperfect as the data shown by shaded areas reflects the physical size of the region irrespective of its population size, so that larger areas assume more visual importance than small ones. How could you reconstruct the map to get away from this physical bias? You would have to ignore the conventional boundaries and create units which would result in a rather strange map of Western Europe! You might try to construct such a map using the ERDF quotas given in Table 3.3 and compare this with another one showing percentage of total EC population.

---

* ECU's are *European Currency Units*, an artificial common unit of account designed by the EC as a means of standardising payments between members. Effectively it establishes an exchange rate that applies to all EC transactions (for example in February 1984 1 ECU = £0.57). This avoids the use of the multiplicity of national currencies or other internationally accepted measures of value such as gold or silver, as each member's currency is translated into an appropriate amount of ECU's.

---

shows that the accession of Greece has resulted in a further large share of the existing pool of aid being diverted to another peripheral region.

Assistance from the other EC investment programmes listed in Table 3.2 has tended to duplicate this distribution of aid. Of the loans given by the European Investment Bank between 1958 and 1977 over 40 per cent went to Italy, predominantly to the Mezzogiorno, and other major beneficiaries have been southern and western France and the assisted areas of the UK. Nevertheless, there is some variation in the spatial patterns of assistance given under the different EC initiatives and Table 3.4 attempts to give some indication of this by showing the 10 recipients of the largest amount of investments under the three major programmes in 1980. EIB grants were again concentrated in the peripheral, predominantly rural, regions of southern Italy and Ireland, but also assisted primarily industrial areas in

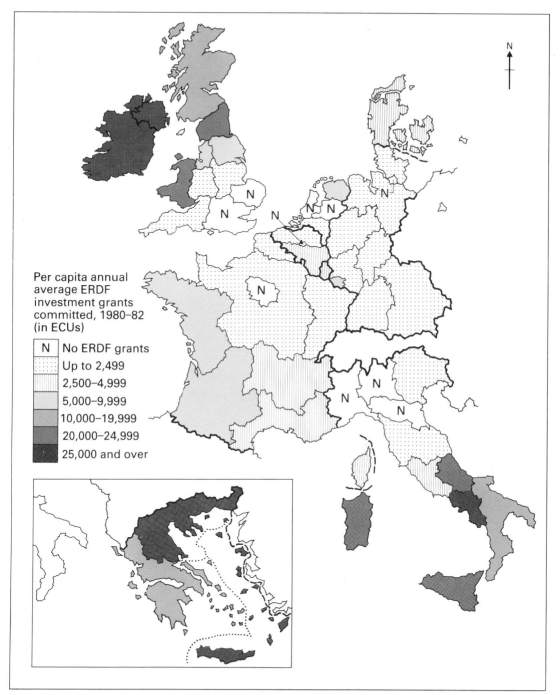

Figure 3.3 Intensity of EC Regional Aid, 1980–82

northern Italy and the UK as well as projects in the northern half of France (in Centre and Pays de la Loire). A rather different geographical pattern is seen in European Coal and Steel Community aid which naturally is isolated to the coal and steel producing

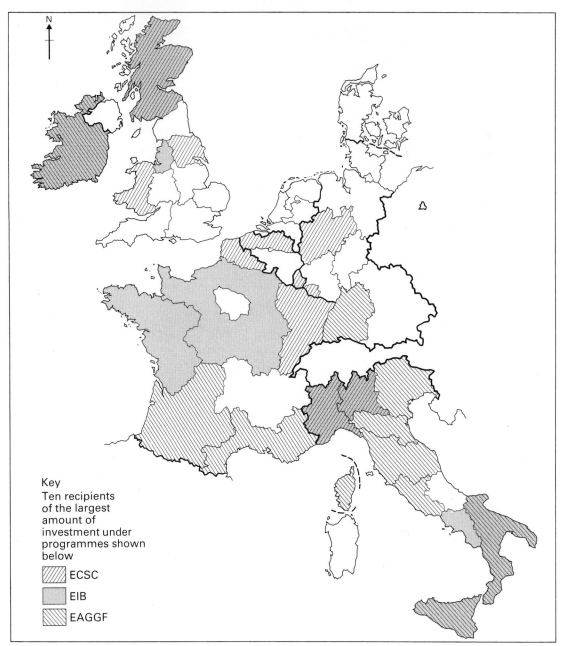

Figure 3.4 Regions gaining from EC investment policies, 1980

Note: The map treats the three programmes as of equal importance but in fact the amounts devoted to them varied markedly. The EIB was the largest fund (2703 million ECUs) followed by the ECSC programme (1017 million ECUs). Expenditure on the EAGGF totalled 299 million ECUs but this is only a small part of the amount spent on the Common Agricultural Policy and is concerned with direct measures rather than general subsidies to farmers. In comparison, 1133 million ECUs was spent on the ERDF in the year in question.

regions of the EC, many of which are in and around the core. West Germany, particularly the Ruhr area of North Rhine-Westphalia, was a major beneficiary during the 1970s but by 1980, as Fig. 3.4 shows, four other centres of assistance emerged: a contiguous area of Lorraine, Saarland and Luxemburg; adjacent areas of north-east France and northern Belgium; northern Italy; and south Wales, central Scotland and Yorkshire in Britain. Conversely, investment in direct measures under the Guidance section of the European Agricultural Fund was primarily found in the peripheral zones of Italy, southern France and Ireland, the exception being the Baden-Württemburg region in southern Germany.

Although much effort appears to have been devoted to regional regeneration by the EC, the various layers of assistance with their differing geographical distributions have merely scratched the surface of the problem. The range of programmes and the large areas qualifying for assistance have ensured that the resources made available have been too thinly and widely scattered. Moreover, the amounts involved are relatively small compared with other elements in the EC budget, especially the Common Agricultural Policy, and are dwarfed in some cases by the regional aid policies of individual countries. Future prospects will merely emphasise the inadequacies of the situation as the progressive enlargement of the Community to include Greece, Spain and Portugal has introduced even wider regional disparities. The admission of the poorer Mediterranean rural regions, in particular, has exposed differences that have made previous regional variations pale into insignificance. Existing arrangements seem totally ill-equipped to cope with this emerging problem of immense dimensions.

II) OTHER COMMUNITY ACTION
The amount spent on regional development by the EC is overshadowed by the massive sum expended on the Common Agricultural Policy (CAP). This is widely distributed throughout the Community, as shown in Table 3.3, but of course much is channelled into the peripheral rural regions rather than the industrial core. Nevertheless, because of the CAP's blanket nature the less prosperous parts of the EC do not benefit to the same extent as in other programmes which are distributed according to need.

Self-assessment exercise

Compare the figures in the three columns of Table 3.3. How far can it be said that the operation of the Common Agricultural Policy is supporting regional policy by assisting those areas in greatest need?

Clearly, CAP expenditure has profound implications for the economic and social well-being of the major agricultural areas and it has arguably a more significant impact than any regional planning policy could have. Money paid to farmers has a multiplier effect helping to support the region's economy at a much higher level than otherwise as goods and services are purchased locally, as well as keeping people on the land and preventing a wholesale drift from the countryside to the cities. Regardless of the arguments about overproduction, 'winelakes', 'butter mountains' and so on, the CAP performs an indirect but crucial geographical redistribution of income which helps to sustain the rural regions of the EC by channelling vast amounts of money into their general economies (unlike the specific job creation projects of regional planning programmes). However, its rather crude approach cannot differentiate sufficiently between the various types of farmer, so that those in most need such as the impoverished hill farmers or those on fragmented holdings in Mediterranean regions do not reap comparable benefits, unlike the already prosperous farmers in and around the Eurocore such as the grain farmers of the Paris Basin.

The impact of the CAP will be taken up in greater depth in Chapter 5, here a rather different case study of indirect action by a supra-national agency and how it affects a region's vitality is discussed.

Case study: Iron and steel in Lorraine

A relative late-comer on the scene of European industrialisation, the iron and steel industry of Lorraine, is in many ways distinct from other areas of production. It only began to develop on a large scale during the last decade of the nineteenth century following technological innovations in the 1880s which permitted the use of low-grade iron ore with a high phosphorus content. Lorraine sits astride one of the largest such ore fields in the world. However, technical problems plus the political uncertainty engendered by its frontier situation had ensured that other producers, notably the Ruhr, northern France, southern and eastern Belgium, and Britain's coalfield sites, were well established before the rise of Lorraine. These major rivals were all primarily based on deposits of coal whereas Lorraine's coal was of poor quality and much had to be imported to the region initially. Consequently, production of iron and steel came to be centred close to the ore field just to the west of the Moselle around Thionville and Gandrange with important outliers to the north-west at Longwy and further south near Nancy (see Fig. 3.5). Lorraine iron ore also helped to spawn the iron and steel industry in the Saarland, its neighbouring region across the German frontier, where coal was to be found, but once again its frontier situation slowed the Saar's growth.

Not only does the nature and development of the Lorraine iron and steel industry make it exceptional but its geographical situation was rather different from its principal rivals. It is the only large producer region to be situated in the heart of the sub-continent. Its competitors had plentiful supplies of coal at hand and many had good locations on navigable waterways which facilitated the transport of raw materials, particularly iron ore. Lorraine on the other hand was in a position of relative isolation, remote from the major

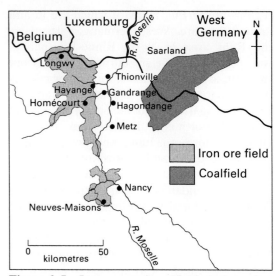

Figure 3.5 Iron ore and coalfields in Lorraine

French markets and main European coalfields, and faced with an immense transport problem. This situation has become critical in recent years as the changing geography of iron and steel production has exposed Lorraine to even greater competitive pressures. The first major change has been that the industry as a whole is no longer tied so closely to its sources of raw materials. Technological advances have resulted in the more efficient use of raw materials and in some cases coal has been replaced altogether. Moreover, low grade ores such as those mined in Lorraine are expensive to transport and compare unfavourably with foreign, high-grade ores which can be imported in vast quantities by bulk carrier. The result has been a significant move of manufacturing throughout Europe to coastal locations with deep-water facilities. In the case of France large, new integrated steel plants were built at Dunkirk and Fos (near Marseilles) – paradoxically with the collaboration of Lorraine producers – and as a result Lorraine's share of French production has dramatically fallen to less than half of its 80 per cent share in the 1950s.

Not only does Lorraine face increased competition from more modern, efficient French steelworks but also the creation of the

ECSC and then the EC brought a second major change which exposed the region to the might of other European producers, particularly from West Germany. But on the other hand, this change meant that Lorraine was no longer peripheral but was at the heart of a very much larger market area. Moreover, these moves towards European economic unity also changed the context within which the iron and steel industry of Lorraine had to operate. No longer did its entrepreneurs have to react solely to market pressures. Now they were also faced with a degree of central planning and control which attempted to fix production quotas, reduce state subsidies, and give assistance to modernise factories as well as removing tariff barriers. They became small fish in a very big pond with little power to influence matters.

How then did the iron and steel producers of Lorraine react to this deteriorating situation? The initial move in the 1960s was to introduce modern techniques, but many firms were too small, lacking the financial resources for large-scale investments that were necessary, and this led to the next stage, that of rationalisation and merger. Under this process the small, spatially-dispersed and often family firms were gradually brought together into two massive combines, SACILOR and USINOR; production was concentrated in the larger, more efficient works, as at Gandrange and Hagondange on the Moselle, and the smaller units in the confined valleys futher west (such as at Homécourt and Hayange) closed down in the rationalisation schemes. This internal restructuring gave the industry a sounder base but still could not make the industry as competitive as the most efficient, integrated iron and steel works found at coastal sites which, ironically, the two giant groups were busy constructing at more or less the same time. Thus, USINOR's works at Dunkirk prospered using iron ore from Mauritania at the expense of its other concerns in Lorraine.

However, the initial restructuring process was relatively painless compared with the crises which were to follow. By the mid 1970s the European iron and steel industry was faced with a grave problem of over-capacity brought on by the impact of recession and intense foreign competition. In the case of the French industry this was made worse by low levels of productivity (30 per cent below those in West Germany, over 50 per cent below Japan) that primarily originated from its older, less-efficient plant predominantly located in Lorraine. In an effort to rectify the situation the French government agreed to a 10 per cent cut in the labour force in 1977, most jobs being lost in Lorraine, but in the following year with the two giant steel corporations continuing to make massive losses the government again stepped in. Virtually taking the corporations over, it ordered a reduction of 22,000 in the workforce, including 16,000 jobs lost in Lorraine alone. The region faced the loss of over a quarter of its steel jobs and several plants in just two years. There was a swift and violent reaction, nowhere more so than in Longwy, a small, one-industry town of only 80,000 people distant from the main area of production, which was faced with the prospect of losing half its steel workforce of 12,700. Here dismay turned into civil unrest and violence. Only state intervention in giving extremely generous redundancy payments and government pressure on other firms (notably Renault and Peugeot-Citröen) to create 6,000 new jobs in the area by 1983 placated the protestors.

The second crisis derived more directly from the activities of the supra-national agency. By the late 1970s the European Commission began to take an increasingly interventionist role in the affairs of the iron and steel industry. Prompted by the fear that member states might revert to protectionist policies in the face of the loss of their home market to foreign competition, they produced a series of plans aimed at cutting back production in an ordered and balanced fashion, each country being given a 'fair' production quota. Moreover, the Commission began to take steps towards the banning of

Figure 3.6 Protest at Longwy

state subsidies to native iron and steel industries which they considered to distort competition. Using considerable powers under the ECSC rules rather than EC treaties, the Commission has enforced a series of cuts in production throughout the EC. These are summarised in Table 3.5 which shows the large scale reductions demanded of France between 1980 and 1983.

France then was faced with the loss of one-fifth of its steel capacity during the first half of the 1980s and this was compounded by the Commission's directive that all state subsidies must end after 1986. In response to these demands from the EC the French government ordered another round of redundancies in early 1984. Once again these centred on the out-dated steel works of Lorraine. In this round of cuts at least a further 20,000 steelworkers will be made redundant by 1987, almost a quarter of Lorraine's already depleted workforce, as part of a further rationalisaton and modernisation programme involving the investment of 15 billion francs. A planned development at Gandrange has been axed and new electric-arc furnaces will be built at Neuves-Maisons (which will lead to a reduction of jobs from 2,200 to 600) and at Longwy where the total cost is 3,000 jobs. Despite early retirement and retraining measures which guarantee redundant workers 70 per cent of their salary for two years, the reaction was perhaps predictable. At Longwy there was a re-run of the events of 1979 with violent clashes between angry workers and the police, throughout Lorraine steelworkers protested, motorways were blocked, and shopkeepers and other workers staged sympathy strikes. All this culminated in a massive protest march in Paris in April 1984 to confront the French government with the problems of the region. Their response was not to preserve jobs in the steel industry (which had lost 10 billion francs in 1983 alone)

Table 3.5 EC steel production figures and quotas, 1980–85

	Production in 1982 Total*	(% EC total)	Closures 1980–3*	Proposed cuts 1984–5*	Total reduction 1980–5*
Belgium	16.0	(9.5)	1.7	1.4	3.1
Denmark	0.9	(0.6)	0.1	—	0.1
France	26.9	(15.9)	4.7	0.6	5.3
Italy	36.3	(21.5)	2.4	3.5	5.8
Luxemburg	5.2	(3.1)	0.6	0.4	1.0
Netherlands	7.3	(4.3)	0.3	0.7	1.0
UK	22.8	(13.5)	4.0	0.5	4.5
West Germany	53.1	(31.6)	4.8	1.2	6.0

* in million tonnes

but to make a series of conciliatory gestures including designating Lorraine as one of the 15 industrial 'poles of attraction' and pressuring state-run industries to develop there.

What then does this case study tell us of the nature of supra-national intervention and the fate of European regions? The power and influence of the EC is shown in two ways; firstly, the *indirect* measures which abolish protective tariff barriers and expose industries to the full glare of international competition, and secondly, the *direct* intervention which controls the output of particular industries. Both of these have proved such potent forces that even powerful national governments are reduced to the role of acting as agents to implement the policies and then attempting to mitigate their worst effects. The end product is of course the social and economic crises experienced by certain regions due to the geographical concentration of vulnerable sectors of the economy there. Often termed industrial restructuring, it is misleading to attribute this solely to market forces. State agencies at supra-national, national, and regional level have increasingly become involved in the painful processes of adjustment. They have acted variously as the architects of change, the agents to implement policies decided elsewhere, and the means by which some of the resultant social hardship can be attacked. The move towards 'giantism' in both industry and government has meant that the fate of regions such as Lorraine is no longer in their own hands. Their future is no longer decided in Nancy and Metz, or even in Paris exclusively. Today decision-makers in Brussels are playing a crucial role in determining the wealth or poverty of regions throughout Europe.

2 Regional planning within Western European states

Virtually every government in Western Europe has felt obliged to intervene in the workings of their national economy with the specific objective of reducing the inequalities between the rich and poor regions, be they Paris and Brittany, Vienna and Bürgenland, Stockholm and Lapland, or Lombardy and Sicily. The mechanism adopted is usually referred to as regional planning but care must be taken to clarify this concept given the plethora of different forms of planning variously termed urban and regional, town and country, urban and rural, national and spatial, and so on. Basically there are two forms of regional planning which operate at different spatial scales:

- **Inter-regional planning.** Often termed regional policy, it is designed and implemented by central governments. It is primarily economic planning which aims to distribute resources (primarily jobs) more fairly *between* regions so that deprived areas get greatest help. Public sector investment is channelled into these regions and private industry is enticed

(and sometimes directed) away from prosperous areas.
- **Intra-regional planning**. Designed and implemented by urban or regional authorities, it is basically concerned with physical planning (i.e. allocating land uses) *within* a particular region using instruments such as green belts and new towns. It has mainly been associated with planning the expansion of large city regions.

The two types of regional planning are quite distinct in terms of their aims, the scale at which they operate, and the processes involved. Here we will be concerned with inter-regional planning, whereby national governments seek to influence the economic prosperity of the different parts of their country at the regional level. The case of intra-regional planning will be taken up later in Chapter 4 where it will be viewed in the context of shaping change in the large city regions of Western Europe. Clearly, a review of the regional planning systems in each country in Western Europe is not possible here but a brief overview will be given and this will be followed by a case study of the classic example of inter-regional planning in the Mezzogiorno region of southern Italy.

a) Variations and trends in regional policy

The overall picture is one of great diversity with each country having its own particular blend of regional policies using different types of measures in response to a variety of regional problems. These range from frontier isolation to rural underdevelopment, from industrial decline to capital city dominance. Given that a country such as France can suffer from a combination of all four of these interlocking problems (in Alsace-Lorraine, Brittany, the Nord, and Paris respectively) this only adds to the complexity. Table 3.6 attempts to isolate the four main factors behind these variations.

Moreover, the situation is not static, the countries of Western Europe began regional planning initiatives at different periods of time in response to the emergence of particular problems. So, for example, high unemployment in the 1930s initially prompted action in Britain, whereas French measures only gained impetus in the early 1950s following the exposure of the dominating role of Paris. On the other hand, agricultural underdevelopment in southern Italy and its associated political, social and economic crisis underpinned the regional development programme for the Mezzogiorno from 1950 onwards. However, Dutch regional planners only began to attack the problems of urban expansion into rural areas on a concerted basis during the 1960s.

As agricultural and industrial change has gathered pace so the nature of the problems perceived by planners and politicians has evolved and accordingly the measures introduced to remedy them and the areas to which they apply have changed also. Nevertheless a number of general points can be made. Normally planners have tried to create growth in the deprived regions rather than encourage people to migrate to areas of greater opportunity. This doctrine of taking work to the workers is justified on social grounds as it seeks to prevent the upheaval of moving home and leaving family and friends. Such migrations gradually drain the poorer region of the younger, more dynamic and skilled elements of its population further isolating the disadvantaged left behind. The measures introduced have generally been a combination of the 'stick and carrot' to try to get the private sector to invest and develop in deprived regions by dangling an attractive incentive in front of them in the form of generous tax relief, low-interest loans, grants and so on. In some cases this approach has proved ineffectual and has been supplemented by tougher measures which prevent further growth in prosperous areas but grant permission for development in areas of need. Furthermore, as governments have become major employers themselves they have tried to decentralise their own office activities,

Table 3.6 Variations in regional planning systems

Nature of political system	• Centralised state	(e.g. UK, France, Sweden) National government designs the policy and administers it under tight control from the centre, usually a capital city that dominates the economy
	• Federal system	(e.g. Switzerland, Austria, West Germany) Policies result from negotiation and bargaining between central government and individual regions, which have considerable power and autonomy
Perceived problems	• Declining industrial regions	(e.g. southern Belgium, north-east France, the Ruhr, the UK coalfields) They suffer from high unemployment as traditional industries fail (such as coal, iron and steel, textiles, ship building, etc.), a polluted environment, and an inability to attract new growth industries
	• Backward agricultural regions	(e.g. the Mezzogiorno, Greece, Ireland, northern Scandinavia) A preponderance of small, marginal farms accompanied by a drift from the countryside, high unemployment, low incomes, and little industrial potential due to remoteness
	• Capital city dominates the provinces	(e.g. Paris, London, Copenhagen) Power and decision-making is centred here, draining the provinces, as activity is concentrated in the core, but at the cost of high prices, congestion, and inner-area decline
	• Isolated frontier regions	(e.g. Saarland, Alsace-Lorraine, eastern parts of West Germany, eastern frontiers of Austria) Peripheral to national economies, frontier acting as barrier to trade and movement
Measures introduced	• *Direct state investment* in particular industries (often propping-up ailing concerns) or in the form of control of public sector employment (e.g. decentralisation of government offices) • *Incentives to private industry*, encouraged by grants, tax rebates, low interest rates, and purpose-built factories to locate in ailing regions which they would otherwise not consider suitable • *Preventive legislation* to discourage further industrial expansion in prosperous regions and to direct it to deprived ones • *Creation of infrastructures* to help improve the accessibility and attractiveness of regions by constructing motorways, ports, airports, and so on	
Areas covered	• Extensive	Often a hierarchy is established with the poorest areas getting most aid and progressively less assistance is given as the severity of problems decreases
	• Concentrated	Investment is concentrated where it is likely to have greatest success. Often associated with the concept of growth poles (geographical foci for development)

invest in facilities such as motorways, and pressurise state-controlled manufacturing industries to help deprived regions by building new factories in places where they might otherwise not do so. All these factors have resulted in increasing and more complex government intervention to try to prop up ailing regional economies.

How then do these factors translate into regional planning programmes? This is best illustrated perhaps by comparing two examples taken from the extremes of a centralised political system and a decentralised federal state. Thus, the following sections briefly outline the origins and nature of regional planning in France and West Germany.

b) France: planning from the centre

The exposure of the dominating role of the

French capital by Gravier in his book *Paris et le désert français* published in 1947 was the primary impetus to the creation of a regional planning system. However, it is not suprising given the central role of Paris dating back to the Napoleonic era that the French chose to stimulate regional development not by giving more power and autonomy to the provinces but by setting up a central government agency to control the whole process. From 1963 DATAR (*Délégation à l'Aménagement du Territorie et à l'Action Régional*) has designed policies to counterbalance Parisian dominance by planning urban growth and encouraging the decentralisation of industry as well as producing strategies for rural redevelopment. Typically this key body was a small 'think tank' comprising a group of less than 100 'technocrats' at the core of the government which devised policies and coordinated their implementation by the various ministeries, which were eventually transmitted downwards to the regions. One of the major elements in DATAR's plans has been the creation of eight *métropoles d'équilibres* strategically located to try to cover the whole country to act as growth poles to siphon off development which otherwise might have gone to Paris (see Fig. 3.7). The metropoles are large provincial cities as it was felt that these had the greatest chance of success. In addition, controls were established on new industrial growth in and around Paris and financial encouragement given to relocate in a wide range of problem regions. Rural renovation regions were also created (mainly in Brittany and the Massif Central). The overhaul of this system was a key element in the programme of the socialist government of President Mitterand, who, in 1982–3, saw the need for greater regional autonomy and introduced wide-ranging reforms to decentralise power to the regions.

c) West Germany: planning in a federal state

Just as the imprint of the Napoleonic era is a crucial factor in the government of France, so historical factors are vital in the understanding of the present West German system. Following the Second World War the occupied country was reconstituted as a federal state comprising 11 *Länder* (provinces) each with considerable powers and autonomy. This was done partly for political reasons, primarily to avoid the formation of a strong central government, but it also fitted in with the long tradition of Germany as a conglomeration of small states. Planning in particular was tainted by its association with the totalitarianism of the Nazi regime, and accordingly it was relegated to purely local matters at first so that it was not until the 1960s that regional initiatives developed. These have not been primarily concerned with the reduction of inter-regional

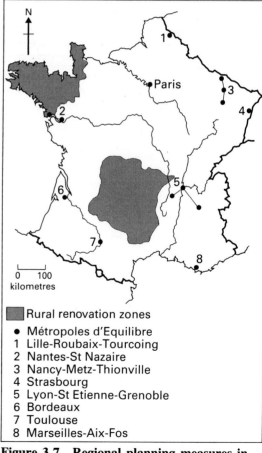

Figure 3.7 Regional planning measures in France

notably the Saarland and northern parts of the Ruhr. In addition, the frontier zone (*Zonenrandgebiet*), a 40 km wide corridor along the entire eastern frontier, also receives investment incentives. Recent co-operation between the Federal government and the *Länder* has produced regional action areas in predominantly the same regions shown in Fig. 3.8. where government assistance to industry is concentrated.

Self assessment exercise

Go back to Table 3.6. and, using categories shown there, draw up a list of similarities and contrasts between the French and German regional planning systems.

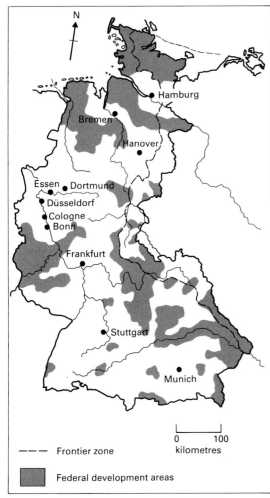

Figure 3.8 Regional planning measures in West Germany

disparities; the *Länder* have made considerable efforts at the co-ordination of development within their regions through intra-regional planning and the Federal government has little power to intervene in the affairs of its member states in this sphere. Nevertheless, a number of programmes have been introduced, as shown in Fig. 3.8. These cover the development areas which are primarily isolated rural areas suffering from underdevelopment such as Schleswig-Holstein, eastern Lower Saxony, the Eifel, a central zone along the Eastern Frontier and Bavarian Forest. Some industrial areas were later added,

The contrasts between the French and West German examples to some extent illustrate the dilemmas governments face when they seek to attack the problem of regional inequalities. They have firstly to identify the nature of the problem and then decide which measures are best suited within the limits imposed by their political and economic systems. A strong system of state intervention would be unpalatable to Germans given their recent history and thriving private sector, whereas the centrist tradition in France and the striving to achieve industrial modernisation made regional planning more acceptable. Moreover, the 'economic miracle' of West Germany made it seem that nothing much needed to be done to help the weaker regional economies, whereas the advanced process of agricultural and industrial decay in parts of France (together perhaps with the stirrings of regional militancy) in contrast with the success of Paris made intervention a priority. Similar differences can be observed throughout much of Western Europe and these help to explain the diversity of regional planning policies.

The most important question concerning regional planning, however, is: have all the efforts of national governments had an impact in reducing regional inequalities? Many millions of francs, marks, lire and guilders have been expended in the cause of regional development and yet as Chapter 1 has illustrated the gap between rich and poor parts of Europe is still disturbingly wide. One way of analysing this issue in detail is to examine one particular region, the Mezzogiorno, which has been the focus of strenuous efforts during the last 30 years. In many ways southern Italy is the classic testing ground for regional planning. It is a region which is peripheral in both economic and geographical terms and which is shunned by private investors, so that state aid and intervention are its only hope for salvation. (Chapters 1 and 2 have already presented an analysis of its problems and their causes in the context of geographical change in Western Europe see pp. 23–27, 51–52.)

Case study: The crusade for regional development in the Mezzogiorno

Regional planning initiatives in Italy have focused on the fundamental division of the country between the prosperous northern areas and the chronically disadvantaged southern half of the nation. It seemed that the unity of the country was threatened during a brief period of unrest in southern Italy following the Second World War and the state responded by instigating urgent action to attack the severe social and economic problems of the south. The cornerstone of the policy was the *Cassa per il Mezzogiorno*, a special government agency set up in 1950 outside the normal ministries given special powers and funds, and charged with reducing the gap between north and south. Its activities were concentrated in the area, christened the Mezzogiorno, which effectively encompassed the poverty-stricken regions of mainland Italy to the south of Rome plus the islands of Sicily and Sardinia (Fig. 3.9).

Initially the *Cassa* concerned itself primarily with agricultural improvements, over three-quarters of its budget was devoted to these schemes to modernise farms and to control the natural environment (irrigation schemes and erosion control). This was not surprising given that agriculture was the Mezzogiorno's dominant economic activity (56 per cent of the active population were in agriculture in 1951) but doubts began to be expressed about the widespread geographical distribution of assistance to areas where it had little likelihood of success and it was increasingly claimed that it should be concentrated in those parts with the greatest potential for development. Concern was also voiced over the political exploitation of funds to win votes, with instances such as the expenditure of half a million pounds on the construction of a football stadium in Naples, and how it was benefiting large landowners rather than the impoverished peasants. Consequently, in 1957 the *Cassa* changed direction and switched its attention to the promotion of industrial development as it was considered that this was more likely to lead to real and permanent increases in the standard of living in the south. In a two-pronged programme generous incentives were given to private industry moving to the south and state-controlled industry was obliged to

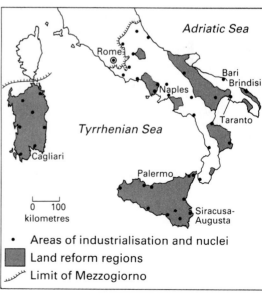

Figure 3.9 The Mezzogiorno

direct at least 60 per cent of new investment to the Mezzogiorno. Industrial development was carefully located in the larger, more accessible urban centres in line with the concept of growth poles, which are shown in Fig. 3.9. Once again, it appears that the 42 industrial areas and nuclei designated were partly a product of political patronage and proved too numerous to succeed. Although in some cases it seemed that some locations would prove successful, particularly the Bari-Brindisi-Taranto triangle, in very few cases did the initial firms create a significant multiplier effect and few local industries developed. Moreover, the majority of new factories were large, capital-intensive concerns which employed comparatively little labour, often branch plants of firms based in the north or multi-national corporations which prove most susceptible to closure in a recession.

Two further initiatives of the *Cassa* should be noted: the improvement of infrastructure, particularly roads, and the promotion of tourism. In a sense these are closely interrelated. For example, the construction of a major motorway, the Autostrada del Sole, broke the isolation of the south, encouraged the location of industry in the northern part of the Mezzogiorno, and opened up the south to tourists from all over Europe. Railways, other roads, and ports have been improved with aid from the *Cassa*, which has also financed in part the construction of 14,000 hotels.

Another form of Italian state intervention must also be noted. The complementary process of land reform was begun in the early 1950s at the same time as the *Cassa* commenced its work. For many centuries large landowners had rented out small parcels of land to peasants on a piecemeal basis with no security of tenure and this had led to small, fragmented holdings of low productivity. Government action was prompted by outbreaks of violence in the south and it involved the forced purchase of land from the estates of the large landowners, who retained some land and received compensation for land that was

Figure 3.10 Autostrada del Sole

taken. The land thus gained was then redistributed to peasants without any land or who farmed the smallest plots in order to create farms large enough to support a family. Land reform was not the only aim; farmhouses were built, irrigation schemes introduced, and co-operatives were formed. This process took place in eight specially designated areas, six of which were in the Mezzogiorno and these are shown in Fig. 3.9. The overall results of the scheme have proved very impressive but at great expense. Over half a million hectares of land have been redistributed in the Mezzogiorno and over 90,000 southern families received improved farms in this way. However, the geographical impact has been patchy with some very successful schemes such as that along the Ionian coast around Taranto contrasting markedly with the failure of reform in central Sicily, where Mafia involvement prevented any progress. Although not strictly part of the regional planning process, these agricultural reforms have played an important role in the regional development of southern Italy. They stabilised a situation of great unrest, helped keep peasants on the land and improved productivity in the more successful areas, thereby complementing the efforts of the regional planners of the *Cassa*.

The quest for regional development in the Mezzogiorno has consumed an immense amount of money, as the Cassa has expended £20,000 million between 1950 and 1975 drawing not only on Italian sources but also on the European Investment Bank and the World Bank. This figure is approximately the equivalent of £1,000 for each person living there. Has this money been well-spent and has any appreciable reduction been seen in the gap between north and south? Superficially much seems to have been achieved. Agricultural and industrial developments have changed the geography of the south favouring the valleys and coastal plains at the expense of the marginal, isolated uplands, new transport networks have markedly improved accessibility, and much of the absolute poverty has disappeared or been confined to small pockets in cities such as Naples. A whole battery of statistics have been put forward to demonstrate the progress made; for example, in 1950 agriculture employed 57 per cent and industry 20 per cent of the active workforce, whereas by 1970 the figures were 33 per cent and 32 per cent respectively. Similarly in terms of wealth, average income per head in the south has risen from 41 per cent of that in the north in 1951 to 59 per cent in 1971. So clearly, significant changes have been brought about and the quality of life has improved markedly. Yet some of these changes are illusory and the deep-seated problems remain. The perennial drift of population leaving the south to find employment elsewhere has continued; 4.5 million people emigrated between 1950 and 1976 mainly to the industrial conurbations in the Eurocore. This makes the situation look better than it is as, for example, although 1.1 million new jobs were created in the south between 1951 and 1969 some 1.7 million workers left the region. The growth rate of various socio-economic indicators also shows that the south had apparently out-performed the north in many instances but the absolute gap has increased because the relative percentage increase is based on a smaller base population. By the 1980s fundamental differences in the quality of life remained. Figure 3.11 shows that in terms of average wealth (measured by GDP per capita) the southern regions are still at a level of approximately half that in the north and that the pattern of unemployment and infant mortality still shows a large gap between north and south.

Such evidence has led some to conclude that all this effort has done little to solve the underlying problems of the Mezzogiorno. They cite the fact that the new industrial plants have been large, capital-intensive, externally-owned enterprises which have failed to generate locally-based industrial growth and which are very vulnerable as branch plants. Similarly, agricultural improvements have produced crops such as citrus fruits which are

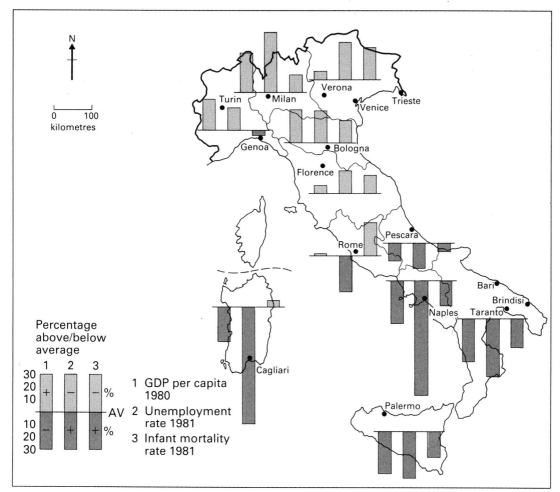

Figure 3.11 Regional inequalities in Italy

Note: The map depicts variations from the national average for three indicators; wealth (**measured by gross domestic product per capita**), employment (**the average unemployment rate**), and social conditions (**the infant mortality rate**). Positive attributes are shown above the line (i.e. higher than average GDP, below average unemployment and infant mortality) and these effectively pick out the five northern regions. The transition is seen around Rome and this grades into the Mezzogiorno where all the regions have overwhelmingly negative characteristics dipping below the line (i.e. lower than average GDP, higher than average unemployment and infant mortality). The overall effect is to highlight the sharp and continuing dichotomy between the two parts of Italy, despite the vast efforts made to reduce such regional inequalities during the last four decades.

not covered by the Common Agricultural Policy and which are in strong competition with other Mediterranean countries such as Spain. One conclusion is that the overall process has been one of 'modernisation without development' whereby the vast inputs of money have helped to improve the outward appearance of life but have failed to change the shaky economic foundations on which it is based. It is indisputable that regional planning has helped to improve the quality of life in the Mezzogiorno but it has failed so far to generate self-sustaining growth and solve the underlying problems despite more than 30 years of strenuous effort. The Mezzogiorno remains as vulnerable today, dependent on outside funding to maintain its livelihood, and the gap between north and south has not closed appreciably.

Self assessment exercise

Given the bleak conclusion arrived at above, can you put forward some suggestions for future measures? Will the continuation or intensification of present policies achieve success or are alternatives needed?

d) The effectiveness of regional planning policies

In trying to reach an overall verdict on the role and effectiveness of regional planning in Western Europe one is confronted with a fundamental problem. In evaluating the policies it is often impossible to say what would have happened without intervention, and yet this is the measure against which the results of regional planning should be compared. Nevertheless, the continuance of wide and deep-seated disparities between the rich and poor parts of Western Europe must cast doubts on its ability to significantly reduce such inequalities despite progressively large areas benefiting from increasingly large amounts of government assistance. Moreover, in some European countries, notably the UK and the Netherlands, there are signs of a down-grading of regional initiatives and a movement towards policies designed to remedy inequalities at a different spatial scale – the inner areas of large cities.

Generally academics put forward one of three diverse verdicts on the value of regional policy. The first is associated with advocates of market forces as a means of equalising disparities. It argues that state intervention distorts the free market and that industry should be allowed to locate where it wishes, arguing that to make it move elsewhere would reduce its profitability so that it becomes susceptible to closure and is ultimately self-defeating. If free market forces lead to concentration in the core it is argued that in time rising costs and congestion will render such core areas uncompetitive, and gradually industry will move out to the cheaper locations in peripheral regions. However, such ideas ignore the social costs of allowing market forces full rein, not simply in terms of unemployment and social security payments in deprived areas, but in terms of disruption involved in migration and the waste of public investment in under-used infrastructure (roads, housing, schools, etc.) in regions lacking development. Such problems are, of course, very difficult for governments to ignore. Moreover, in economies where market forces have been allowed to operate in a more or less unrestricted fashion (such as the USA) there is little evidence of new industry moving out of the metropolitan cores to recolonise depressed industrial areas, rather it tends to relocate in new situations, often semi-rural with environmental attractions. The growth of cities such as Grenoble (see Chapter 2) would tend to support this argument in Western Europe.

The second verdict is more 'middle-of-the-road' and stresses the improvements that have been made in terms of the creation of new jobs, transport improvements, agricultural development and so on in peripheral regions, whilst at the same time recognising that the problems have not been solved. It is frequently associated with the idea of 'running in order to stand still', in other words the underlying job loss in traditional economic activities has continued apace and all that new development has done is to top up this situation. Nevertheless fundamental to this idea is that in a mixed economy the state can intervene in a worthwhile and effective manner to mitigate the worst effects of market forces. Its proponents argue that the situation in the disadvantaged regions would have been much worse without such remedial action and that far from denigrating such achievements people should praise the efforts. Nevertheless, they are criticised from both extremes of the political-economic spectrum; from the right for interfering with market forces and reducing efficiency, and from the left for failing to change the fundamental inequalities in society and being unable to control the activities of large multi-national corporations.

The final verdict is a more radical one which regards regional planning in Western Europe as merely 'window-dressing' to mask the continuing inequality in society. It is a sop, if you like, designed to appease the working classes, to convince them that something is being done to remedy their plight, but at the same time doing nothing to challenge the dominance of capitalists, the ruling classes. Central to this verdict is the concept of the reproduction of labour power, whereby peripheral regions with government assistance in the form of regional planning, housing, education, and health programmes reproduce vast amounts of labour not in the physical sense but in the sense of providing a large, able-bodied, skilled and amenable labour force at low cost (see Box 3.1). It is argued that regional planning can only be effective in a completely different form of society where the state plans centrally along socialist principles so that each part of the nation gets an equal share of resources. There is, however, a paradox that emerges. In recent years right-wing governments in Western Europe have begun to cut back on programmes such as regional planning in a process termed 'rolling back the state' aimed at revitalising the private sector. Yet these governments are the very ones that would usually be considered closest to the ruling classes and most supportive of the needs of capitalists. The apparently contradictory behaviour appears to raise the issue that public expenditure cuts would not be made by right-wing governments if they consciously only acted in the interests of capitalists.

Box 3.1

The *reproduction of labour power* interprets government action in terms of its effects. By providing state housing, education, transport networks and health care facilities governments meet the social and some of the economic costs of industry – if firms had to provide housing and schooling for their workers and if they had to build road and rail links to move their products it would greatly increase their costs. As well as providing a subsidy to industry this has a further benefit as it provides a healthy, educated and contented work force. A geographical dimension can be added as peripheral underdeveloped areas (such as Portugal or southern Italy) or declining industrial regions (such as the Nord region of France and parts of Belgium) can be seen to provide a large reserve of such workers available either to incoming firms or for migration to the prosperous areas of the Eurocore. Thus, regional planning policies can be seen to be part and parcel of this process as they effectively subsidise firms who move to regions designated for assistance either directly through grants and loans or by povidjng public facilities such as motorways, universities or airports.

Rather than prescribing which verdict is correct we leave it to you to evaluate each of them bearing in mind the material presented in this chapter, particularly the two detailed case studies. It may be that you can find evidence in them to support each of the three verdicts and at the end of the day it will be your own values and political ideology that will lead you to conclude that one explanation is better than the others.

3 Other forms of state intervention

As with supra-national agencies, national governments find it increasingly necessary to intervene in the workings of the economy in all sorts of ways and many of these have a powerful effect on individual regions although they fall outside the sphere of regional planning. The case study of the Mezzogiorno showed how a land reform policy has a significant regional impact in a positive way, whereas in the case of Lorraine its economic well-being was adversely affected by decisions made in Paris concerning the development and later restructuring of the iron and steel industry. Virtually all governments are major employers as the public sector has grown and in recent years many have been forced to take over or nationalise their ailing industries (for example, car manufacturers such as British Leyland and Renault) which has the effect of sustaining employment in the regions where they are localised. Increasingly governments have themselves become more decentralised and have also pressurised state-controlled industries to expand in ailing regions, especially when the private sector has shown a distinct reluctance to accept the regional planning incentives on offer. In these ways governments can act in both a positive and negative way and the fortunes of regions can be affected to a much greater extent than by the policies strictly concerned with regional planning.

Moreover, the seemingly-neutral activities of government such as tax gathering can also have a significant geographical impact. Most tax is raised from the wealthy and those in work who tend to be concentrated in the prosperous core areas, whereas those dependent on unemployment and social security payments frequently live in the ailing peripheral zones. In effect, the government can gather more money than it spends in one part of its territory and transfer to another region where its net receipts are negative. The result is to bolster up an ailing regional economy through the operation of a welfare state. These activities are known as transfer payments and are part of a complex subject known as regional accounting. We cannot give a full breakdown of such transfers in a national economy here but the following brief case study illustrates how the process operates.

A striking example of transfer of public funds between regions is provided by Belgium, a nation of two distinct halves where Dutch-speaking Flanders has coexisted uneasily with French-speaking Wallonia ever since the formation of the nation state in 1830. As Fig. 3.12 underlines, Wallonia (which includes the South Belgium coalfield, see Chapter 1 pp. 28) is economically much the weaker of the two. Yet this automatically means that Wallonia pays less tax (due to lower incomes)

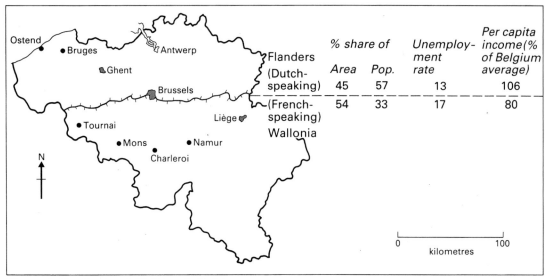

	% share of		Unemployment rate	Per capita income (% of Belgium average)
	Area	Pop.		
Flanders (Dutch-speaking)	45	57	13	106
(French-speaking) Wallonia	54	33	17	80

Figure 3.12 Regional divisions in Belgium

and receives more in social benefits (because of higher unemployment). In a country where average unemployment pay was over £60 per week in 1980, this last item alone represents a very sizeable inter-regional transfer. According to economists at the University of Louvain, inter-regional transfer payments exceeded £1,500 million per year in the mid-1970s, almost £5,000 for every man, woman and child in Wallonia. In Belgium, these regional 'subsidies' are used as political ammunition by those who would like to see the country split into a two-state federation. Even so, they are by no means unique to Belgium but are universal throughout Western Europe, part and parcel of the operation of welfare capitalism.

4 Conclusion: state intervention and regional planning in retrospect

What conclusions can be drawn from this examination of state intervention and regional planning? As Western Europe is moving from an industrial to a post-industrial society so the need for government to intervene has increased but so has the difficulty in formulating effective policies. Agricultural and manufacturing employment has fallen dramatically and given its uneven geographical distribution, this has left many regions over-dependent on these declining sectors. Private industry does not face the social responsibility for its actions and so the state steps in and regional planning is one weapon in its armoury. In most people's eyes, the issue is not really whether there should be regional planning but rather how to make it more effective. The move to office and service employment, to 'high-tech' industry, and to 'giantism' in the economy (with the dominance of multi-national corporations) are all presenting new challenges to regional planners and governments who are in most cases still grappling with the problems of the previous age. Planners simply do not have the powers to bring about their wishes in many cases and governments frequently indulge in actions that make matters worse when faced with international pressures. The pragmatic conclusion is that in the mixed economies of Western Europe the state will continue to intervene to patch up the inequalities that arise. Even the most right-wing governments have maintained some form of regional aid and nowhere in Western Europe has moved to a strictly centrally-planned economy. The uneasy relationship between the economic imperatives of the free market and the social responsibility of government will continue to produce ailing regions that can only look for assistance to state intervention in its many forms.

4 Problems of Urban Development

In September 1984 the *Guardian* newspaper reported that tourists in central Amsterdam had suffered repeated attacks by the squatter community of that city – homeless people, mostly young, who illegally occupy inner city houses and flats, often awaiting conversion into offices or other commercial uses. Squatters were responsible for the ambushing of a pleasure barge, the discharge of a smokebomb in a hotel lobby and the slashing of the tyres of foreign coaches. The purpose of these demonstrations was to publicise the squatters' message that too many luxury hotels were being built in a city with a huge housing shortage.

Amsterdam is no isolated case of social disorder. Copenhagen has seen similar events, the takeover of unoccupied housing is widespread in Italian towns, while in Frankfurt the early 1970s saw badly-housed immigrant workers squatting on an extensive scale. In France, action groups of all kinds protest against the quality of life in inner city areas. In Brussels, the residents of Marolles, a working-class district threatened with extensive demolition, campaigned successfully to persuade the authority to revise its redevelopment plans.

Figure 4.1 Street demonstration in Amsterdam

1 Characteristics of the modern Western European city

The obvious connection between all these events is political protest. In all cases, ordinary citizens have organised together against authority in an attempt to redress a grievance. The other less obvious common factor is their *urban* character. It is no coincidence that they all occurred close to the heart of some of Western Europe's largest cities. When we consider some of the characteristics of the typical modern city – a congested space containing a huge population often rapidly growing in numbers and made cosmopolitan by newcomers and visitors from foreign countries and provincial regions – then we might wonder how life continues to function with some degree of harmony. The potential for social conflict seems to be increased by three factors.

a) High population density

Urban land is an extremely scarce commodity. By definition, the city is a crowded geographical space. Because urban land is heavily in demand from many competing uses it is expensive. Expensive land means expensive housing. There can be little doubt that housing constitutes the major social issue of modern urban life. The problem of supplying decent housing at a price or rent everyone can afford has so far proved insuperable in almost all Western European countries (see Box 4.1). This housing crisis first reared its head in the

---

Box 4.1 Housing policy

Because housing is such a vital commodity it is almost always given high priority by the governments of modern states. Until recently Western governments have played a steadily increasing role in housing provision. This was not necessary simply because of war damage but also because of population increase, improving living standards and rising expectations. Private builders alone were unable or unwilling to meet the demand and government intervention has taken two main forms:

- *Direct intervention*, whereby the state itself becomes a landlord, building and renting out housing of acceptable quality at rents affordable by low income groups (social housing). Few European countries have gone as far as Britain, where almost one in three households lives in local authority-owned dwellings. Many nations have set up special 'mixed' housing agencies, funded by both public and private capital.

- *Indirect stimulus*, through the dispensation of grants and favourable loans to private builders and developers as well as tax relief for owner-occupiers.

Housing in modern Western society is a politically sensitive issue on which opinions are sharply divided. On the left of the political spectrum the belief is that the state has the duty to ensure that *all* its members are decently housed. The provision of large quantities of good cheap housing is unprofitable for the private sector as it is expensive to build and takes a long time to repay the initial investment, a situation made even worse in large urban areas by the high price of land. Consequently, there has been a chronic shortage of working-class housing, a crisis which would have been even more acute but for government intervention.

At the opposite end of the spectrum the belief is that housing should be supplied principally by private enterprise. The duty of the state is simply to provide a safety net only for those in extreme need through no fault of their own, such as the chronically sick and invalid who may need sheltered accommodation or migrant workers needing temporary homes until they can find settled housing. Some Western European governments, notably in Britain, have cut back their housing programmes in recent years in line with this philosophy.

100 GEOGRAPHICAL ISSUES IN WESTERN EUROPE

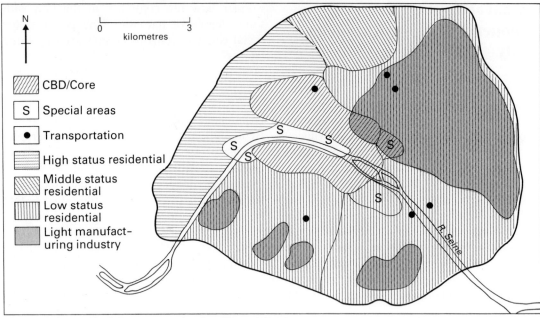

Figure 4.2 a) Social contrasts in the City of Paris

growing industrial towns of north-west Europe in the nineteenth century and since then it has been ever-present over an expanding geographical orbit.

The housing crisis is very much an urban phenomenon and it is most acute in those countries with the highest proportion living in towns and cities. In France the bulk of bad housing is located in Paris where 300,000 dwellings (25 per cent of the total) lack bathrooms or interior W.C.s and where a further 100,000 are actually homeless. In the Netherlands and West Germany the problem is greatest in the older, most heavily built-up areas of the Randstad and Rhine-Ruhr conurbations. At the other, more fortunate, extreme is Sweden. Despite considerable urbanisation (83 per cent of Swedes now live in urban settlements), the population is generally better housed than other Western European nations. Many attribute this to committed government support for housebuilding, though this belief is not shared by all. Whatever the explanation, the Swedish case proves that there is no natural law which dictates that cities must inevitably contain squalid and inadequate housing. Densely populated cities are the arena for social conflict but not the cause of social problems.

b) Social inequality

Because urban dwellers are packed in close proximity to one another, the contrasts between the 'haves' and 'have-nots' are particularly visible. In the West European city, class divisions tend to be geographically localised with specific neighbourhoods bearing the brunt of poverty and bad housing, while others display all the outward signs of affluence. Though interestingly, in the cities in the south of the sub-continent segregation takes a different form. Here, the apartment blocks often display vertical segregation – the poor being relegated to the worst housing on the upper floors (see Box 4.2). As Figure 4.2 suggests, privilege, deprivation and all shades in between are reflected in spatial contrasts across the city. The inner-city slum dweller or shanty-town immigrant would find it very difficult not to notice the comfortable housing conditions enjoyed by residents in more

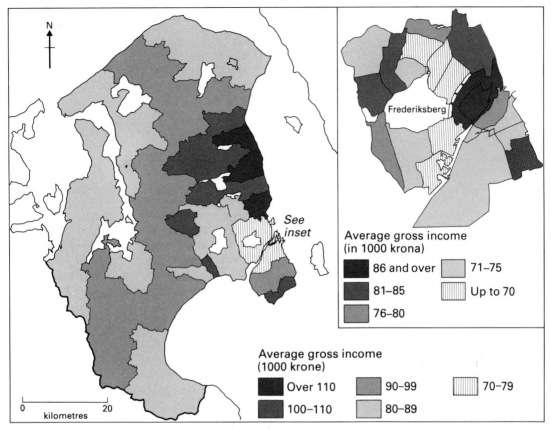

Figure 4.2 b) Social contrasts in Copenhagen

fortunate (usually suburban) parts of the city. Such vivid contrasts serve as constant reminders of exclusion from the fruits of a generally affluent society. This being the case, it is hardly surprisingly that discontent occasionally bursts forth as direct action. The squatters of Amsterdam, for example, see themselves as unjustly victimised in a city where no one need go homeless if the available housing stock was more equitably distributed. In these circumstances, people driven by desperation may take illegal action to claim a share of what they see as rightfully theirs.

One important lesson to be drawn from Fig. 4.2 is that urban poverty tends to be a widespread occurrence even inside regions which we normally regard as rich (see Fig. 1.5, p. 17). It is not confined to the notoriously poor cities such as Naples where, for example, an estimated one in three families are without a regular means of livelihood. Prosperous core-region cities such as Copenhagen and Paris betray no less internal inequality than peripheral Naples. Although overall material standards may be far higher in the core cities, these nevertheless contain considerable minorities who are disqualified from a satisfactory share of the city's rewards and resources. It is no consolation to the Copenhagen poor that they may actually be better off than an inhabitant of a shanty town in Lisbon.

c) Growth and change

Throughout this book we have repeatedly underlined the *dynamic* nature of life in modern Europe, a society undergoing

> **Box 4.2 Urban residential segregation**
>
> One of the classic themes in urban geography is the relationship between social and spatial structure, whereby each distinct social group is generally concentrated in specific areas of the city separate from other groups. You will probably be aware of a series of models (such as that of Burgess or Hoyt) which attempt to explain (arguably inadequately) such spatial patterns.
>
> *Class* segregation is undoubtedly the most basic geographical divide within the city. In effect each layer of the social hierarchy is concentrated in those areas of housing which it can afford. Thus the geographical distribution of the urban population into prosperous (usually suburban), intermediate, and modest (usually inner city) areas reflects income differences and the operation of the housing market. A major cause for concern here is the existence of slum neighbourhoods inhabited by people who are too poor or discriminated against to afford decent housing of any kind.
>
> *Age and family status* is another important line of division. In recent years young married couples with children have tended to move out to the edge of cities in search of houses with gardens, more open space, freedom from traffic and air pollution and a generally more suitable living environment for children. This process of selective migration has left the central city as more and more the preserve of single people and the aged.
>
> *Ethnic origin* also complicates the picture. In those cities where there has been a major influx of foreign migrants, this has caused a further geographical split, with foreigners (especially Arabs, Africans, Asians, and other non-whites) concentrated in neighbourhoods where few if any native-born Europeans will live (see pp. 121–123). Immigrant ghettos have long been a depressing feature of North American life but now they have become replicated in the Western European city, whether in the form of inner city slum clusters or outer city shanty towns.
>
> Geographical segregation is thought by many to be a critical social issue because it potentially threatens the unity of society. When classes, races and age groups are cut off from one another in their own local territories there is very little opportunity for social mixing. Urban children no longer grow up as part of one community. Their experience is formed in schools and communities populated by only a very limited range of classes and cultures. This can breed ignorance about other sections of society and reinforce class and racial prejudices. Isolation from the elderly can widen the generation gap. Segregation can also consign the less fortunate members of society to oblivion: out of sight in their slum ghettos, they are also conveniently out of mind. All in all, segregation is the very negation of community.

ceaseless change. Inevitably, of course, it is the large city which is at the heart of these changes – new technological innovations, novel forms of economic activity and fresh styles of social behaviour tend to be experienced first and most deeply here. This is only to be expected since the very decisions that cause change are taken in the city itself at the headquarters of business, finance and government.

As we have seen in Chapters 1 and 2, this brings substantial gains to many urban dwellers as in the case of Grenoble (Chapter 2, pp. 65–66). Yet we have also seen (Chapter 1, pp. 28–30) the reverse side of the coin in Southern Belgium. In creating new opportunities, economic growth frequently destroys established arrangements, sweeping away not only old industries but also settled communities, homes and familiar landmarks. Progress is potentially an environmentally and socially destructive force.

The restricted and densely built-up space of the large city bears the brunt of this upheaval. At the city centre, where land values reach their peak, only the highest bidders with enormous financial resources can afford to buy land. In effect, the ownership of central city land is open only to large organisations such as business corporations and government.

Consequently, over the past century or so city centres have become more and more the exclusive preserve of offices, government buildings, civic and cultural monuments and the largest retail stores. Apart from the most expensive luxury apartments, housing has been pushed further and further away from the centre. This process of residential displacement has accelerated sharply in the post-war period. Throughout Western Europe, this has been the age of the bulldozer, as old central city neighbourhoods of housing and small businesses have been flattened to make way for multi-storey offices, car parks, modern shopping precincts and the new traffic schemes without which none of these activities could function. The process has varied in intensity from city to city but few of the larger places have entirely escaped the traumas of modernisation. The pace of change has slackened since 1970, as conservation and renovation have now become major elements in urban policy.

The question of change is closely bound up with the problem of inequality, as the reshaping of the city creates new patterns of deprivation. Often, urban redevelopment encroaches on the lives of previously well-settled and contented communities, as in the case of Marolles, a stable working-class area in Brussels, provoking a sense of outrage. Moreover, the recent wave of urban protest has not been confined to the desperate but has extended to relatively comfortable neighbourhoods under threat from motorway construction or some other drastic form of redevelopment.

Not unexpectedly the problems described here are most acute of all in Paris, where pressure on limited space frequently produces near-intolerable conditions for local residents. Already Western Europe's largest single population concentration in 1945, Paris has subsequently experienced continuous growth, together with rapid expansion of its tertiary economy resulting in a mighty wave of office

Figure 4.3   Urban redevelopment in central Paris: la defense

construction. On top of this, it has attracted an ever-growing stream of tourists for whom hotels, restaurants and other facilities have sprung up. The net result of all this has been large scale redevelopment, as urban space has been reshaped to cater for the mounting demands being made on it. This has involved quite unprecedented upheaval and has inevitably sparked off bitter reactions from local inhabitants. Various pressure groups have sprung up to protest over housing issues (the clearance of settled inner city communities and rehousing in unsatisfactory flats on the outskirts), environmental issues (the threat posed by office skyscrapers in particular to the unique historical heritage of Paris) and transport issues (the cost of getting to work from increasingly distant suburbs).

Although the sheer magnitude of its problems puts Paris in a virtual class of its own, it does not make it unique. On a smaller scale, growth and change are widespread features of urban life from Oslo to Naples, claiming similar victims and evoking similar reactions. The next section probes more deeply into the characteristics and problems of urban growth and examines attempts to plan change in Western European cities.

Self-assessment exercise

Examine the article from *The Economist* opposite, concerning the 'Nouveaux Pauvres' (new poor) in France. This raises the other side of the problem – not urban protest but poverty.

i) Isolate the urban characteristics of the problem – which cities are involved and how are they affected?
ii) Why do you think that the 'nouveaux pauvres' are to be found in the major French cities? Is it anything to do with the urban areas themselves or is it the result of wider processes?
iii) List the actions that are being carried out to alleviate the problem. Are any of them likely to provide a long-term solution?

2 Urban growth and its consequences

Prior to the Industrial Revolution, the predominantly agricultural economy of Western Europe dictated a dispersed rural pattern of human settlement. Since the great majority of workers were engaged in agriculture, so, in the words of Rousseau, they were 'scattered over the earth to till it'. They lived in villages, hamlets and isolated farmsteads, with towns occupying a very minor place in the landscape. In comparison with today, urban places were small in size and contained only a small fraction of the total population. At the beginning of the nineteenth century, on the threshold of the industrial age, Paris was by far the largest city in continental Europe, with a population of a little over 500,000. It was followed by Naples (440,000), and by Amsterdam, Vienna and Madrid, none of which had many more than 250,000 inhabitants.

The obvious reason for the comparative absence of urbanisation was that the pre-industrial economy had relatively little use for towns and cities. Their main function was that of trading points, places where the rural population could gather to exchange their surplus produce. In certain cases, where a strategic geographical location encouraged international trade, this function was greatly enlarged. One of the great developments in the three centuries preceding the Industrial Revolution was the emergence of the great mercantile cities – Amsterdam, Lisbon, Antwerp, Cadiz, Seville, Bordeaux, Nantes, Hamburg – whose fortunes were buoyed up by the opening up of trans-Atlantic and colonial trade routes. Apart from these great ocean ports, however, it is fair to say that the importance of the pre-industrial city was primarily political. Ever since the Middle Ages, the European city had been a military stronghold and a centre of administration (both civil and ecclesiastical). The great modern functions of the city – mass manufacturing, high finance and banking, informatio

France

Nouveaux pauvres

FROM OUR PARIS CORRESPONDENT

Left and right in France have at last found something they can agree on: the country has too many poor people. Everybody from the Socialist prime minister, Mr Laurent Fabius, to the neo-Gaullist leader, Mr Jacques Chirac, is suddenly worrying about the steep rise in the number of the "new poor" — the people who have been out of a job for more than a year. For the first twelve months, unemployed Frenchmen get 70% of their previous salary. After a year, that drops to between FFr40 ($4.30) a day for people under 40 and FFr80 ($8.60) for those over 55, and the recipients' eligibility has to be renewed every six months.

When he became president, in 1981, Mr François Mitterrand pledged to hold French unemployment below 2m. It is now 2.3m, almost 10% of the workforce. The number of long-term unemployed has increased accordingly. The effect can be seen in the growing number of young and middle-aged people begging for money in the streets and in Paris underground stations. In Lyons, since 1982, there has been a tenfold increase in the number of young people receiving subsidised meals from the city. City mayors from all over France met in Paris on October 9th to draw attention to the problem, and the meeting was remarkably non-partisan.

The government insists that the problem is not new. However, the plight of the long-term jobless has been worsened by measures taken over the past two years to bring France's social security budget under control. Previous conservative governments had failed to do so. The Socialists, by contrast, managed to produce a surplus on the social security account, and saved the separate unemployment fund from going bust. But they achieved this by cutting benefits, as part of Mr Mitterrand's plan to curb state spending.

The president and his ministers take pride in the rigour of their new economic self-discipline. The man in Marseilles or Lille with no job for the past year, and little prospect of one, can hardly be expected to cheer, especially with winter approaching. The mayors who met in Paris called for urgent financial help for families in distress. They want local committees to be set up to provide speedy assistance for the worst-off, with local authorities bearing 65% of the cost and the central government 35%.

Nobody knows exactly how many people need help. Mr Chirac claims that 600,000 jobless people no longer get the full unemployment benefit, and that the number is swelling by 100,000 a month. The government estimates that there are 100,000 unemployed without enough to eat or a place to live. Whatever the exact figure, welfare payments to the long-term unemployed to supplement their 40–80 francs a day have been skyrocketing. In Dunkirk they have tripled since 1982. Mayors from Toulon, Angers and Amiens report a 75% rise in payments to the poor. In Paris, aid to the jobless and homeless rose by 141% between 1981 and 1983 to FFr30.8m ($3.3m).

The goverments's policy of economic austerity and industrial reconversion means that unemployment will go on rising. The prime minister, Mr Fabius, does not have much room for manoeuvre; he must hold down government spending if he is to meet his 1985 budget targets.

While the government considers its response to the mayors' call, France's leading cut-price grocer Mr Edouard Leclerc, has taken matters into his own hands. He has announced that his supermarkets will provide food to be given free, or cheap, to poor people. For those who have some money in their pockets, Mr Leclerc is offering a full meal for FFr10. His only concern is that he will be swamped by the "old poor" — the tramps who consider France's streets their home.

Figure 4.4 *The Economist, 13 October 1984*

dissemination, decision-taking – were either in their infancy or had not yet been invented.

Since the Industrial Revolution, the entire role of the city has been transformed to one of overwhelming importance in all spheres of life. The age of industrialisation has also been the age of urbanisation. With its emphasis on geographical concentration, the modern economy is by definition carried on in cities and the past 150 years have seen an immense expansion in urban-based activities and a decline in the scattered rural economy. The traditional trading functions of the city have become progressively enlarged whilst the city has also taken on the additional functions of manufacturing and services. Hence, the term *urbanisation* denotes a progressive shift in emphasis from the rural to the urban sector.

The great majority of Western Europeans are now dwellers in towns and cities. At the upper extremity many of these urban places have grown to quite formidable dimensions. Paris, the largest city in our study area, contains no less than 6 million people within its built-up area and a further 4 million living in neighbouring satellite settlements, that is, places *physically* separate from the central city but dependent on it as a place of work and a provider of services. Satellites are said to be *functionally* linked to the central or parent city. They are linked to it chiefly by daily flows of commuters, residents of the satellite but working in the central city and hence part of its economy. They are also linked by periodic flows of shoppers, leisure and pleasure seekers and others dependent upon what only the large city can provide. Thus one can live in a small village or even a hamlet and still be regarded as urbanised especially if one is reliant on the city for a livelihood.

In saying this we have identified a further key characteristic of the modern city – its geographical form is extremely complex. Unlike its ancient predecessor it is no longer an easily identifiable built-up object standing amid open country and bounded by a wall. It is an extensive sprawling mass which is often unclearly differentiated from the surrounding countryside. Furthermore, it often incorporates a multitude of formerly separate settlements into a single body. Because of this it is no longer possible to use such terms as town or city with utter certainty. Indeed, the geographer or planner frequently prefers expressions such as *conurbation, agglomeration, city-region*, or *metropolitan region* (see Box 4.3).

These concepts recognise that the larger urban settlement now tends to consist of several formerly distinct centres which have subsequently coalesced to form a single physical and functional unit.

When we consider these conurbations in more detail, we find that they take on a variety of geographical forms. The three largest conurbations of continental Western Europe – Paris, Rhine-Ruhr in West Germany, and the Dutch Randstad – embrace between them a total of 25 million inhabitants. They dominate the settlement geography of their nations, containing no less than 19 per cent, 18 per cent and 28 per cent of their respective national populations. This is urbanisation on a mammoth scale familiar to British readers, the continental equivalent of Greater London, the West Midlands and Greater Manchester.

a) A central city

Of the three, Paris possesses the simplest spatial structure since it is based upon only one centre. Its present shape results from outward expansion from a single nucleus on the River Seine, an expansion which has incorporated successive rings of outlying settlements and which has subsequently spread over a radius of 18 km from the historic central point of Notre Dame Cathedral (see Fig. 4.5). Even so, the gigantic territorial expanse of the Parisian conurbation poses awkward problems of definition. Where does this metropolis begin and end and what is to be included in it? Peter Hall has argued that there are at least four distinct 'Parises' defined on four different scales (see Fig. 4.6).

- **The historic city (Zone 1):** This consists of the 10 central administrative divisions (*arrondissements*, of which there are 20) and represents the area which had been built-up by the mid-nineteenth century and was at that time still confined within defensive walls. Nowadays this is the tourist's Paris, containing most of the best-known architectural monuments, cultural

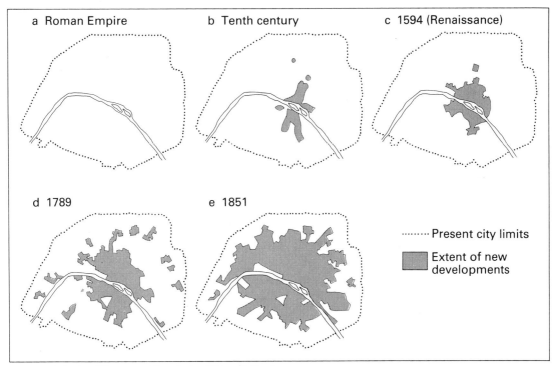

Figure 4.5 a) The growth of the City of Paris

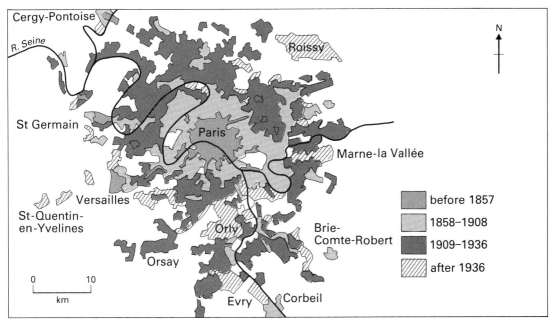

Figure 4.5 b) The growth of the Paris agglomeration

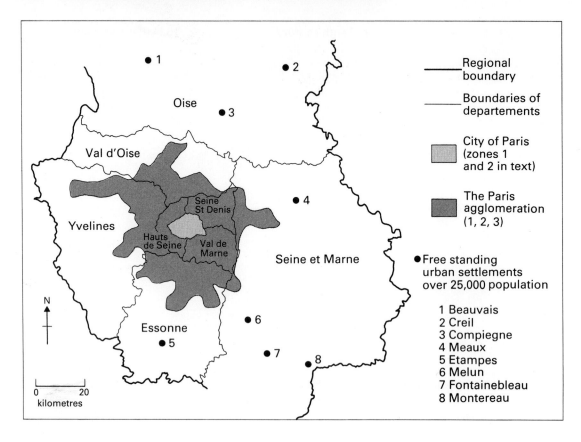
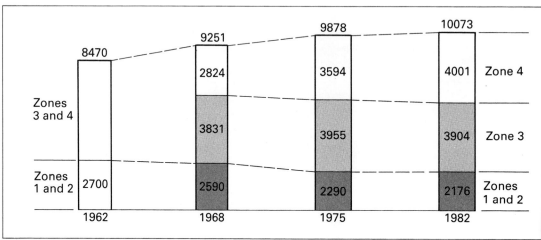

Figure 4.6 The Paris region (the population figures, above, are in thousands)

Note: The diagram illustrates the simultaneous growth of the city region and the decline of its core. The city itself has been undergoing population decline since the end of the First World War. Now the area of depopulation has expanded outwards to Zone III (which for the purpose of this diagram is defined as the départements of Val de Marne, Seine St Denis and Hauts de Seine).

Box 4.3 Urban forms

For some years geographers have realised that the terms 'town' and 'city' are inadequate to express the increasing diversity of urban forms. As urban settlements expanded new terminology was required to differentiate them from the conventional idea of a single centre city.

Conurbation was first introduced in 1915 and is now widely used. It refers to a city which has expanded to such an extent that it has swallowed up previously free-standing small towns and villages to form one large urban mass. Usually it is taken to refer to the physical extent of the urban area. Conurbations are defined as the continuously urbanised built-up area, and the term urban agglomeration is often used interchangeably with it. Another term used is *metropolis* which is also a large, continuously built-up area but focused on more than one centre.

City region also refers to the expanded city but in a functional and not simply a physical sense. It is concerned with the more extensive area to which the central city provides services and from which it draws labour (this is sometimes referred to as a daily urban system). Hence, it is not measured in terms of the continuous built-up area of the core and its suburbs but extends beyond these physical boundaries to small towns and villages standing in the countryside of the surrounding region. The largest city regions are often termed *metropolitan regions*.

A major problem for urban geographers and planners has been that the administrative boundaries of the city do not coincide with the wider conurbation or city region. Often they are historical relics which have not kept pace with the rapid progress of urban growth. Thus, the 'city' authority often has control of the declining (in population terms) heart of the city whereas the suburban areas and peripheral zones (sometimes termed *exurban*) are governed by a fragmented group of local authorities formerly more concerned with rural matters. As a result, statistical information is sometimes hard to come by and difficult to match to the realities of the geography of the conurbation or city region. In most Western European countries the census undertakings have had to devise special combinations of local authority units to roughly coincide with the conurbation/city region boundaries.

---

facilities and leisure activities. It is also the heart of the business economy and the location of the great French educational and governmental institutions. It is an area in which few people are now able to live: a place where for over a century residential land has been progressively forced to give way to the demands of expanding commercial enterprise.

- **The administrative city (Zone 2)**: This consists of the historic city together with a surrounding ring of 10 arrondissements whose housing and industry were developed mainly between 1850 and 1914. Originally the outer 10 districts acted as new suburban living quarters for people forced out of the overcrowded heart of the city. Now, however, they themselves have become surrounded by later suburbs. The continual expansion of the metropolis has converted them from suburb to inner city in less than a century. Taken together the 20 arrondissements of the central and inner areas make up the Département of Seine – this is the administrative definition of Paris, a legally constituted local government unit. The point to grasp here, however, is that local administrative boundaries cannot contain urbanisation and since these boundaries were fixed Paris has burst through them to spread far and wide over the neighbouring territory.

- **The agglomeration (Zone 3)**: There is a third Paris consisting of the administrative city plus the extensive outer area created by twentieth century urbanisation. This third ring, commonly referred to as Seine-

Banlieu, is a physical continuation of the city but is quite independent of it in local government affairs. It is essentially a sprawling mass of low density suburban housing.

- **The functional city region (Zone 4)**: Finally and beyond all this we must take account of the scattered communities – towns such as Beauvais, Etampes and Fontainebleu and new towns such as Cergy-Pontoise which are quite separate physically but which have been drawn into the commuting, shopping and service orbit of central Paris.

b) Polycentric conurbations

Unlike Paris, Rhine-Ruhr and Randstad are *polycentric* conurbations – instead of spreading outwards from a single dominant nucleus they have each been formed by the running together of closely neighbouring settlements. Rhine-Ruhr consists of a large number of urban centres, of which only Cologne was of more than moderate size or local importance before the 1850s. Since that time, vigorous industrialisation has caused all these towns and cities to grow outwards, in some places merging together, and the whole region has become functionally intertwined. Rhine-Ruhr is now a highly complex unit encompassing numerous and quite distinct sub-units. Major cities such as Dortmund, Bochum, Essen, Duisburg, Düsseldorf and Cologne exist as separate entities within a built-up area forming an inverted L-shape kept apart by green wedges of open land (Fig. 4.7).

The polycentric mass of the Randstad contains many of the leading cities of the Netherlands and houses the majority of the citizens of North and South Holland provinces. Its unique feature is its shape. Instead of an unbroken expanse, urban development in Holland has taken the form of a girdle surrounding a large open space, much of it agricultural. As we shall see, Dutch town planners have been much preoccupied with preventing urban encroachment into this green heart.

c) Megalopolis Europe

Paris, Rhine-Ruhr and the Randstad are the three great conurbations or metropolitan regions of continental Europe. Yet, for all its vastness, the conurbation is not the ultimate urban form. In the 1960s the word *megalopolis* entered the urban geographer's vocabulary. Whereas the conurbation is a fusion of several settlements, the megalopolis is a fusion of several conurbations.

The three giant European conurbations are adjacent to and intimately connected with numerous other city clusters, smaller in scale but still major centres in their own right. These include the agglomerations of the Belgium coalfield, Brussels, north-east France, eastern Netherlands and the Rhine-Main and Rhine-Neckar regions. There are grounds for arguing that all these should be classified as a single super-settlement, Megalopolis Europe, truly formidable in scale and with tentacles reaching over several international boundaries. Though there are many substantial patches of rural land contained within this zone, few of its inhabitants are entirely immune from the influence of its urban centres. Furthermore, many of these open spaces are in imminent danger of being eaten away by the ceaseless outward expansion of the conurbations.

d) The problems of urban decentralisation

In the eyes of many of its residents, urban Europe has become an increasingly unpleasant place to live in. This has led to a growing urge to escape to more spacious conditions outside the central city. Hence, one of the most marked social trends of recent times has been the mass exodus of urban dwellers seeking new homes and hopefully a better life beyond the city limits. This process of *suburbanisation* – the formation of extensive rings of low density residential areas on the periphery of

PROBLEMS OF URBAN DEVELOPMENT 111

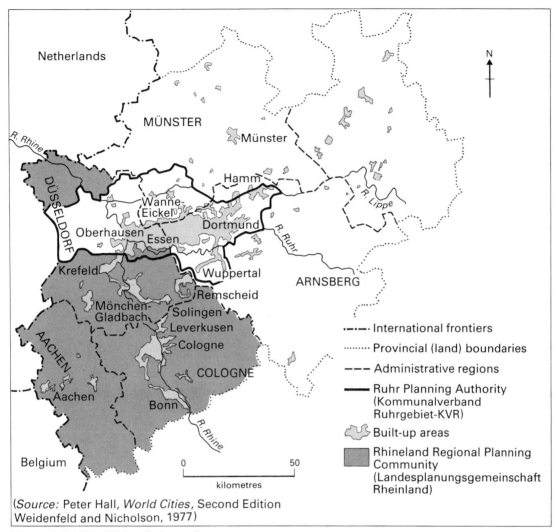

Figure 4.7 The Rhine-Ruhr region

Rhine-Ruhr is a complex multi-centred urban agglomeration extending over five of the administrative regions of North-Rhine Westphalia. Within this agglomeration two regional planning authorities – the Ruhr planning authority and the Rhineland regional planning community – are concerned with the major part of it.

urban areas at increasing distance from the city centre with a high level of commuting to central work locations – had been operating throughout most of the present century but has accelerated considerably in the post-war period. Urban population growth must give rise to territorial expansion in order to accommodate the additional housing and employment. What is new about the post-war era is that territorial expansion has proceeded much more rapidly than population growth. Throughout much of Western Europe cities have now entered a phase of *decentralisation* with population falling at the heart of cities and growth taking place at ever increasing distances outwards (see Fig. 4.8).

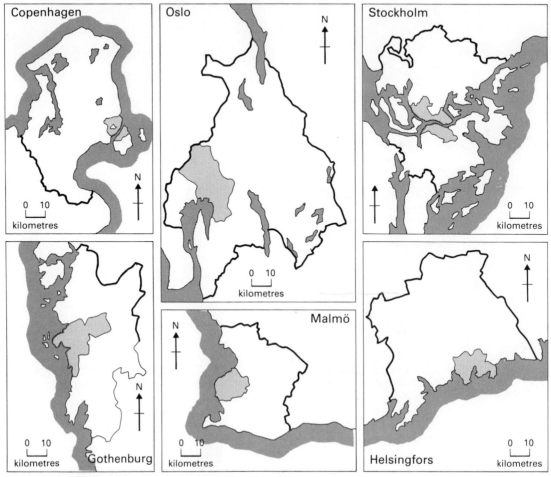

Figure 4.8 Population change in Scandinavian city regions

The maps depict the boundaries of the *central cities* (defined by the administrative boundaries of the historic city cores) within the wider *city regions* which reflect the area served by the city's services and its commuting zone.

The following population characteristics are seen:

	Copenhagen	*Stockholm*	*Oslo*	*Helsingfors*	*Göteborg*	*Malmö*
% population in central city	27.9	46.1	54.0	50.4	60.9	50.5
Population change in central city 1980–3	−15,913	+1,568	−7,562	+716	−10,513	−5,731
Population change in remainder of region 1980–3	−2,455	+23,160	+13,532	+28,686	+11,663	+6,360

PROBLEMS OF URBAN DEVELOPMENT

The turn-around from central area growth to inner city decline varied but by 1984 most cities showed population loss from the central city:

	Total population in the central city (in thousands)					
	Copenhagen	*Stockholm*	*Oslo*	*Helsingfors*	*Göteborg*	*Malmö*
1950	765	738	434	369	349	189
1960	728	805	475	448	401	225
1970	637	752	477	524	447	258
1980	499	649	455	484	435	235
1984	483	651	447	484	424	229

Self-assessment exercise

Figure 4.8 illustrates the population dynamics of six Scandinavian cities which can be regarded as advanced urban systems. All have undergone important population changes in the recent past. Examine the maps and statistics, then answer the following questions:

- What are the major population changes that have occurred?
- Are there any variations in the pattern? Which city can be regarded as the most advanced?
- All the maps are drawn to the same scale but some of the central cities and city regions are larger than others. What impact do you think this will have? If boundaries are tightly drawn could this explain variations in the pattern of change – consider the case of Copenhagen?

Data usually have greater impact if they are presented visually rather than in tabular form. Using the six maps as a base, illustrate population change in the central cities and city regions between 1980 and 1984 by the cartographic method which you consider most appropriate. You need to give a visual indication of the absolute size of population change and whether it increased or decreased.

Decentralised metropolitan development in north-west Europe is coming increasingly to resemble that in Britain and North America. Until quite recently, it was possible to depict the continental attitude to city life as being very different to that of Anglo-Saxon societies. Britons and Americans have in general been seen as having anti-urban attitudes favouring low-density suburban homes, the ultimate result being the vast sprawling suburbs which are the hallmark of Los Angeles, New York and London. By contrast, the continental tradition (especially in Latin countries) can be seen as revelling in the vigorous and closely-packed life of cities. As a result continental cities are much more compact and less given to suburban overspill. Moreover, continental Europeans are generally more likely to be apartment dwellers.

Recent evidence suggests that a growing number of continental Europeans would now prefer their own home in the suburbs rather than an apartment in the central city. All over Western Europe, these new aspirations have led to great changes in residential patterns. In Megalopolis Europe, the flight from the urban heart is proceeding on much the same scale as in the UK and creating similar 'grey' areas of low density sprawl, neither town nor country. Even societies such as the Mediterranean countries, which have traditionally been identified as natural city-dwellers are changing their residence patterns (as in Marseilles, Milan and Venice).

e) Urban sprawl: the consequence of decentralisation

Ironically, the quest for a promised land

beyond the urban margins has brought fresh strains in its wake. The social problems generated by urban decentralisation include:

- The loss of time and the personal wear and tear suffered by long distance commuters; the strain on family life entailed by the lengthy absence of the breadwinner(s);
- Overloading of public transport systems and roads;
- Rush hour congestion especially at the city centre;
- Environmental damage – increased pollution, destruction of buildings to make way for additions to the transport system;
- The loss of agricultural and amenity land and the swallowing up of rural settlements by urban sprawl;
- The waste of precious energy resources consumed in the ever-lengthening journey to work.

Decentralisation is not simply a question of adding on successive rings of suburbs. In the most advanced countries of Western Europe it appears that the long-term trend of urbanisation has come to an end because those leaving the city to find greener pastures beyond its limits now out-number those new arrivals who tend to settle in the core areas. This process has been termed *counter-urbanisation* and it has two elements:

- *decentralisation* i.e. the migration of people from the central areas of the city to its adjoining suburbs;
- *deconcentration* i.e. the movement out of the conurbation altogether to smaller towns and villages some distance away. The rural impact of this will be examined in the next chapter.

First noted in the United States and Great Britain, counter-urbanisation has recently become the dominant trend in many cities of north-western Europe with large-scale decline in the population of the urban cores and fastest growth in small towns and rural areas surrounding the large urban complexes. As the following case study shows this is now having a profound effect upon the population geography of many West European countries.

Case study: Counter-urbanisation in the Paris region

Firstly, it is necessary to view demographic change in Paris in the context of that in the country as a whole. For most of recent history Paris has proved a magnet to migrants from all over France and other countries (see Chapter 2). As a consequence, urban growth expanded at the cost of rural depopulation. By the 1970s, however, it became clear that this was no longer the case. French population decline was no longer simply a case of rural depopulation (although significant falls were still being recorded in the Massif Central and Pyrénées regions) but it was also to be found in the industrial areas of north-east France. Of particular relevance here is the previously unseen phenomenon of population decrease in the inner Paris region. In contrast, some of the areas of most dynamic growth were to be found in the outer ring of *départements* of the Paris region, as well as in the south-east of the country. This reversal of previous urbanisation trends did not apply solely to Paris – during the same period all the major French cities suffered a fall in population from their inner areas (see Table 4.1), although in half of these cases this was not significant enough to cause an overall decline. Nevertheless, there was a considerable turnaround between the two time periods and even dynamic growth centres such as Toulouse saw their expansion considerably curtailed.

In examining population change in Paris in more detail, a clear pattern emerges of decline in the core and continued suburban and peripheral growth. The Ile de France region is divided into three basic units (see Fig. 4.6); in the City of Paris at the heart of the region population fell by 0.8 per cent between 1975 and 1982, in the three inner suburban departments (the 'petite couronne') decline was slightly lower at 0.3 per cent, whereas the four

PROBLEMS OF URBAN DEVELOPMENT 115

Table 4.1 Population change in major French cities, 1968–82

	Overall urban region percentage change		Urban commune (inner core) percentage change	
	1968–75	1976–82	1968–75	1976–82
Paris	3.6	−0.5	−11.1	−5.8
Lyons	7.5	−0.1	−13.5	−9.6
Marseilles	5.9	0.9	2.3	−3.9
Lille	5.1	−0.1	−8.8	−10.7
Bordeaux	7.5	2.6	17.5	−6.7
Toulouse	14.1	2.5	0.7	−8.1
Nantes	11.8	2.5	−0.9	−6.2
Nice	11.4	2.7	6.5	−2.3

other departments (the 'grande couronne') grew by 1.5 per cent during the same period. When the components of these overall changes are examined, it is clear that although there are variations in the rate of natural increase it is the impact of migration that is most significant. Figure 4.9 clearly shows that an overwhelming majority of *arrondissements* in the City of Paris and *communes* in the 'petite couronne' registered a net loss of population between 1975 and 1982, whereas in the 'grande couronne', except for a number of *cantons* bordering the inner areas, there was a significant influx of people. In particular, the new towns of Marne-la-Vallée, Saint Quentin-en-Yveline, Cergy-Pontoise, Evry and Melun-Senart have acted as major foci of growth (see Fig. 4.10). It is important to note that not all these migrants will have originated in the central areas. Indeed, some will have come from the peripheral areas of other French cities to the fringes of Paris or even from rural areas. Counter-urbanisation is not a simple outward movement from one centre but a complex amalgam of many changes.

One consequence of counter-urbanisation has been severe inner city decline in demographic terms. Some 18 out of the 20 arrondissements of the City of Paris lost population between 1975 and 1982 during which the

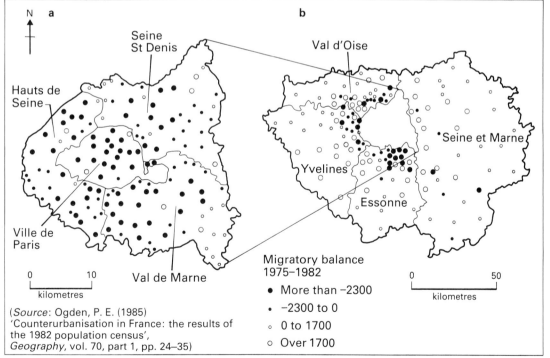

Figure 4.9 Migratory movements, Paris 1975–82 (after Ogden)
a) Arrondissements of Paris and communes of the 'Petite Couronne'
b) Cantons of the 'Grande Couronne'

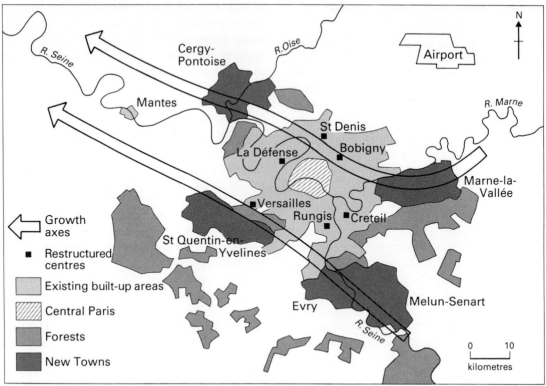

Figure 4.10 Planning proposals for the Paris region

Planning proposals for the Paris region were introduced in a major planning strategy published in 1965 – the *Schema Directeur*. The Schema's principal strategy aimed to concentrate urban expansion within two major axes: to the north a linear belt incorporating the new towns of Cergy-Pontoise and Marne-la-Vallée; to the south a parallel belt encompassing St Quentin-en-Yvelines, Evry and Mélun-Senart. Strict controls were imposed for parts of the Seine valley, thereby conserving much of the best quality open and forested land.

overall population fell from 2.3 million to 2.17 million. Losses were particularly severe in the most central districts which have lost half their population during the last 20 years. Large-scale urban renewal has been a major factor in this upheaval as public buildings have replaced housing, as in the redevelopment of the Les Halles quarter and the Maine-Montparnasse area.

The experience of Paris is by no means an isolated one. It seems that a turning point has been reached in the experience of many Western European cities. The urban influxes and progressive suburbanisation which have characterised the past century are being replaced by a counter-urbanisation trend with inner city decline and continuing suburbanisation supplemented by the growth of smaller towns in peripheral zones and the repopulation of the most accessible rural areas. These changes are fashioned as much by social change (such as the perception of inner city areas as hostile environments for family life and the acceptance of longer journeys to work) as by economic factors such as the decentralisation of some employment and the general rise in personal mobility. Whatever their causes, counter-urbanisation is creating increasing social polarisation as those most able to leave the city (usually this means the most wealthy) are 'voting with their feet' leaving behind the most disadvantaged and deprived sections of the community (the poor, the unemployed, the aged, ethnic minorities and single-parent families). Moreover, it poses severe problems to planners who are seeking both to contain urban sprawl and improve conditions in the inner city.

3 Urban planning

a) Planned overspill

Regardless of what label we use to describe it, we can now appreciate that the seemingly unstoppable outward march of the metropolis can be seen as a potentially destructive process. This menace was officially recognised very early in the post-war period when, taking a lead from British town planning initiatives, several Western European governments announced policies for the control of urban development. It was considered that the level and location of development could no longer be dictated by developers, builders, land speculators, property owners and other private profit-making interests. Intervention by central and local government was necessary to ensure that the interests of society as a whole were protected. In particular, public intervention was used to ensure that ugly formless sprawl would be checked and to protect agricultural land and open spaces around the major cities.

The early planning exercises were announced amid an atmosphere of optimism. It was thought that *urban land use planning* – the reshaping of city regions by the rational location of work, housing, services and transport lines – would be the key to a better life for all. It was hoped that the quality of the physical environment would be protected, people would live closer to their employment and other needs, and new traffic systems would make it easier for all to move about. A new profession of town planners (many of whose members were trained geographers) came into being to supply the expert advice upon which the 'better tomorrow' would be built. The aims, methods and achievements of urban planning can be illustrated by a case study of the Dutch experience in shaping the swelling metropolis of Randstad in Holland.

Case study: The Randstad

At the heart of the Eurocore, the Western Netherlands has perhaps benefited more than any other part of Western Europe from the advantages of centrality and economic integration. Its pivotal position in the Western European economy has ensured immense advances in social and economic well-being but this increased affluence has created physical, social and economic problems which Dutch planners have grappled with in a variety of ways. Their responses have been strongly conditioned by the physical form of the urban area of the Western Netherlands because here, unlike most other European conurbations, it is not centred around a single major urban core (such as London or Paris) but consists of a series of towns and cities loosely grouped together.

This multi-centred conurbation has been termed the Randstad (Dutch for rim city) because of its unique form. It consists of a roughly horse-shoe shaped belt of towns and cities grouped around a central, primarily rural zone (see Fig. 4.11). The northern wing of the Randstad stretches from Utrecht through Amsterdam to Haarlem, and the southern wing extends from the Hague-Leiden agglomeration through Rotterdam to Dordrecht. These encompass a core zone with smaller towns and villages which have largely retained their traditional character within a low-lying landscape of market gardening, bulb-growing, livestock and arable farming. This contrast has resulted in the Randstad being labelled the 'green-heart metropolis' and it is the clash between urban expansion and preservation of the green heart that planners have sought to manage.

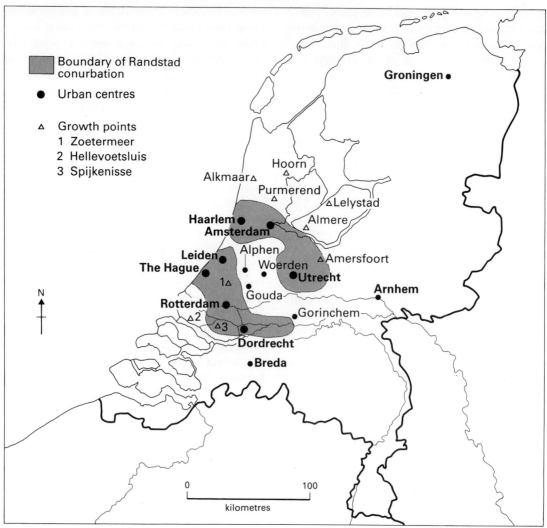

Figure 4.11 The Randstad

Clearly, the Randstad is very different from the conventional Western European city. Its population together with that of the green heart is approaching 6 million and yet it contains two major conurbations of around 1 million inhabitants (Amsterdam and Rotterdam) and two others (the Hague and Utrecht) with approximately half a million people, as well as numerous sizeable towns all of which have their own distinctive characteristics. Moreover, the functions of these urban centres tend to be distinct; Rotterdam concentrates on heavy port-related industry and wholesaling, Amsterdam mixes a range of port and lighter industries with important commercial activities (primarily offices and retailing), whilst the Hague is overwhelmingly the centre of administration and government.

Essentially, the problems of this urban region were bound up with the spiralling demand for land. The Netherlands is the most densely populated country in Western Europe with the bulk of its population concentrated in the Randstad. It has also experienced the highest rate of natural increase of any Western European country and its economic success since 1945 has attracted many migrants not only from within the Netherlands but also 'guest' workers from Mediterranean countries as well as immigrants from former Dutch colonies. This demographic growth together with rising prosperity, increased car ownership, improved transport networks and the Dutch propensity for low density, single family houses with gardens (along British lines rather than continental preferences) created a massive demand for housing land on the edges of the Randstad cities, with grave danger of large-scale encroachment into the treasured green heart. Competing claims for this land came from industry seeking to locate close to this large and prosperous urban market, from the intensive market gardening industry (some of it under glass and benefiting from Dutch natural gas supplies) and from the recreational demands of millions of increasingly affluent and more mobile urban dwellers in the Randstad. But it was the housing needs that posed the most serious dilemma. The major cities, especially Amsterdam and Rotterdam, were beginning to experience population losses as the more wealthy moved out to suburban zones beyond their boundaries. This raised the spectre of a continuous linear city over 100 kilometres long as the towns and cities of the Randstad began to coalesce and, even worse in the eyes of many planners, that its unique character would be lost as the green heart would begin to fill in to become one vast urban sprawl – a nightmarish 'Dutch Los Angeles' in the eyes of one geographer.

Faced with these pressures and utilising the Randstad's unique properties to best advantage, Dutch planners put forward a series of policies during the 1960s and 1970s to deal with the major issue as they perceived it at that time – that of growth. A four-fold programme was put forward. Firstly, in an attempt to defuse the pressure on the Randstad, regional development would be encouraged in the peripheral parts of the Netherlands in places such as Groningen, Arnhem and Breda. But it was acknowledged that even so growth in the Randstad would continue and the other three initiatives were designed to cope with this. So, secondly, priority was given to the preservation of the rural heart of the Randstad to maintain agricultural uses and restrict any limited expansion to the small historic towns such as Alphen, Gorinchen, Gouda and Woerden (Fig. 4.11). The third policy was designed to prevent the towns and cities of the Randstad ring merging together. Buffer zones were established so as to keep them physically separate as, for example, at Spaarnwoude between Amsterdam and Haarlem where a massive recreational area has been developed.

It was the fourth strategy, however, that was the most innovative. Given the unique shape of the Randstad, if it was not allowed to grow inwards or along the existing axes of development, then it had to expand outwards. But in seeking to prevent chaotic urban sprawl Dutch planners rejected the option of a green belt (as adopted in many European cities) in favour of growth along a restricted number of radial routes, with development zones separated by green wedges of agricultural land or recreational areas. Two obvious growth areas presented themselves; one to the north of Amsterdam in Kennemerland in existing towns such as Alkmaar, Hoorn and Purmerend, and the other to the east of the northern wing of the Randstad at Amersfoort. The other two options were not so straightforward and built upon the Dutch expertise in reclaiming land from the sea. In order to relieve the problems of the Randstad *new* land was to be created to the north-east of Amsterdam by reclaiming a large part of the Zuider Zee. In a bold engineering scheme the Flevoland polders were created and, although primarily for agricultural and recreational use, they contain two new towns at Almere and Lelystad designed to attract over a quarter of a million people from the northern Randstad, especially from Amsterdam (see Fig. 4.11). A similarly ambitious scheme was put forward to relieve pressure in the southern Randstad by opening up land to the south of Rotterdam in the isolated region of the Rhine-Maas-Scheldt delta. The Delta Plan has a dual purpose with the main aim the prevention of large-scale flooding as occurred in 1953, combined with the development of areas for housing, industry and recreation, principally at the growth points of Spijkenisse and Hellevoetsluis (see Fig. 4.11). Only one area of the Randstad could not be helped by outward expansion; the Hague, Delft and Leiden are hemmed in by the North Sea and New Waterway, and here Dutch planners have resorted to the construction of a new town at the existing village of Zoetermeer to the east of the Hague, but this is rigorously surrounded by buffer zones to prevent further encroachment into the green heart.

Overall then, the growth-oriented policies of the 1960s were designed to guide development to where it could do least harm and to prevent growth in the most sensitive areas. But it was becoming evident by the early 1970s that the assumptions upon which these plans were based were no longer applicable. A new set of priorities began to develop and this aroused planners into a reappraisal, which will be examined later in the context of inner city decline.

The above experience is generally representative of events elsewhere in Megalopolis Europe. There have been several inevitable differences in approach in Belgium, France and West Germany but in broad outline they have faced comparable problems and have made similar responses. Certainly, all have been forced to react to the seemingly unstoppable force of urban expansion.

Paris is a particularly instructive case in point. The original plan for the region envisaged that almost all growth would be contained within the city limits for the foreseeable future. Yet by 1965 this strategy had been abandoned, when it was discovered that population was growing at an alarming rate, which was causing the city to burst out of its planned limits. Consequently, the revised strategy (Schema Directeur, first introduced in 1963 and continually updated ever since) has contented itself with regulating overspill rather than with containment (see Fig. 4.10). Like the Dutch, the plan has relied on new towns (initially eight but later cut to five) and radical improvements to transport networks. There is also an important new ring of large scale suburban centres (*poles restructurateurs*), such complexes of offices, flats, shops and hypermarkets are designed to divert employment away from the intolerably over-strained central business district, in effect to give Paris a more polycentric form similar to that of the Randstad and Rhine-Ruhr. Planning developments in Rhine-Ruhr concentrated on the prevention of urban coalescence and the improvement of transport networks. But, here, a new element was a concerted effort to improve the badly polluted environment of this heavily industrialised region.

b) The regeneration of the urban heart

The need to control and direct decentralisation has taxed the ingenuity of the planning profession to (or even beyond) its limits. In theory, the rapid outward movement of population should act as a kind of safety valve to relieve pressure on the overcrowded living space at the city's heart. Yet paradoxically, living conditions in most large urban centres have actually deteriorated during the course of the process. The *inner city* has increasingly come to be regarded as a 'problem area', a geographical concentration of physical decay, social deprivation and social discontent. Two factors go some way to explaining this paradox.

New pressures on the inner city have come mainly from commercial redevelopment, the expansion and modernisation of the central business district to accomodate the growth of the office sector. They have been felt most acutely in the largest metropolitan centres – Brussels, whose rise as a centre for international organisations, especially EC institutions, has triggered off an office explosion; Paris, where employment in government,

Box 4.4 Immigrant housing in the city

Foreign migrants in the Western European city tend to be residentially *segregated* – concentrated in neighbourhoods separate from those of the native-born population. Moreover, immigrant communities are almost always obliged to occupy the worst housing and environments in the city, with conditions of overcrowding, poor quality and lack of amenity below the normally accepted standards of the host community in accommodation rejected by the native population. Among the various forms of immigrant housing are:

- The barrack-like *hostels* in which many large companies (especially in West Germany) house their workers. These are usually close to or even part of the work site. They represent the ultimate in regimentation.
- *Shanty towns* whose makeshift dwellings have been flung up without official permission on waste land, often with few normal services or facilities. The problem has been most acute in France where by the early 1970s it was estimated that well over 20 000 people, lived in these shack encampments (*bidonvilles*). Since then official demolition programmes have eliminated most of the larger bidonvilles but large numbers of the displaced occupants have simply set up smaller clusters of shacks dotted about the suburbs of Barcelona, Paris, Lyons and Marseilles.
- *Foyers* In the larger French cities a species of racketeer called *marchand des sommeilles* (sleep sellers) has emerged to exploit the poverty, ignorance and insecurity of immigrants. The marchand rents beds by the hour or shift. Closely connected with this system is the *foyer*, often a converted garage or cellar divided into tiny cubicles, so as to realise very high returns from the kind of rents that poorly paid immigrants can afford.
- *Inner city housing* For the longer-established migrant workers, especially those who have brought in their families, the classic form of occupancy is inner city housing, often in the oldest, most densely crowded areas with the least desirable housing. Usually these are the areas which have been abandoned by the former population, who have moved out to better housing or been rehoused by the public sector.

In explaining the existence of these ghettos three basic causes can be put forward:

- *Poor jobs and low pay*, driving migrants into the cheapest housing.
- *Racism* on the part of local residents and landlords, who attempt to exclude foreigners from all but the least desirable areas.
- *The migrant's own preference* to live in their own ethnic community among people from their own national origin, with familiar language, religion and customs.

Though literally true, this last point may be thought invidious in that it usually portrays migrant groups as *preferring* to live in slum conditions beyond the pale of society and as wilfully refusing to become assimilated into the host community. It would be truer to say that it stems from the failure to provide enough proper housing. As one German observer puts it, 'We called for workers and there came human beings'.

---

financial and commercial sectors has grown spectacularly since the 1950s; and cities such as Düsseldorf and Frankfurt where the growth of finance and banking is also reflected in modern city centre reconstruction.

The second source of social tension in the inner city has been the great influx of *foreign migrant workers*, many of whom have moved into the oldest and most delapidated neighbourhoods in the search for cheap accommodation (see Fig. 4.12). Inevitably this has led to a serious conflict of interests with local residents, who see the newcomers as competitors for housing and other local resources. A particularly sharp edge is placed on this conflict when housing, education, environmental conditions and local services are seriously lacking. Consequently, the newcomers

122 GEOGRAPHICAL ISSUES IN WESTERN EUROPE

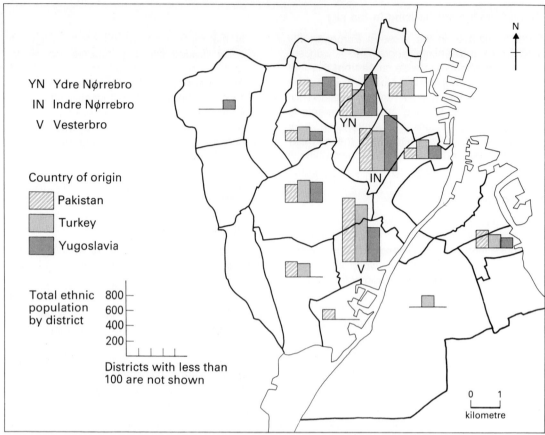

Figure 4.12 a) Distribution of ethnic minorities in Copenhagen, 1984

Figure 4.12 b) Distribution of immigrants in Paris

COMMENTARY

Pakistanis (3,709 residents), Yugoslavs (3,460) and Turks (3,604) are the most populous of the various alien nationalities resident in Copenhagen and Frederiksberg. Note however that it is only in 9 of the 21 districts of the city that all three groups are present in greater numbers than 100. This fact suggests a high degree of residential concentration. In fact no less than 47 per cent of the Pakistani-Turkish-Yugoslav population is contained within just three districts – Indre Nørrebrö, Ydre Nørrebrö and Vesterbrö. These districts contain only 19 per cent of the total population, indicating that the distribution of immigrants follows a different pattern from that of the population as a whole.

Note: In each of the shaded areas in Figure 4.12b the immigrant population comprises over 16 per cent of the total, once again indicating a marked degree of spatial concentration and segregation.

find themselves in a particularly odious position, blamed for the very conditions which affect them worse than any other section of the population (see Box 4.4).

These two factors were an important part of the new perception of urban planning problems that began to filter through in the 1970s, as the following case study shows.

Case study: The Randstad revisited

In most Western European city regions the changed circumstances of the 1970s hastened planners into an urgent reappraisal of their policies, and the Randstad was no exception. The direction of growth was no longer the overriding priority as other problems came to be perceived as more significant. There was a rapid decline in the birth rate and in-migration fell significantly.

Accordingly, there was no longer a pressing need to decant large numbers of people from the Randstad cities. To some extent decentralisation was, and still is, taking place in a spontaneous manner as the more affluent continued to move out to the fringes of the city region. Together with other concerns which began to surface at this time, such as the pressure for conservation of the urban environment (historic buildings, open space and so on) and worries over the future supply of energy resources, there was a growing realisation that the new problem to be addressed was at the heart of the major towns and cities. The renovation (both physical, social and economic) of the inner areas of the Randstad cities became to be accorded top priority as it was argued that planned decentralisation had gone too far and was in fact threatening the continued vitality of cities such as Amsterdam and Rotterdam.

Attitudes changed not only to the problems to be tackled, but also to the way in which they were to be approached. Gone were the masterplans setting out a fixed, detailed blueprint of some future utopia; instead the plans of the 1970s were to be less detailed, more flexible and more socially aware, so that they were capable of coping with the fluctuations of a rapidly changing world. Predictably, the large scale, planned decentralisation schemes were the first victims; though not abandoned completely, they were cut back considerably. The policies of preservation of the green heart and the maintenance of buffer zones would continue to be followed vigorously, as despite past efforts population was still growing in the green heart as local authorities were either unwilling or unable to implement the advisory plans of the regional planners. But the new element was to be a strenuous effort to revitalise the inner areas of the major Randstad cities to arrest the process of decay. Nowhere is this better seen than in Amsterdam where policies have come full circle.

AMSTERDAM

Throughout most of its history Amsterdam has proved to be a magnet attracting many migrants as its port and industries expanded and it developed as the commercial and administrative capital of the Netherlands. The city achieved its peak population of 867,000 in 1967 but since 1945 it has not been *demographic growth* that has been considered problematic so much as *physical expansion* caused mainly by the development of suburbs, but also arising from commercial uses, open space and extensive facilities such as Schiphol Airport (Fig. 4.13). As housing shortages developed, population density fell and outward pressure increased, it was feared that there would be serious incursions into the vulnerable green heart of the Randstad. Consequently, Amsterdam became part and parcel of the planners' attempts to disperse surplus population to growth points and new towns to the north of the Randstad during the 1960s and 1970s (as described earlier, pp. 117–120). It was considered that the city had reached its optimum size.

Subsequent events have cast doubts on the wisdom of this policy. Amsterdam's population has fallen to 687,000 in 1984, partly as a result of planned decentralisation but mainly due to continuing spontaneous suburbanisation around the city's fringes. This decline in population has now come to be considered by the city planners and politicians as a danger to the future viability of Amsterdam's culture and economy. They fear that the continuing outflow will drain away demand for its services and facilities, imperil the tax base of the city, as well as imposing severe strains on the transport networks as commuters undertake lengthy journeys in to Amsterdam (industry has not tended to follow people outwards). Moreover, despite the splendid (and more dubious) attractions of its city centre, dubbed the 'Venice of the North', Amsterdam has been unable to avoid the fate that has befallen most Western cities – inner city decline. Within the inner districts the most deprived section of the population is to be found living in the worst housing, experiencing the greatest difficulty in finding work, and confined in an urban environment with the poorest services, few recreational facilities and severe traffic congestion.

These changed circumstances and the realisation of the existence of an inner city problem in the midst of general affluence brought about a reappraisal of planning policy. Instead of large scale decentralisation schemes, planners now favour redevelopment and growth *within* Amsterdam, so much so that their new Structure Plan published in 1984 is entitled 'Focus on the City'. It puts forward the concept of a compact city with a complete range of facilities, rather than a central city with a collection of far-flung satellites. A major component of their strategy is a vigorous programme of urban renewal (Fig. 4.13). First started in the 1960s, a considerable initiative has been mounted to deal with the decaying residential areas around the city centre. Built around 100 years ago these districts of small, inadequate houses at high density with confined internal courtyards, narrow streets and few open spaces have attracted large numbers of new migrants – foreign workers and young people in particular. They suffer from structural instability also, as houses were constructed with poor foundations on reclaimed marshland. In an attempt to improve the poor standards of accommodation, a phased programme of housing renovation and redevelopment has begun. In some areas this has been met with open hostility from many residents, especially the squatters in the redevelopment zones, sparking off riots and street disturbances which have become a regular occurrence in Amsterdam as compulsory purchase orders are used to acquire whole neighbourhoods. Nevertheless the initial results of the renewal strategy are impressive; to date some twelve urban renewal areas have been created, in which around 12,000 dwellings have been renovated and 7,000 new units constructed, much of the housing built to high standards blending in to its historic setting, and popu-

PROBLEMS OF URBAN DEVELOPMENT 125

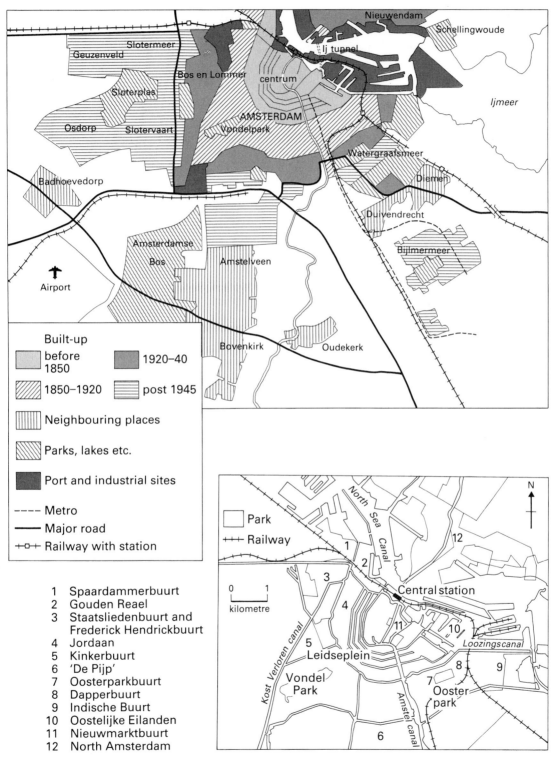

1 Spaardammerbuurt
2 Gouden Reael
3 Staatsliedenbuurt and Frederick Hendrickbuurt
4 Jordaan
5 Kinkerbuurt
6 'De Pijp'
7 Oosterparkbuurt
8 Dapperbuurt
9 Indische Buurt
10 Oostelijke Eilanden
11 Nieuwmarktbuurt
12 North Amsterdam

Figure 4.13 Amsterdam (inset showing areas of urban renewal)

lation has been retained in the inner city. One side-effect has been the 'gentrification' of some inner city areas, where more affluent families have moved back to recolonise certain fashionable districts such as the Jordaan.

However, the fall in population densities and the continuing housing shortage is such that these inner city programmes by themselves are not enough and the conventional solution of building large scale peripheral housing estates has been adopted. Like many other cities, the response initially was to use high rise developments, indeed, Amsterdam contains one of the 'best' examples of this approach at Bijlmermeer to the south-east (Fig. 4.14). This futuristic complex of high-rise, densely-packed flats with elevated highways strictly segregating people from traffic (built along the principles of Le Corbusier) has suffered from all the problems commonly associated with high-rise living. Such an approach has now been abandoned in favour of low-rise housing on a more human scale. It is envisaged that this housing will be built within the city boundaries and until it is completed the planned decentralisation schemes to new towns such as Almere and Lelystad will continue but on a much reduced scale (Fig. 4.11).

The experience of Amsterdam illustrates the changing priorities of urban planning and how it has refocused its attention on the internal problems of the city rather than urban sprawl. It shows that many people still wish to live within the city provided that a suitable living and working environment can be created. The big question remains, however, whether planners can persuade or compel the free market to slow down the decentralisation process and thereby ensure a future for the inner city areas of our large cities by attracting people and jobs back to them.

Self-assessment exercise

Throughout these first three sections we have continually returned to Paris to illustrate the points that we wished to make. Given its position as the largest urban complex in continental Europe this was a natural choice and we now want you to go back to this linking theme and bring together the Parisian material.

i) What are the principal physical and functional characteristics of Paris and how do they differ from other major city regions?
ii) What are the major problems of the Paris region?
iii) How is the region changing?
iv) What solutions have been put forward to counter the problems of the Paris region? Do you think that they are likely to be successful?

4 Urban change in the southern periphery

Urban development in southern Europe has two aspects. Firstly, for most towns and cities in Iberia, peninsular Italy and Greece large-scale urban growth began relatively recently. In particular, there has been an urban explosion fuelled by rural-urban migration during the last 30 years. Table 4.2 shows there has been a rapid expansion not only in the largest cities but also of the smaller urban settlements. Moreover, the urban hierarchy in both Greece and Portugal tends to be dominated by just two metropolitan areas (Athens and Thessalonika in Greece, Lisbon and Oporto in Portugal) whereas in Italy and Spain the growth of the largest cities has tended to be slow in comparison with medium-sized and smaller urban settlements, such as the growth poles created by regional policy in the Mezzogiorno (see pp. 90–94). Hence, for most of southern Europe population is still centralising in its urban growth areas.

The second feature worth noting is that there are significant exceptions to this trend. The largest cities whose modern growth has been active the longest have begun to show all the signs of urban decentralisation. In

PROBLEMS OF URBAN DEVELOPMENT 127

Figure 4.14 Bijlmarmeer

Table 4.2 Urban growth in southern Europe

	Urban population as a percentage of total population		Percentage of urban population			
			In the largest city		In cities over 500,000 persons	
	1960	1980	1960	1980	1960	1980
Greece	43	62	51	57	51	70
Italy	59	69	13	17	46	52
Portugal	23	31	47	44	47	44
Spain	57	74	13	17	37	44

particular the industrial cities of northern Italy – Turin, Milan and Genoa – have long passed their saturation point and are now showing population loss at the centre with overspill to sprawling suburbs. Similar trends can now be discerned in the largest cities in Spain. Indeed, southern European cities are an amalgam of those in their urban infancy together with others which seem to be embarked on a similar course to their northern counterparts.

a) Stages of urbanisation

Some recent writers have suggested that the regions of Western Europe should be classified according to their *stage* of urban development. Three major stages have been suggested: 1 *Urbanisation*; 2 *Suburbanisation*; and 3 *Desuburbanisation*. Stage 1 areas are those like the Mezzogiorno, where urban growth and centralisation are still the order of the day. Stage 2 are those city regions where urban cores are declining and suburban population rising, such as in Austria and Sweden. Stage 3 countries have moved beyond mere suburbanisation to a point where the most rapid growth is now taking place in medium-sized settlements up to 100 km away from the conurbation centres as in Paris. Is this sequence somehow inevitable? Is there no choice for the newly modernising cities of the Mediterranean but to repeat the history of the north and suffer the same consequences? Or can their decision-makers learn from the experiences of others?

All the signs suggest that the answer to the last question is negative. The example of northern Italy is particularly instructive here, since this is an area which has crammed much of its modern urbanisation into the period since the 1950s. During Italy's era of rapid industrialisation in the 1950s and 1960s migration from countryside to town was extremely heavy. But the newcomers were not provided with the necessary housing, schools and medical services. Consequently, the largest cities – Turin, Milan and Genoa – were faced with completely new problems of overcrowding, homelessness, crime and social unrest. The situation was very reminiscent of Paris, London or the coalfield areas in the nineteenth century: a different age but an identical stage of urbanisation with similar social stresses.

We should note here that similar events are now being re-enacted in Iberia, where in Lisbon, Madrid and other cities thousands of recent rural migrants and workers returning from overseas are forced into shanty towns. In respect of housing conditions these cities are closer to the Third World than to the rest of Europe. The pressures on urban resources are similar and there is the same lack of official planning. In Greece, it is only in the present decade that direct government action has been taken to combat the major problems of Athens, its ugly sprawling suburbs, traffic chaos and air pollution.

In many other ways Athens typifies the worsening plight of the south. It has been transformed out of all recognition in a very short space of time. The rise of tourism, new manufacturing industry and the flight from the countryside has created an urban monster from what was a city on a human scale just four decades ago. In 1940 its population was less than half a million. This is now almost double and if the surrounding suburbs are included the entire agglomeration numbers 2 million people. All this has happened largely without planning controls.

(b) Planning problems

The absence of strong official policies to regulate the use of scarce urban land is the main cause for pessimism about the future of the southern European city. As we shall see in the following case study, however, not every Mediterranean city has been content to leave its development to the free market.

Case study: Bologna – an alternative planning approach

Here, we have chosen not a typical example but a city which has attempted to break out of the stranglehold of deteriorating living conditions and chaotic flight to the suburbs. The factor that distinguishes Bologna from almost all other Western European cities is political. Since 1945 the city council (*municipio*) has been controlled by the Communist Party, which has used this unbroken spell of power to introduce highly radical policies on conservation, housing, land use and transport. Consequently, this is an extremely controversial case study. For socialists, Bologna is a model of what can be achieved by committed government by and for the people. In the eyes of its detractors, however, this is a spendthrift council which has penalised property owners and driven private investment away from the city. Whatever our view of this argument, there is no doubt that Bologna has refused to passively accept the kind of environmental and social disruption which plagues many other large European cities. It offers a complete contrast both to the laissez-faire chaos which prevails elsewhere in southern Europe and to rule by professional experts in the north.

The first unusual feature of local politics here is that there has been a genuine attempt to encourage popular involvement in the planning process through the setting-up in 1962 of 15 local ward councils. The *municipio* regards this as a genuine act of decentralisation. It maintains that no new building or change in land use can occur in any neighbourhood without the approval of the local residents – using the ward council as their mouthpiece. Urban planning is thus carried out in consultation with the people most affected by the proposed development, so that ordinary people are active participants rather than passive recipients of bureaucratic decisions. The minority opposition parties (notably the Christian Democrats) take a more cynical view of matters. They argue that the ward councils are not really democratic, since until 1981 they were appointed by the Communist mayor rather than elected directly. They also find it hard to believe that a communist leopard which until quite recently owed allegiance to Moscow should change its spots so completely on the question of democracy.

Be that as it may, Bologna's ruling party has won considerable approval in the highest circles for some of its planning measures. Honours have arrived in the shape of a gold medal from UNESCO in recognition of its conservation policy and a prize from the World Traffic Conference for its transport policy. Since 1962 much of the *municipio's* planning strategy has hinged upon the need to conserve the walled medieval heart of the city (*Centro Storico*), a treasure house of architectural and cultural monuments, a medieval complex second only to Venice. But conservation here has been seen as a *social* as well as an environmental process, the creation of a city centre for living as well as admiring. Furthermore, it should be a living space for all sections of the community, not simply the very privileged. Methods used to achieve this ambitious aim include:

- A complete ban on high rise office development, the destroyer of so many other ancient centres. Economic growth and modernisation have been sacrificed to conservation, the idea has been promoted under the slogan 'Conservation is Revolution'.
- A complete ban on traffic other than buses, taxis and residents' own vehicles. Public transport is heavily subsidised by the *municipio*, running free at peak hours and at any time for pensioners.
- Renovation of the ancient housing of the *centro*, much of it in appalling squalid conditions. But, whereas the practice is often to re-let such improved accommodation at luxury rents, in Bologna landlords are subsidised to carry out renovations then obliged to re-let to the original tenants. In this way, the central working class communities are preserved rather than displaced to far flung municipal estates.
- Outside the centre, green belts have been imposed, (notably around the opulent southern suburbs which climb into the Appennine hills) in order to prevent

further sprawl into the most desirable amenity land.

Whether all this is to be judged a success depends on the measures used. Certainly Bologna has better housing standards than much of urban Italy, excellent transport links between the central area and suburbs, and a conservation record which is the envy of many other places. Most outside observers remark on the community feeling and local identity of the Bolognese people, a great social resource helped by the retention of people in the inner city and involvement in planning affairs.

On the minus side, however, the anti-communist critics have ample ammunition. There is no doubt that planning of this type is a vastly expensive exercise. On the one hand, there is the lost tax revenue incurred by excluding large offices from the centre. On the other, there is the increased municipal expenditure incurred by subsidising public transport and housing rehabilitation, together with the cost of ward councils with their numerous social schemes such as the provision of free nursery education. The city is heavily in debt and the *municipio* makes no bones about the fact that its second largest budget item is the servicing of bank loans. Rather like a Third World country Bologna must borrow ever more money to pay the interest on existing loans.

We leave it to the reader to deliver the final verdict. Whatever the decision, one thing is certain – Bologna is different.

5 Conclusion

We began this chapter by examining some examples of social unrest in specific urban areas and progressively broadened out to look at the sweeping changes occurring in urban Europe. What can geography contribute to the explanation of such developments? Some academics argue that geography is of little or no use in helping us to explain why such changes are happening. They would see large-scale, inner city unemployment, for example, as being caused by broad economic changes on a national scale (perhaps resulting from intense international competition) and the geographical concentration in inner urban areas as merely the end-product of these other processes, arguing that the geography of the city and the characteristics of its various areas do not cause the problems that occur there. We hope that by now, you will be questioning this. After reading our discussion you should realise that cities *are* different (for example, at various stages of development) with different problems. In a sense, all places are *unique* and this in turn conditions how planners seek to intervene to change their geography. Clearly, geography is not everything, but to understand the major issues in urban areas you need to combine an appreciation of the broad social, economic and political changes affecting society with an understanding of the effect of geography in translating these processes in different ways from place to place.

5 Rural Western Europe

Despite the pervasive influence of urbanisation since 1945, we must not ignore the fact that urban land covers only a fraction of the land surface of Western Europe – by far the greatest area of land being taken up by some form of rural use. Whether it be agriculture or forestry, even bare mountain, heathland or wetland the landscape of the whole continent has been imprinted with the mark of rural activities which, for many centuries, have exerted a considerable influence upon the physical environment, local culture, politics and the pattern of settlement. Although the relative importance of many rural regions – in terms of population and employment – has declined in recent years, the predominant position of agriculture as the most extensive user of rural land in Western Europe remains largely unchallenged. The visible impact of farming with its distinct cropping patterns, field systems and spatial organisation of settlement has, itself, played a fundamental role in the evolution of the landscape of Western Europe.

Although a gradual transfer of land from rural to urban uses has taken place, in 1985 well over 40 per cent of the land surface of Western Europe was still devoted to agricultural production. Since 1945, despite the loss of some farmland, agricultural productivity has risen many times. Not surprisingly, then, agriculture still remains the key to rural development. Yet despite the knowledge that rural land will be as extensive as ever in the twenty-first century major internal changes have taken place since the end of the Second World War which have had a major impact upon rural regions throughout Western Europe. The purpose of this chapter is to identify these major changes and to investigate the many issues which have arisen as a consequence.

1 Agricultural development

The phrase *agricultural development* calls many images to mind: streamlined efficiency as tractors, harvesters and milking machines carry out the back-breaking tasks which formerly fell to the lot of human labour and draught animals; limitless bounty, with ever increasing yields of crops and animals from each patch of land; increasing profitability, with large, modern farms run according to all the best principles of scientific management. Up to a point these images are an accurate description of events which have actually been occurring on the farms of Western Europe. Change has been in the air ever since the eighteenth century when the new crops and rotations of the agricultural revolution brought about the first 'great leap forward' in agriculture since Roman times. Change quickened in the nineteenth century when industrialisation began to create new opportunities for commercial agriculture. The expanding population of north-western Europe represented a huge market for food, a demand which farmers in Denmark, the Netherlands and elsewhere were not slow to respond to. At the same time, the new engineering and chemical industries began to supply the machines and fertilisers necessary to achieve greater output.

The real surge, however, has come since 1945. As in all other walks of life, the postwar era has been a time of constant innovation in the agrarian sphere. Nevertheless,

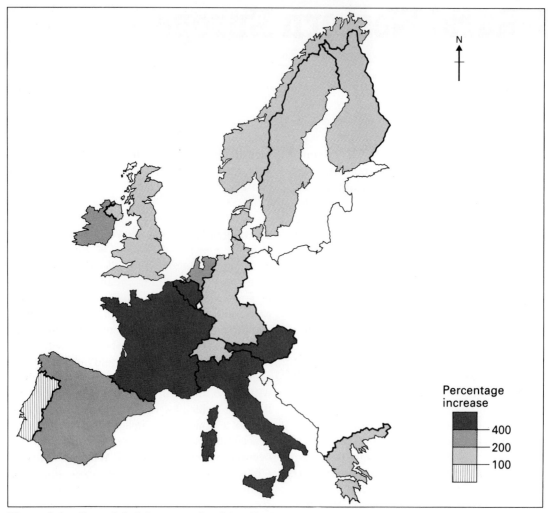

Figure 5.1 Growth in labour productivity (food output per worker) 1950–83

what distinguishes this period from all former ages is the degree to which governments themselves have intervened in the farming industry. Without exception, the post-war governments of Western Europe, as well as supra-national bodies such as the EC have consistently pursued policies to stimulate farm production, to update attitudes and working practices and to give farmers a better deal than they would have enjoyed under open market conditions. One important aspect of policy has been the maintenance of the selling price of food at an artificially high level, in order to protect farm incomes (see pp. 154–160). Apart from this, governments have also aided the industry by investing in agricultural research (into new methods of production, improved strains of seeds, better breeds of livestock); by programmes of farm consolidation (inducing small, inefficient farmers to sell to large concerns, thereby creating larger more streamlined farm units); by the establishment of expert advisory services to inform and educate farmers in modern methods.

Since the formation of the EC in 1958, official intervention has stepped up dramatically. The EC system of guaranteed high prices to farmers has been perhaps the greatest single influence on agricultural development of the modern era. One of its more obvious effects has been to stimulate a massive – some might say grossly excessive – increase in food production. The vexed question of the Common Agricultural Policy and its attendant food mountains and wine lakes is taken up more fully later in the present chapter.

Whatever the problems of EC intervention, it is unquestionably true that West European agriculture has made giant strides since 1945. The watchwords of the post-war era have been productivity, modernisation and growth. Fig. 5.1 charts the impressive gains in output achieved since the early post-war years. Two trends are especially noteworthy.

a) Rising productivity of land

The huge increases in output between 1950 and 1980 have occurred on a virtually unchanged area of cultivable land. Inevitably a certain amount of land reclamation has taken place. In the Netherlands, for example, over 6,000 square kilometres have been taken from the sea with the help of a number of large-scale civil engineering projects. On the whole, however, reclamation has contributed only a marginal increase in the cultivable land available to Western Europe's farmers. Consequently, the key to rising production has been an *intensified use* of the existing space rather than an outward extension of the arable frontier. Each unit of land has been induced to yield up a greater product than before: or to put it another way, land productivity (output per hectare) has steadily increased. On average, each hectare now yields a far larger cereal harvest and carries a far larger livestock population than at the start of the period (averages are deceptive though – Fig. 5.2 shows that considerable geographical variations exist in cereal yields, for example). Output per hectare of oil seeds, fruits, vegetables and almost every other conceivable farm commodity has also increased.

In part, this great burgeoning is due to a shift in land use, with a greater proportion of land now used for crops rather than grass. On the most modern commercial farms, animals tend to be fed on fodder crops and imported grain, rather than left to forage at will on grassland. Even more, however, productivity growth is due to the application of scientific innovations. Increased yields reflect increased inputs of artificial fertilisers, weedkillers and insecticides together with improved varieties of plants and animals, which mature faster, grow bigger and reproduce more energetically than nature ever intended.

To complete the picture, we should also note the importance of improved management. Farming, for so long a traditional way of life, is gradually being transformed into a commercial activity. On the modern farm enterprise, decisions about what to plant, how to raise it and where to sell it are taken on rational grounds of profitability. Furthermore the rise of corporate involvement in modern agriculture is increasing all the time with multi-national companies such as Nestlés, CIBA-Geigy, Rank-Hovis-McDougall and Birds Eye transforming farming into a business – an 'agribusiness'. In such circumstances farmers act out the role of caretakers having forfeited their autonomy as decision makers to produce foodstuffs under contract at fixed prices to the major firms who then handle all the processing, marketing and distribution. This business-like philosophy contrasts vividly with that of the traditional European rural life, where many families were rooted to the land for reasons of tradition, sentiment or sheer ignorance of any alternative way of life. This is not to say that all farmers and peasants lacked business sense or that they were wasteful and utterly inefficient. Indeed many of the ancient husbandry practices handed down from one generation to the next made an extremely good use of the soil and of the limited tools and machines available. The point is that the traditional agriculturalists did not measure success entirely in terms of profits, productivity and growth. Farming was

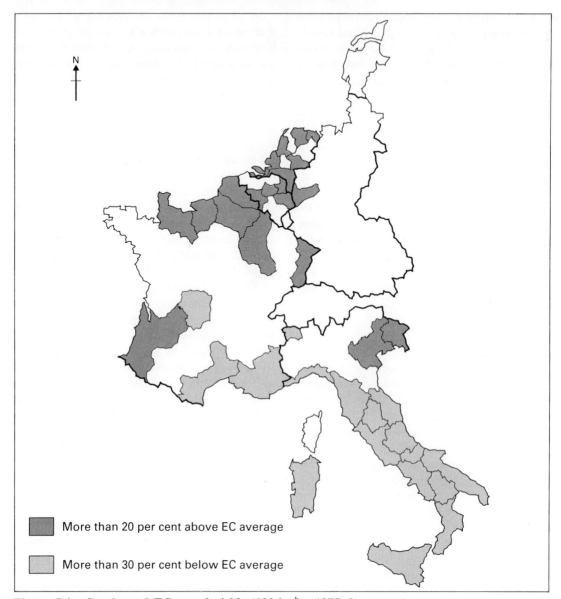

Figure 5.2 Continental EC cereal yields (100 kg/ha 1975–8 average)

Note: Cereal yields are obviously not a comprehensive measure of land productivity. Given the importance of cereals in the European diet and agricultural economy, however, they probably offer a better guide than any other single crop.

more than simply a livelihood; it was part of an entire way of life, a measure of social standing in the rural community. The rise of the modern commercial farm represents nothing less than a complete cultural change – and culture is one of the important conditions upon which post-war agricultural development has been based.

Together, technical advance and modern business attitudes have brought about a growing mastery over natural forces. It is now impossible to explain the agricultural geography of Western Europe as largely the outcome of climatic, soil and topographical variation. Commercial profitability is the greatest single factor determining how land is used. If market prices favour milk, then much land will be used for the production of grass and fodder for dairy cattle. If the relative price of cereal crops rises, then much of the same land will be turned over to wheat, barley, maize and other grains. National government and EC policy exert such decisive influences over prices that we are justified in identifying government among the 'human factors' determining present day agricultural geography.

b) Labour productivity

Modernisation has also been reflected in a growing output per agricultural worker. Indeed, it might seem contradictory that Western Europe has experienced its immense expansion in land yields at the same time as the number of active workers has dwindled rapidly. On reflection of course we see that there is no contradiction between the two trends. The loss of manpower is yet another consequence of progress, this time in the form of mechanisation. Agriculture is an industry in which the replacement of human labour by machines has been proceeding for a century or more in some regions. In the most advanced nations, for example, workers are now outnumbered by tractors. Aside from tractors and other field machinery, the post-war situation has also been transformed by the rapid spread of electrification into rural areas. In some countries this now takes in virtually every farm holding and makes possible the use of a complete range of labour saving devices.

The agrarian labour force has also been trimmed by the elimination of many of the smallest farmers along with their holdings (Fig. 5.3). As a consequence over twenty three million agricultural jobs have been lost in Western Europe during the period since 1950. For every three agricultural workers (farmers plus hired labour) in 1950, there is now approximately one. The map (Fig. 5.3) confirms that this very substantial replacement of labour by capital has occurred throughout the entire continent, though a smaller number of countries – the United Kingdom, Portugal and Greece – show a significantly lower decrease. In the case of the United Kingdom this is because the process started much earlier than elsewhere in Western Europe and was virtually complete by the early post-war period. By contrast, it has only recently taken off in Portugal and Greece.

Self-assessment exercise

Figure 5.2 shows a clear geographical pattern between the high yielding nations of north-western Europe and the low productivity of nations in southern Europe. Make a list of some of the reasons why such differences should exist.

You may like to consider factors such as climate, relief, land tenure, level of technology, etc.

Since the 1950s, many governments have pursued policies aimed at combining small holdings into larger units for the sake of greater efficiency. As in many other branches of the modern economy, so too in agriculture is *scale of production* a crucial consideration. Large units produce more cheaply than smaller ones. They can justify the use of machines which, on a smaller farm, would lie idle for much of the time; they have scope for specialised, expert workers such as cowmen, shepherds and drivers, rather than the general worker typical of the small holding; their returns are sufficiently large to create surplus capital for investment in improved plant and equipment. In the increasingly competitive and businesslike world of post-war agriculture, many of the smaller operators have gone to the wall. Still more have been encouraged to retire or leave the industry by

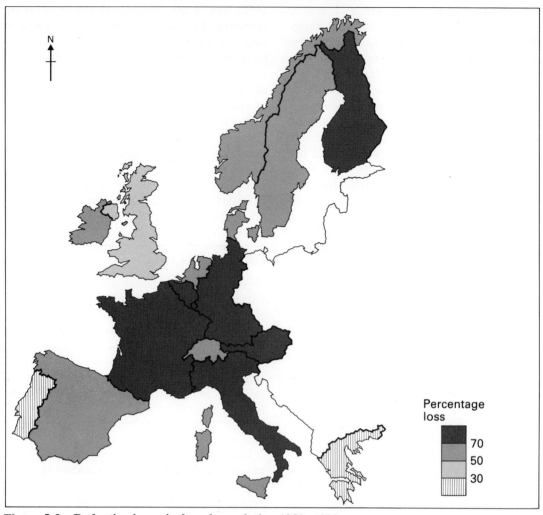

Figure 5.3 Reduction in agricultural population 1950–1983

government inducements and compensation. The net result has been a consistent rise in the average size of the West European farm and a steady diminution of the farm population.

Quite clearly, the people of Western Europe have made immeasurable gains from agricultural development. Abundant supplies of cheap food have freed them from scarcity and famine, supported a growing population (up to the 1970s) and made a major contribution to the general raising of living standards. There are, however, a number of drawbacks which cloud this rather rosy picture. Firstly, not all regions of the continent have participated equally in the onward march of progress. In particular the lagging rural regions on the outer rim of the EC and the peripheral regions elsewhere in Western Europe are still carrying many of the economic problems of traditional agriculture. Secondly, as with any other form of economic progress, there are damaging side effects. Both these issues are sufficiently important to merit extended discussion.

c) Leading and lagging regions

We cannot assume that all farmers in every

part of the sub-continent have benefited equally from the kind of progress so far described. As with the other great changes discussed elsewhere agricultural development has taken the form of a spatial diffusion process. Modernisation began and has progressed much further in the countries and regions of the Eurocore: it has lagged behind in the outlying areas of the south and the far north and west. The highest productivity has been achieved in regions such as the West Netherlands, Denmark, Flanders, North Germany and the Paris Basin. All these are regions where the benefits of mechanisation, intensive land use and scientific management have been realised to the fullest extent possible. By contrast, the remoter peripheral areas of the EC, Southern Italy, Spain, Portugal and Greece, fall well short of these high standards, lagging well behind the technological progress of the Eurocore and in consequence failing to share in the general rise in farming prosperity. The extent of this geographical lag can best be appreciated by a comparison of two regions at the opposite end of the spectrum – West Netherlands and Abruzzi-Molise in Southern Italy.

Case study: West Netherlands versus Abruzzi-Molise

On most measures of efficiency and prosperity, it is the West Netherlands which appears as the agricultural leader of the EC. Consisting of four provinces this is the portion of the Netherlands correctly referred to as Holland. To most Britons, Holland conveys an immensely romantic image: flat, lush pastoral landscapes, teeming with flowers (it is always spring!), dotted with windmills and tended by amiable rustics clad in national dress. Somewhat regrettably for those of a poetic disposition, the modern reality bears little resemblance to this fantasy. Windmills are now heavily outnumbered by electric pumps draining the reclaimed sub-sea level polderlands (which are themselves the result of massive civil engineering projects). Bulbs still flourish by the billion but their munificence owes more to heated glasshouses, fertilisers and pesticides of the giant chemical corporations than to the tender loving care of apple-cheeked maidens in clogs and aprons. In fact the farm population is now outnumbered almost 25 to 1 by non-agriculturalists, chiefly workers in the great Randstad conurbation (see Chapter 4) which encircles the farmland heart. Those who remain are anything but quaint in outlook. The typical farmer is likely to possess a degree or some formal qualification in agriculture or management and to look with some scepticism on time-honoured traditional farming lore.

The keynote, then, is intensive exploitation of the land for profit, large amounts of capital invested in applying the latest technological innovations. Consequently, Dutch farming achieves huge returns, whether measured in yields per hectare or output per worker. In addition to technology, returns are boosted by: 1) the high soil quality of the reclaimed land; (note that even here we are discussing an artificially created agro-environment rather than a purely natural bounty, in which horticultural crops in particular are grown under glass and heated during the coldest months with natural gas obtained from the vast Dutch reserves) and 2) specialisation lines such as horticulture and dairying, which make an extremely intensive use of the space available.

In many senses the situation in Abruzzi-Molise is the inverse of that in Holland (Table 5.1). Here, no less than one in four workers are economically dependent on the land, a proportion more than six times the Holland level and four times the EC average. Despite this immense input of human effort (or perhaps because of it) returns are low; pitifully so in comparison with Holland or almost any other region of the Eurocore. Each worker in West Netherlands produces over three times as much (by value) as his counterpart in Abruzzi-Molise.

Low output certainly has nothing to do with the idleness and fecklessness of workers. On the contrary, the backward agricultural

Table 5.1 Comparative agricultural profiles of West Netherlands and Abruzzi-Molise

	West Netherlands	Abruzzi-Molise
% total area utilised	50	57
Wheat yield (100 kg/ha)	64	24
Sugar beet yield (100 kg/ha)	455	410
Cattle per 100 ha	169	29
Population engaged in agriculture	90,000	128,000
% total workforce	4	24
Value of output per worker (EC = 100)	186	49

systems of Mediterranean Europe demand much more of the traditional back-breaking toil than do the super efficient systems which now dominate the Atlantic side of the continent. The crux of the matter is that the southern Italian worker must struggle to produce by hand tools and draught animals what the northern European can achieve much more easily by mechanised aids and electric power. As Table 5.1 confirms, the picture in Abruzzi-Molise is one of too many workers confined on too little land.

At this point we should emphasise that agricultural backwardness is very much part and parcel of a general economic retardation which affects every kind of economic activity in the peripheral rural regions. In progressive regions like Holland, the greatest stimulus to agricultural advance has come from outside the industry: from manufacturing and urbanisation. Not only has this provided farmers with all the benefits of huge markets and advanced technology: it has also provided an alternative source of employment for displaced farm workers. But in Abruzzi-Molise (as in much of the southern and western periphery), a lack of industrial development has meant that few such opportunities have been created. The Abruzzi-Molise region is remote, mountainous and much of it is far removed from the main lines of Italian communications. It has therefore failed to attract major industrial development and has been by-passed by the mainstream of tourist activity. Peasants and landless workers are therefore trapped on the land. For the larger landowners and farmers there is little incentive to modernise, when there is such an oversupply of cheap, able-bodied labour. The main burden of agricultural underdevelopment is thus borne by the workers and peasants themselves, who earn among the lowest incomes in the EC. The lack of opportunities in the region are revealed by an unemployment rate of 29 per cent amongst young people under 24 years of age (1979) (see also the case study of Sicily, in Chapter 1).

One of the many lessons to emerge from this case study is that agricultural development is very largely determined by events *outside* the agricultural sector itself. In general the regions with the most advanced farming are also those with the highest levels of industrial and urban development. Proximity to the city is a key factor determining modern patterns of land use and productivity. In saying this we should not lose sight of traditional geographical considerations such as variations in the natural environment. When we compare Holland with Abruzzi-Molise, we must not forget that we are comparing a region of cool, temperate lowland with one of drought-ridden Mediterranean mountains. Even so, the key difference between the two is not so much the nature of the land but the way the land is used.

Quite clearly the lagging areas of the EC have a considerable amount of catching up to do – regions such as Iberia which are still a world apart from the mainstream of the subcontinent. Despite strenuous efforts by the modern governments of Spain and Portugal to pull themselves into the twentieth century by industrialisation, there is still a huge gulf to be bridged. In no sector is this more true than in agriculture and in no country is the problem more acute than in Portugal.

Case study: Portuguese agricultural underdevelopment

Until 1974 Portugal had long been ruled by a fascist-style dictatorship, a one-party state run by a tight-knit alliance between the landowning, military and business-owning classes. By-passed by the Industrial Revolution, the country had remained heavily dependent upon agriculture. In 1970, agriculture claimed no less than 33 per cent of the active working population and in 1984 the figure was still over 20 per cent, a level of agrarian dependency which marks it out vividly from other countries of the EC (with the exception of Greece) or Scandinavia.

Without doubt part of the agricultural problem was, and still is, environmental in nature. In common with the rest of southern Europe, the southern portion of the country suffers from inadequate and unreliable rainfall and there are a host of drought-related problems such as soil erosion. Little more than one quarter of the national space is cultivated. However, as we have seen, agricultural land is amenable to improvement (notably, in this case, through irrigation). Thus the key questions to ask are not simply about climate and soils but about the lack of modernisation and the persistence of age-old, labour intensive forms of husbandry.

When, in 1974, the military regime was replaced by an elected government, the communist party became the dominant force. Their approach to agricultural development is interesting for us, since it represents one of the few socialist rural programmes west of the Soviet Bloc. For the communists, the root cause of agrarian backwardness was not environmental or technical, but *social* – the medieval class structure of Portuguese rural society. Broadly speaking the rural population had, for centuries, been divided into three principal social layers:

- *The upper stratum* – landowners of large estates (*latifundia*), in many cases aristocrats with a very lengthy pedigree;
- *The middle stratum* – of peasants (small landowners and tenants);
- *The lower stratum* – landless labourers, often hired on a seasonal or even a daily basis by the large estates.

This system was by no means unique to Portugal, being widespread throughout much of Mediterranean Europe and even further north. Wherever its geographical location, this ancient rural class structure gives rise to two very serious social consequences.

INEQUALITY AND POVERTY

Because so much land is held by a minority of people in large (sometimes vast) estates, there is insufficient land for the rest to gain a decent living, however hardworking, businesslike or innovative they may be. In Portugal, as elsewhere in southern Europe, the situation had grown worse over generations as a result of high birth rates which caused smallholdings to be further split between many inheritors. In Northern Portugal, over half the farm units are still less than two hectares in size.

From the viewpoint of the upper-class landlords, this was an excellent arrangement. Large number of peasants were forced to seek part-time employment to augment their incomes and, together with the landless, they formed a great army of workers prepared to labour on the estates for near-starvation wages. To say the least, this was an exploitive system, enabling a powerful minority to gain wealth at the expense of the poverty of the majority.

RESISTANCE TO CHANGE

In this situation progress and innovation are extremely unlikely, since any change is completely unnecessary for the ruling class. For the landowners, their estates were so

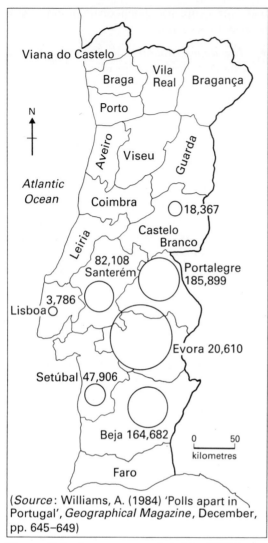

(*Source*: Williams, A. (1984) 'Polls apart in Portugal', *Geographical Magazine*, December, pp. 645–649)

Figure 5.4 Areas expropriated in 1975 by collective farms (hectares)

and church ensured that there would be little grassroots pressure for reform in the system.

After 1974 all this changed. The seizure of land by communist-inspired peasants prompted the Land Reform Law of 1975 which formally initiated a series of rural measures aimed at terminating the autocratic, latifundian system of capitalism. As a result, hopes were high in the countryside that this was the dawn of a new age. Over one million hectares, 14 per cent of all farmland in Portugal, was confiscated from the large estates in the Alentejo, the south of the country (33 per cent of the region's farmland) and redistributed (Fig. 5.4). From this small farms were grouped together into 550 'collectives' for greater scale and efficiency, employing some 59,000 workers – 7 per cent of the national and 35 per cent of the regionally active rural population. In retrospect, however, we find that in many ways Portuguese agriculture is weaker now than previously and although the rural workers in the Alentejo now have greater employment opportunities, improved social welfare and are living in a transformed society, the impetus of land reform elsewhere in the country soon weakened. In some cases a return to the 'old system' has occurred with land, previously seized from estate owners, being returned to them. Whereas in the mid 1960s the nation was self-sufficient in food, it now imports half its needs. Agricultural exports (mainly vine products including port wine and olive oil) are worth only one quarter the value of imports. It is feared that following Portugal's entry into the EC in 1986, the open competition with more efficient producers will force many branches of farming out of business. Olive growing is the most likely candidate for extinction.

There can be little question that politics lies behind this dashing of hopes. Since 1976 the communists have been replaced by a series of short-lived governments, with much switching of policies. One major victim of this has been the collectivisation programme. The landowning classes have recovered some of their former influence with government suffi-

large and labour so cheap that they were able to earn great wealth even by using the most backward technology. Moreover, any kind of change would tend to threaten their position, since it would be likely to raise the expectations of a generally passive rural population. For the peasants and labourers, no progress was possible either. A combination of ignorance, apathy and blind loyalty to lord

ciently enough to claw back much of its confiscated land in the Alentejo. Over 200 of the collectives have now disappeared.

Rural development has also been set back by development elsewhere in the economy. The government's drive for industrialisation and for tourist growth has led to rapid employment expansion and urbanisation in Lisbon, Oporto and various other coastal locations. This has not only drained away government investment from agriculture but it has also caused a migration of many of the younger able-bodied rural workers.

In summary, we can take the Portuguese experience as a model for many of the crucial issues in agricultural development. It provides us with an insight into the nature of the pre-modern rural economy; it reminds us of the stresses and strains of modernisation; it demonstrates that social well-being depends upon wealth distribution as well as wealth creation; it emphasises that modernisation is not just a technical exercise but also a matter of politics; it shows the way in which the agricultural sector is dependent upon trends in the rest of the national economy and indeed the world economy.

d) Land fragmentation and consolidation

Although there remains a vast gulf between the lagging and leading regions this is not to say that the former are not experiencing any progress. Throughout the peripheral regions of Western Europe there are many examples of recent agricultural transformations stimulated, often, and guided by, government policy.

Some of the earliest state intervention in Western Europe occurred during the nineteenth century when the enclosure movement in Britain was adopted in northern Germany, Switzerland and Austria, whilst in Scandinavia early attempts were made to replace fragmented open fields with consolidated holdings. In other countries of Western Europe, however, land consolidation has not proceeded so swiftly and in many parts of Spain, northern Portugal and Crete the spatial structure of many farm holdings is still characterised by many small, non-contiguous strips or pockets of land scattered over a wide area. Such a pattern has developed for any one of a number of reasons – in France, for example, it was very much a legacy of land inheritance laws introduced during the Napoleonic era in which farmland was divided up between heirs with the passing of successive generations. Elsewhere land fragmentation has been seen as a response to piecemeal reclamation as at Vriezenveen in the Netherlands and in other areas where low-lying, marshy land has been reclaimed, as in the Hunte Valley near Bremen in West Germany.

Inevitably, such a structure contains both economic and social disadvantages being not only time-consuming, uneconomic and inefficient but also expensive to maintain in terms of providing a basic farming infrastructure of fences, buildings and water supply. Social disadvantages accrue as a result of the problems involved in often trying to determine the exact pattern of ownership and the social tensions which often arise as a result of disputes over ownership, access and damage. For local and regional government too, land use planning is hampered by this outmoded structure in which high population densities often make official decision making extremely unpopular.

Despite the largely negative consequences of land fragmentation, in some parts of Western Europe, most notably in Alpine regions, farms divided into small parcels of land remain closely related to local environmental conditions. The system of transhumance, based upon the seasonal movement of farmers and their livestock from mountain pastures to valley floor meadows, remains a satisfactory solution to overcoming environmental constraints.

Today, land consolidation has taken place in most countries of the sub-continent varying in scope from being a package of integrated agricultural reforms to the simple, but effective, transfer of small parcels of land. Some of the most extensive schemes have occurred in Spain, France and West Germany. In the latter two countries consolidation programmes have involved approximately ten million hectares of land, taking until the end of this century to complete.

Case study: Land consolidation in Spain

Despite the size of the schemes in Spain, there is still a pressing need for further development. In the depressed agricultural region of Galicia, Spain's most north-westerly region, some of Western Europe's highest population densities are recorded – with over 400 persons per square kilometre of farmed land. The farms themselves are extremely small, seldom exceeding five hectares and often being subdivided into 30 or more separate fragmented strips giving the distinct rural social structure so often associated with *minifundia*. Such a system is quite the opposite to latifundia and has been created over two or more centuries of tireless sub-division of rented holdings as tenant farmers and share croppers have gained title to their land. The new generations of peasant landowners, apart from being fiercely conservative, remain materially deprived and poorly educated lacking the funds, technology and market expertise to compete effectively. Instead supremely interdependent farming families have evolved channeling their total energy into maintaining the title of the land regardless of the economic and social consequences whilst at the same time retarding worthwhile agricultural development.

Elsewhere in Spain the agricultural landscape has similar characteristics. In Andalucia, a region comprising 17 per cent of Spain's territory and providing nearly 20 per cent of her agricultural output, the great estates around Huelva, Cadiz, Seville and Cordoba contrast vividly with the multitude of smallholdings, often backward and inefficient, stretching from the depressed eastern highlands to the more prosperous 'vegetable patches' of the south-eastern coastal belt.

Throughout Spain, despite a number of attempts at land reform, solutions to many of the problems have remained elusive within an agricultural system which has, over the centuries, become so firmly entrenched. Of the attempts which have been made during the past 60 years many have floundered as a result of the intense political pressure exerted upon national and regional government by the large estate owners who, not surprisingly, have wished to see the status quo maintained. The years immediately prior to the start of the Spanish Civil War in 1936 heralded just such an attempt with the Republican government of the day proposing to introduce extensive land reforms by expropriating land with the sound intention of redistributing it among poorer peasant farmers. Unfortunately, the nationalist victory in the Civil War under the leadership of General Franco, who was supported by powerful interest groups such as the latifundists, ensured that the Law of Agrarian Reform (which had been introduced in 1932) was soon repealed. The status quo was to remain unchanged for another thirty years.

In the 1960s, however, some of the first positive steps were taken though more as a result of market forces than government initiatives. Firstly, the situation was helped by the massive exodus from agriculture during the late 1960s and early 1970s as well as significant emigration of population to the countries of north-western Europe. As a result the proportion of the population employed in agriculture was halved from 48.9 per cent in 1950 to just 24.7 per cent in 1974. Reduction of the labour force allowed a smooth transition to the second major change – mechanisation. Thirdly, changing consumer habits, both at home and abroad led to a spectacular rise in demand for fruit, vegetables, meat and dairy products. To meet these rapidly changing circumstances large wealthy estate farmers adapted swiftly to new crops – but so too did smaller family farms who, in many cases, because of the scale of their operation were more responsive to changing

RURAL WESTERN EUROPE 143

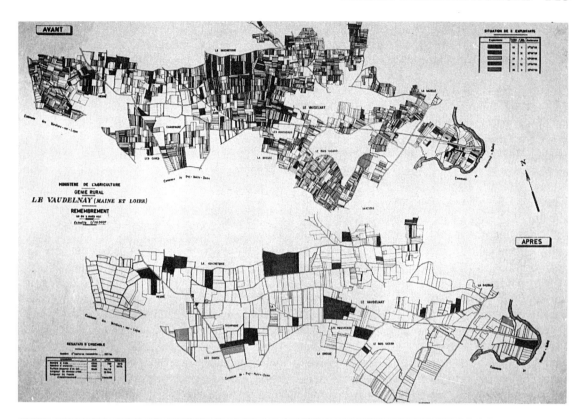

Figure 5.5 Land consolidation schemes in France

market forces. In Catalonia, Murcia and Valencia, adjacent to the Mediterranean coast, small and medium sized farms intensified their fruit and vegetable production techniques. In the coastal parts of Cantabria and in the regions bordering the Pyrenees meat and dairy production was expanded to meet the growing demand.

Thus through the processes of emigration, mechanisation, specialisation and intensification Spanish agriculture has been transformed. With fewer people dependent upon the land for a source of income the pressure for the reform subsided. More recently, however, many of the old problems have started to resurface as a result of the decline in industrial growth which has had the effect of reducing out-migration from areas such as Andalucia. Inevitably, high unemployment levels – nearly 23 per cent in Andalucia in 1983 – further exacerbated by the return of migrants from northern Europe have started to undermine the smooth path of agricultural change. Once more demands for land reform are being voiced in regions such as Andalucia and as a response the socialist government in 1983 legislated for land reform in the region by means of expropriating or leasing many of the more inefficient large estates. Whether such a policy will follow a smooth path still remains to be seen.

Land consolidation can provide a solution to many of the worst problems in areas like Andalucia and Galicia and can chart the way for farm amalgamation. Additionally, a number of countries have introduced legislation to ensure that when farmland falls vacant it is used for enlarging neighbouring holdings – such policies have been pursued in Austria, France, the Netherlands, Sweden and West Germany. Elsewhere, pensions or grants towards retraining are offered to farmers who are prepared to sell their land for amalgamation. In France and Sweden public agencies have been established to purchase land in order to create 'banks' which can be used for farm amalgamation at a later date.

In addition to these and other schemes, farmers in many countries are encouraged to set up joint farming groups for the purpose of pooling machinery and equipment. This enables small operators to have greater access to labour saving technology which they would otherwise have been unable to afford. A further innovation since the early 1960s has been the widespread formation of co-operatives for collecting, processing and marketing crops and livestock.

Case study: Land consolidation policies in France

Land consolidation was initiated in France after the First World War, and placed upon an official footing in 1941 when the policy of *remembrement* was adopted in which a special regional commission was appointed to investigate the possibility of land consolidation in any rural area where 75 per cent of farmers requested it. The need to consolidate land holdings in France was quite desperate with over 151 million parcels of land (with an average size of 0.36 hectares) supporting a massive labour force, impeding the spread of mechanisation. Consequently, 15 million hectares of land were eventually earmarked for reorganisation, comprising some 40 per cent of all farmland in France. By 1981, over 7 million hectares had been consolidated under the scheme, the majority of which had taken place in north-east France and the Paris Basin – which alone accounted for 50 per cent of all schemes. In these areas the advantages of an open field system, highly fertile and productive land and essentially progressive tenant farmers, soon paid dividends. Elsewhere, however, progress has been retarded – particularly in western, south-western and central districts of the nation. In areas such as Brittany, Aquitaine and the Massif Central, traditional conservatism amongst owner occupiers and their greater reluctance to pay their share of the costs have hindered developments. In addition the 'bocage' landscape of central and western France with its small enclosed fields have proved more expensive to enlarge than the open field systems found elsewhere, and the poorer physical environment has dampened the enthusiasm shown by small farmers in areas such as the Paris Basin.

Unfortunately the spatial expression of remembrement has only highlighted many of the regional differences which have traditionally existed in French agriculture, with the richer, more productive regions of the north and east growing at the expense of the south and west. In order to rectify these disparities the French government introduced in 1960 a scheme whereby 29 regional bodies (SAFER[1]) were set up to supplement remembrement schemes. Their task was to purchase land which came on to the open market and to use it to enlarge those farms too small for efficient production. This time, however, the cheaper price of land in southern France benefited this region rather than areas such as Brittany or northern France where land prices were more expensive. In 1962 another approach saw the establishment of another scheme (under the auspices of FASASA[2]) which was entitled to give further incentives, this time by offering retraining schemes for disillusioned farmers and promoting early retirement for all farmers over the age of 60. In turn, the land vacated by this scheme would then be used to enlarge adjacent farms.

Unfortunately, the lack of funds has prevented both of these schemes from reaching fruition. Despite this, however, the past 45 years have witnessed a major restructuring of agricultural land in France whilst increasing the viability of each farm. It has been suggested though, that from an economic viewpoint farms are still too small to provide a reasonable income.

1. SAFER – Sociétés d'Aménagement Foncier et d'Établissement Rural
2. FASASA – Fonds d'Action Sociale pour l'Aménagement des Structures Agraires

Case study: Land consolidation in West Germany: the policy of 'Flurbereinigung'

The impact of unfavourable physical conditions combined with the effects of a detrimental historical legacy left farming in West Germany in need of drastic improvement and modernisation after 1945. As a consequence of inheritance rights land holdings in many parts of the nation, but especially in Bavaria and parts of Baden-Würtemburg, were excessively fragmented, preventing the post-war generation of farmers from benefiting from the major advances that were occurring in terms of new technology. To combat these structural limitations the policy of 'flurbereinigung' was adopted with the purpose of amalgamating the scattered parcels of farmland and consolidating holdings. In addition to field enlargement and the eradication of the problems most commonly associated with fragmentation the policy went further, promoting and supporting the construction of new access roads, field drainage and water supply systems dramatically improving the rural infrastructure. Allied to these developments in the 1950s and 1960s field reorganisation often occurred in tandem with agricultural resettlement, involving the removal of farmsteads from their central position within small farming communities to the village periphery or, often as not, to sites within the newly reorganised landholdings. Since the early 1950s the Agricultural Resettlement Programme relocated 23,000 farms throughout West Germany. During the last 10–15 years, however, the high cost of such a policy has led to a significant change in direction away from settlement *relocation* to settlement *renovation* improving the existing buildings within villages and hamlets. Nevertheless, since 1945 60 per cent of West Germany's agricultural area has been so treated accounting for more than 7 million hectares, making it one of the largest land consolidation policies in Western Europe.

The visual impact of such policy in West

(*Source*: Thiene, G. (1983) 'Agricultural change and its impact in rural areas' in Wild, T. (ed.) *Urban and Rural Change in West Germany*, Croom Helm, London)

Figure 5.6 Flurbereinigung in part of Gemeinde Krombach, Spessart

Germany can be seen in Fig. 5.6 which considers the district of Krombach in Hessen. Before 'Flurbereinigung' the 1,065 hectares of farmland in the district was divided into 8,800 strips with an average size of 0.12 hectares. With the implementation of land consolidation this number has been reduced twelve-fold with the original field pattern being all but eliminated.

2 The transformation of rural life

For the past 40 years we have been experiencing a new agricultural revolution – a time of rapid change and development. For many rural people, however, such modernisation is a mixed blessing. The benefits of the new are achieved by the abolition of the old, a painful experience for many of the individuals involved. For the larger, more progressive farmers able to take advantage of it, development has meant a rising living standard. For urban consumers it has meant cheap and diversified food.

Socially desirable or not, rural change is irreversible. The countryside has ceased to play the role which it has occupied for centuries. No longer is it populated mainly by agriculturalists and the traditional rural crafts and services which supported them. Increasingly its traditional population is melting away, to be replaced by urban newcomers – long distance commuters, retired people, second home owners. Increasingly rural society develops new activities such as tourism, geared to the needs of the urban leisure seekers. Although many of these developments are clearly beneficial to many rural areas, preventing a check on depopulation, the changing composition of rural society can cause additional problems in its own right.

a) Depopulation

The major impact of the structural changes which have taken place in agriculture since 1945 have been felt, most noticeably, in terms of rural employment where significant reductions have been made in the labour force. As a consequence labourers, as well as many farmers, have been forced to leave the land and seek employment elsewhere. In most cases this has meant a drift to the towns and cities to find work in the manufacturing or tertiary sectors of the economy (see Chapter 6). In 1980, only 10 per cent of the total workforce of Western Europe was gainfully employed in agriculture and its associated industries (forestry, hunting and fishing). Furthermore, this figure had fallen dramatically throughout the sub-continent and every nation had experienced a sustained decline in their agricultural workforce since 1960 (see Fig. 5.3). By far the greatest decline had occurred in Spain, Finland and Italy whilst only Greece and, as we have already seen, Portugal retained over 25 per cent of their total workforce in agriculture in 1980. Elsewhere, in Belgium and the United Kingdom under 3 per cent of the total workforce was directly employed in agriculture.

Apart from job losses, there are much wider implications for rural society in general. Where agriculture is the main employer migration from the land can, and often does, result in a decline of economic activity contributing to a vicious downward spiral of rural deprivation experienced by those people who remain in the area. With a declining population it becomes increasingly difficult to meet the thresholds which are required to maintain a satisfactory range of essential services. As demand continues to fall, services such as health and educational facilities, public transport and shopping facilities are progressively withdrawn which, in turn, prompts further outmigration (see Fig. 5.7).

Whilst modern agricultural practices rely less upon labour in the more favoured farming regions of the sub-continent, such as the west Netherlands, some of the most substantial losses have occurred, however, in the remoter, peripheral parts of the region. Four provinces in Portugal (Guarda, Braganca, Vila Real and Viseu) lost between 1960 and 1970, 26.8 per cent, 24.4 per cent, 18.9 per cent and 15.4 per cent of their population respectively, amounting to over a quarter of

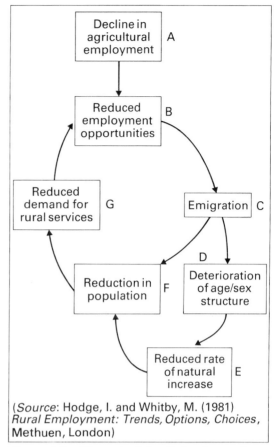

(*Source*: Hodge, I. and Whitby, M. (1981) *Rural Employment: Trends, Options, Choices*, Methuen, London)

Figure 5.7 The principle of circular and cumulative causation as a result of the decline in agricultural employment

Self-assessment exercise

Answer the following questions which relate to the diagram, Fig. 5.7
  i) Box A: Why has there been a decline in agricultural employment throughout Western Europe since 1950?
  ii) Box B: What other sorts of employment activities are affected by a decline in agricultural employment?
  iii) Box C: Where do people migrate to, and for what reasons? What are their particular socio-economic characteristics?
  iv) Box D: Construct a theoretical population pyramid to illustrate the visual character-

istics of a declining rural population.
v) Box E: From the population pyramid you have drawn outline the main demographic features of the population and suggest reasons why there is a reduced rate of natural increase
vi) Box F: Are there any possible advantages of a reduced rural population?
vii) Box G: List the the types of rural service which might be affected by a decline in population. Is there any distinction to be made between a rural region in north western Europe and one in southern Europe?
viii) Is it possible to control or even reverse the basic trends illustrated in Fig. 5.7. If so, list ways in which this may be achieved.

a million people. Elsewhere, in the greater part of Scandinavia, much of Western France and the Massif Central, the Irish Republic, parts of Austria as well as areas in West Germany, southern Italy and much of Greece, depopulation has reached similar levels.

Rural depopulation is not, however, a new phenomenon. It has been a major characteristic of population movement in north-western Europe since the first period of industrialisation in the early nineteenth century, when the emerging industrial regions were in desperate need of additional workers. The unlimited opportunity for employment in these blossoming areas allowed rural peasants to reject, once and for all, the traditional hardships associated with rural life, and gave them the chance to obtain a better standard of living and improve their quality of life far removed from the land (although in reality the hardships associated with the new urban living were often more severe than life had been in the countryside). Nevertheless, the expanding urban areas were a magnetic attraction to hundreds of thousands of country dwellers who migrated in their droves to find, in their belief, a better way of life. Today, although the same motives for rural migration prevail, the chance of finding satisfactory employment in the urban areas of Western Europe has been significantly reduced as the impact of industrial restructuring throughout the region has led to a significant fall in labour requirements.

Nowhere can return migration be more vividly demonstrated than in Portugal. The attraction of the industrialised nations of north-western Europe was mainly responsible for 1.5 million Portuguese citizens leaving their homeland between 1960 and 1974. Of these many were peasant farmers or landless labourers determined to leave their impoverished rural background in order to eke out a better standard of living in the prosperous cities of West Germany and, more importantly, France.

However, for many, the notion of untold prosperity was short-lived with the unannounced arrival of the industrial recession of the mid 1970s severely curtailing the demand for such unskilled labour. With little help or opportunity of bettering themselves, return migration became the only option for many hundreds of thousands of peasants whose numbers were swelled additionally by 800,000 migrants from Portugal's former African colonies of Angola and Mozambique, following their independence in 1975. Whereas the nation's population had fallen by 2.6 per cent between 1960 and 1970, the next decade experienced a growth of 15 per cent.

For many return migrants who had experienced life and tasted the rewards of the advanced urbanised societies of France and West Germany, there was a general reluctance to return to the land. Consequently, rural regions in Portugal such as Beja (-8 per cent), Braganca (-2 per cent) and Guarda (-3 per cent) experienced considerably lower rates than the metropolitan regions of Lisbon-Setúbal (+32 per cent), Oporto (+19 per cent) and Faro (+21 per cent) at the centre of the Algarve tourist industry.

For the residual population problems are exacerbated not least by an unbalanced age and sex structure characterised by those physically or socially less mobile than their more dynamic counterparts. As a consequence many depressed rural areas are over represented by an elderly population as well as an above average proportion of females, the result of which has serious repercussions on the whole vitality of the region. In Italy progressive rural depopulation has resulted in a situation whereby 50 per cent of all farmers are over the age of 55. Inevitably, the over abundance of elderly farmers in regions such as Abruzzi-Molise remains a major obstacle to the technical and capital improvement of the whole agricultural sector of the Italian economy.

Having been with us for so long it is hardly surprising to find that a number of attempts have been made to stem the tide of rural depopulation in certain parts of Western Europe. Unfortunately, experience has shown that there is no universal solution. In most cases schemes have been promoted with the help of government backing, such developments seen as an integral part of regional policy. As a result a variety of development bodies and organisations have been established throughout Western Europe to promote some of the most severely depressed rural regions – such as the Languedoc, the Massif Central and Brittany in France, Northern Norway, Galway in the Irish Republic, the Highlands and Islands of Scotland and the Italian Mezzogiorno which still remains the largest single rural problem region in Western Europe (see Chapter 3). Without exception, limited success (albeit at considerable cost) has been recorded in most of these and other regions – job losses in agriculture have been absorbed into other sectors of the rural economy and the outflow of population has been halted or reduced. Additionally, certain growth points have actually experienced an increase in population as employment opportunities have expanded. Although the most notable growth has taken place in the service sector, in particular the tourist industry, developments have occurred elsewhere including rural manufacturing which has managed to increase its labour force since 1945. However, despite the availability of greenfield sites, local authority co-operation and generous government incentives, all too often firms are deterred from locating in such remote areas, particularly when regional aid is available in more accessible parts of the country. Not all industries are 'footloose' (see Chapter 2). For many there are restrictions on their freedom – the time and cost of transporting both raw materials and finished products to national, regional or even local markets may often be prohibitive, as may be the cost of energy supplies. Additionally, the availability, cost and quality of labour can also be a detrimental factor discouraging new industry.

Not surprisingly then, tourist and other service industries have been heralded as the greatest opportunity for the regeneration and revitalisation of the remotest parts of the sub-continent where, in reality, it represents the only possible source of new employment. In some cases tourism has been regarded as the panacea for development. After all, examples abound in Western Europe of remote and inaccessible rural regions which have been transformed in a couple of decades from obscurity into centres of mass tourism. In some Alpine areas, such as Ticino canton in southern Switzerland, remoter parts of Carinthia and Styria in Austria, the Greek islands from Corfu and Crete to Rhodes, distinct topographical and climatic advantages have ensured their survival – elsewhere, however, results have not been so successful. Obviously, not all regions are blessed with such a high quality environment and have also to face the limitations imposed by the impact of seasonality which can reduce a tourist season to little more than four months every year. Even then the incidence of adverse weather conditions – especially in peripheral regions adjacent to the influence of the Atlantic Ocean such as Western Ireland and Northern Norway – can seriously hamper widespread tourist development. Experience suggests, then, that the relationship between tourism and development is not as clear cut as it might at first seem. Furthermore, like all other industries, it is susceptible to changes in the economic climate as well as the less predictable, but no less serious, changes in fashion and taste. Perhaps more worrying is that the natural beauty of many parts of Western Europe has had, in many cases, to be traded off against the social and cultural disruption of holidaymakers whilst over expansion of facilities often has a negative impact upon the resort or region. In a study of the traditional landscape of Ibiza and Formentera in the Balearic Islands it was found that tourist development not only dislocated traditional industries and deskilled the local labour force but it also inflated land and property values and led to further rural depopulation as small farmers from the surrounding countryside left their smallholdings to find better employment in the fast growing resort towns. Tourism, then, may be seen as a double-edged sword, the long term effects of which are largely unpredictable, although it has been suggested that one consequence is to weaken the position of many rural communities and, in so doing, increase their dependency upon the core areas.

Of all the groups in rural society affected by rural depopulation, none have been so hard hit as the traditional peasant and peasant farmer – people for whom the family and the family farm are the basic units of social and economic life. Although their numbers have declined significantly in recent years it is estimated that anywhere between 8 and 30 million rural inhabitants in Western Europe can be so classified, comprising a distinctive group in the social geography of the region. Although it is difficult to be more precise about the actual numbers it is fair to say that many rural people in Greece, southern Italy, Iberia, southern and central France, the Alpine regions and the remoter parts of the British Isles and Scandinavia are dependent, to a degree, upon peasant agriculture. In all these areas certain characteristics acknowledge its existence. Firstly, there are those whose farms are located on marginal land, where the harshness of the environment has forced away the majority of inhabitants. Those peasants that remain do so because of inertia, lacking the motivation, initiative or means by which to migrate. Next there are those peasant farmers who own slightly more land, achieving a greater productivity which allows them to eke out a full-time living from their meagre resources. Finally, there are part-time farmers. This group above all has expanded rapidly since the Second World War and has started to characterise the agricultural landscape throughout much of Western Europe, particularly in West Germany, Italy and the Irish Republic. In this case peasant farmers either operate small businesses from or on the farm or else work outside the farm in some form of non-agricultural employment. In the latter case such workers have become known as 'five o'clock farmers' carrying out their necessary farmwork in the evenings or at weekends. In some areas, especially in mountain regions, the development of tourism has often provided a lucrative opportunity for farm-based accommodation and recreational activities such as camping, walking, fishing and riding.

The structural changes which have taken place in agriculture across Western Europe since 1945 have been instrumental in shedding labour from the land. With little or no alternative employment the redundant workforce has been compelled to look elsewhere for jobs. Consequently, rural depopulation still remains a potent and significant trend throughout Western Europe and although various remedies to stop the outmigration have been tried and tested – in particular the growth of tourism and the expansion of recreational facilities – results have not been totally successful.

b) 'Counter-urbanisation' – the repopulation of the countryside

Despite the variety of problems discussed in the previous section, depopulation is very much a process of the past in many rural regions of Western Europe. Countryside, not only within the sphere of urban influence but also in the more remote rural areas, has attracted people who had previously spent their whole lives working and living in the city. Increasing affluence, rising rates of car ownership and the growing desire to distance themselves from the noise and congestion of modern city living have drawn more and more people to the notionally romantic and tranquil countryside. Viewed from an urban perspective, the process of counter-urbanisation has affected the more accessible rural regions of most countries in Western Europe since the 1960s, extending suburbanisation, redistributing population from the central city to the periphery of the urban system (see Chapter 4 for a detailed case study of Paris, pp. 114–117). The birth of the new 'dispersed city' within 50 kilometres of major metropolitan centres had led to the growth of a society of pseudo-rural inhabitants who, despite being 'physically remote' remain 'mentally urbanised'. The majority of people who have settled in the metropolitan fringes and villages treat them, often, as dormitories, relying, to a greater or lesser extent, upon the city for employment, shopping and other essential services.

In this instance the overwhelming 'push' factors from urban areas tend to override the 'pull' factors of the countryside. More recently, however, areas well beyond expanding commuter zones have started to show a resurgence of growth suggesting that rural areas themselves are acting independently of urban systems and attracting population in their own right.

The urbanisation of the countryside, which has increased in momentum since the 1960s, has brought with it a number of serious issues directly related to the expanding population. In most areas, for instance, there has been a steady replacement of the indigenous rural population (those invariably displaced by the structural changes in agriculture) by middle class, managerial and professional people commuting to the city, as well as elderly people seeking rural homes for retirement. The influx of both these groups has significantly altered the social composition and power structure of many rural areas creating a new environment in which middle class values and a non-rural way of life predominate. Additionally, the increasing demand to live in these dormitory villages has tended to inflate house prices, causing a rural gentrification to occur which works to the detriment of the indigenous population.

Nevertheless, there can be some benefits to the local community – especially in the provision of some types of service and the generation of employment opportunities – but research has shown that improvements such as these should not be over estimated. Many newcomers, being highly mobile, travelling to town often, still retain important links outside the village and have little use for local facilities which might be inferior or more costly. Although village schools might benefit from a more youthful population, other services – in particular shopping and public transport facilities – may still continue to decline.

Repopulation of rural regions is not only confined to the more accessible parts of the countryside – a second major trend has developed since 1945 beyond the traditional commuting hinterlands in which some of the more remote and attractive parts of Western Europe have also been resettled. In this case 'rural suburbanisation' has been replaced by 'seasonal suburbanisation' in which city dwellers have invested in a 'second home'. Although rural retreats like these are not universally common throughout Western Europe there is an increasing incidence of ownership as sufficient income and growing amounts of leisure time has become the norm for many people. Ownership is most common in France, where 15–20 per cent of all households have some kind of second home, whilst in Scandinavia between 10–15 per cent of people own a rural retreat in addition to their main dwelling. Elsewhere in Western Europe, second home ownership remains below 5 per cent although increasing all the time. Although the majority of second homes tend to be older properties, often purchased cheaply in a state of disrepair, recent years have witnessed the

Figure 5.8 Second homes in Scandinavia

growth of purpose built accommodation for rent or 'time sharing', particularly in countries bordering the Mediterranean. Inevitably, the occupation of second homes is mostly seasonal with local populations swelled intermittently at peak times of the year. As a consequence positive benefits to the local community are often strictly limited, with the mobile, second home owners bringing the supplies that they need with them, putting little money into the local economy. Additionally, and more worrying however, is the impact that second home ownership has upon the market value of local property, inflating prices in some of the more attractive areas, well beyond the means of the local population.

c) Rural deprivation – growing concern

Although we have seen in the previous section that depopulation remains a central issue to the rural geography of Western Europe increasing concern has been shown more recently towards the disadvantaged groups in rural society for whom access to the basic necessities of life is a severe hardship. It may come as some surprise to learn that problems of rural deprivation are not confined to the remotest parts of Western Europe but can, and often do occur within 10 kilometres of a major city. Obviously, deprivation is a relative phenomenon, and the inequalities experienced by residents in the villages of southern Italy are far greater, in comparison to their counterparts in the metropolitan villages in north-western Europe, but nevertheless, in both cases, the quality of life of many residents can differ markedly from their neighbours.

In the remotest parts of Western Europe,

Figure 5.9 *The Sunday Times*, 16 June 1985

problems, albeit not at the same scale, are also evident in some of the more 'accessible' parts of the sub-continent, in areas of sustained population growth. In many countries in north-western Europe, proximity to an urban area does not exempt people from hardship. Although the process of counter-urbanisation has evolved as a response to a mixture of factors, certain sectors of this 'new' rural society remain largely untouched by the growing trend. Whilst car ownership in France or West Germany gives the best of all worlds to those households fortunate enough to possess one, it gives the worst of all worlds to those who, for varying reasons do not.

Despite the higher cost of fuel, the longer distances that have to be travelled and, often, lower wages, car ownership levels in many rural areas in north-western Europe are significantly higher than in surrounding towns and cities. For many people living in such areas, the ownership of a car is forced upon them as a response to the inadequacy of public transport. In such cases sacrifices are often made elsewhere in the household budget so that a car can be maintained at the expense of other necessities such as food and clothing. Furthermore, even within car owning households – especially where there is only one car – constraints are placed upon those who either cannot drive, do not wish to drive, or do not have access to the car for certain periods of the day. As a result there remains a residual population who are dependent upon other means of transport. In many villages the solution may be public transport or lifts offered by acquaintances, but elsewhere, where public transport is less frequent or non-existent, many people are deprived of the access to essential services which those with constant access to a car take for granted. Although elderly people form a significant proportion of those disadvantaged, others include people on low incomes, school children, the unemployed, the infirm and people in single car households.

such as northern Portugal, Sicily and much of Greece, progressive outmigration has left an elderly and largely immobile population deprived of essential services. But similar

It has been repeatedly argued in this book

that the most potent forces working for change in Western Europe derive not from local or regional sources but, increasingly, from national and supra-national agencies. In this respect we can include not only governments and other legislative assemblies such as the EC, but also the influence of major corporations which have both a direct or indirect involvement in the agricultural industry. Nevertheless, the role of such agencies is no better illustrated than by the part played by the EC and, in particular, the Common Agricultural Policy.

3 The Common Agricultural Policy (CAP)

Of all the Community programmes, the Common Agricultural Policy is the one that has generated the most conflict. In many ways it is the foundation stone of the EC today accounting for 70 per cent of its annual budget. On the one hand its provisions are jealously safeguarded by farmers, many of whom are increasingly dependent on it as a source of income and consequently they are prone to take to the streets in vociferous protest whenever changes are being contemplated. Yet on the other hand the CAP is greatly abused by many politicians and much of the public who cannot understand why they should finance the production of vast quantities of unwanted food through a policy which increasingly appears to be out of control. Not only does this seem a mis-use of scarce resources in a Western Europe suffering from widespread and severe unemployment, but the vast surpluses are seen as an obscenity when vast numbers of people in the Third World are starving, as the crises in the Sahel and especially Ethiopia since 1984 have highlighted. Needless to say it is also a potent source of political disagreement between the nations which make up the

Figure 5.10 Danish farmers' protest

European Community with the principal beneficiaries (notably France with her large farming population but also smaller countries like Denmark and Ireland) often coming into conflict with the major paymasters (the urban-industrial nations of West Germany and Britain).

There is, however, a more fundamental issue masked beneath these well-publicised and sometimes emotive debates, in that the CAP acts as the major mechanism for redistributing wealth within the EC. It does so in two ways which are closely interrelated. Firstly, there is a *structural* redistribution from the consumers (the EC is financed in part by a levy on VAT) who predominantly live in urban-industrial areas to the producers – the 8.5 million farmers who receive various payments from the CAP. In effect, it transfers income from the rest of the economy into a very privileged sector as in 1985 only 8.6 per cent of total employment in the ten EC states was in agriculture – although some countries benefit relatively more than others due to their larger agricultural sectors – see Table 5.2. Secondly, a *geographical* redistribution results from this process and this can be viewed on a variety of scales. Some indication of the national impact has already been given in Chapter 3, Table 3.3, which showed that France and Italy were the major beneficiaries. But, of course, there is also a significant regional impact whereby the main agricultural areas (both rich and poor) within each country receive substantial aid under a variety of programmes. This is not a simple urban to rural movement of funds, but a more complex pattern of certain agricultural regions benefiting more than others because some of the crops they produce are favoured with greater payments under the CAP.

a) Development and operation of the CAP

Agricultural policies in individual Western European countries had been framed in the light of the harrowing experience of the Second World War and the perceived need to avoid food shortages by protecting home markets and encouraging a rapid expansion of production. As a result there was a very restricted interchange of agricultural produce between countries pursuing protectionist policies and the newly-established EC saw the establishment of a common market in agricultural goods as an essential accompaniment to that for industrial products. The Treaty of Rome laid down five objectives for a common agricultural policy: to increase productivity; to provide a fair standard of living for farmers; to stabilise markets; to guarantee food supplies; and to provide food for consumers at reasonable prices. The six original members took this set of ambitious (and to some extent contradictory) aims and by the 1960s fashioned a variety of measures which together comprised the CAP.

The need for action was reinforced by a critical review carried out by Sicco Mansholt, Vice-President of the EC Commission for Agriculture, which argued that some parts of Western Europe had too many farmers working parcels of land that were too small, that too much land was in agricultural use, and that farming needed to be modernised so that out-dated, traditional practices (and ways of life) were changed onto a more business-like footing. Mansholt proposed that the

Table 5.2 Active population in agriculture in the European Community, 1985

	Agriculture as % of total national employment, 1985	Total number employed ('000) 1985
Belgium	3.0	107
Denmark	7.1	181
France	7.6	1,590
Greece	28.9	1,037
Ireland	16.0	169
Italy	11.2	2,297
Luxemburg	4.2	7
Netherlands	4.9	250
United Kingdom	2.6	626
West Germany	5.6	1,400
EC	8.6	10,403

Source: EEC Commission, 1987

drift of population from the countryside should be accelerated by encouraging elderly farmers and those with smallholdings to leave the land; that farm sizes should be markedly increased by replacing small peasant holdings with larger, more modern units; and that instead of guaranteeing prices for surplus products, efforts should be made to take land out of cultivation and use it for recreation and afforestation. Some of these controversial proposals were taken up and embodied in the CAP during the early 1970s.

Today's CAP is an amalgamation of all these various policies together with the measures introduced when the Community was enlarged to include Britain, Denmark and Ireland. It is based on three principles: a single market, Community preference, and a joint financial responsibility. Moves towards the abolition of customs duties and tariff barriers within the EC plus the harmonisation of regulations have gone a great way towards the *creation of a uniform market* for agricultural produce throughout the Community. Over 95 per cent of agricultural goods are now covered by a variety of uniform marketing mechanisms. Seventy-two per cent of products (including most cereals, sugar, dairy products, beef, lamb, pork, and some fruit, vegetables, and table wine) are supported by prices which are guaranteed to farmers; when the market price falls below a certain level the Community intervenes, buys up the produce and tries to sell it (for example cheap butter to Russia) or places it into storage (such as milk/wine 'lakes', beef/butter 'mountains'). A further 25 per cent of produce (eggs, poultry, and some cereals, fruit, vegetables and wine) are protected against low-price imports. Each year the Community sets common guaranteed prices for all member states but because of fluctuations in their exchange rates it also establishes 'monetary compensatory amounts' – extra subsidies paid to farmers to protect them from a drop in their income should the value of their currency change, relative to a special 'green' or farming rate of exchange.

The second principle, that of *Community preference*, is designed to protect the European farmer from low-priced imports and consumers from fluctuations in the world market. If the price of imported produce is lower than EC levels, then an import levy is imposed closing the gap between the two prices, whereas if world prices are higher, an export levy is imposed to try to prevent Community farmers diverting goods abroad. Finally, *joint financial responsibility* occurs as member states' contributions to the Community's overall budget are then channelled to the CAP's two sections of the European Agricultural Guidance and Guarantee Fund (EAGGF) with no attempt to balance them according to the size of their agricultural sectors: the Guarantee Section is that part dealing with all the price support measures, subsidies and levies. It is by far the largest component and has grown extremely rapidly in recent years, so much so that it has been considered as being out of control. In 1984, it accounted for 16,500 million ECUs (nearly £10,000 million).

The Guidance Section finances more positive, direct measures aimed at improving the agricultural situation such as funding for the improvement of farms, rural facilities, processing and marketing. It differentiates in favour of the less favoured regions, notably hill-farming and Mediterranean areas. In 1983 it had a five-yearly budget of 3,800 million ECUs which gives an annual budget of around £450 million per annum, a mere fraction of the Guarantee Fund.

The European Commission claims impressive results for its agricultural policies; it cites the increase in productivity of 6.7 per cent per annum between 1968 and 1973, and 2.5 per cent a year thereafter, and the fall of 55 per cent in the number of farmers in the Community since 1960 as tangible successes. Claims are also made that it has safeguarded farmers' incomes, ensured that Europe has not suffered any food shortages, maintained market stability and helped to achieve reasonable consumer prices. Yet to accept this would be to gloss over some alarming developments that have arisen as a consequence of the operation of the CAP. Doubts are mainly expressed about the escalating cost of the whole exercise, whether the measures used are effective, and over their side-effects (notably the huge food surpluses generated), as well as being a seemingly-endless source of disagreement between the member states as they try to safeguard national interests. Some have argued that the CAP is both economically inefficient as well as being socially unjust.

b) The economic and social impact of the CAP – who benefits?

In economic terms the impact of the CAP has been summed up as a means of paying more to maintain fewer farmers most of them in a state of dissatisfaction. Its spiralling cost, which periodically threatens to bankrupt the EC, is largely a consequence of the nature of the Guarantee Section of the EAGGF, the rapid expansion of which in the 1970s is illustrated in Fig. 5.11, and although trimmed back briefly in 1981, it has expanded sharply again. To guarantee prices regardless of the

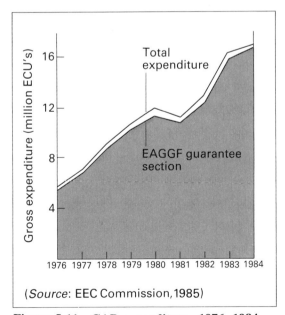

(*Source*: EEC Commission, 1985)

Figure 5.11 CAP expenditure, 1976–1984

Lupins set to flower as feed

**by Toby Moore
Agriculture Correspondent**

LUPINS might be about to supplant oilseed rape as the fashionable, Euro-subsidised feed crop in Britain. French farmers love the lupin, for it is ideally suited to their climate, but it might be less successful in Britain, and its introduction is likely to annoy Australian growers who now serve the British market.

French lobbyists have persuaded Brussels that lupinus albus (a cousin of the suburban lupin) makes an excellent feed compound. As a result – a price of £196 a tonne has been guaranteed – about twice the current world price.

The rush for the lupin this year is a direct result of farmers' fears that the EEC commission will finally force farmers to cut their highly-subsidised production of wheat and oilseed rape this year to reduce Common Market food mountains.

David Kay of the Writtle agricultural college which has spent nine years developing the white lupin, cautions against expecting too much too soon. "Circumstances have pushed things farther than current development mainly due to French influence," he says.

Even so, the subsidy will ensure that crops will be sown in many arable parts of the country. "Concern at what could happen to oilseed rape and cereal support," Kay says, "has led to enormous interest in other crops, with farmers saying: 'Let's give anything a go that can pass through a combine harvester!'"

Sidney Chapman, Conservative MP for Chipping Barnet, describes the lupin subsidy as a further example of the "riduculous common agricultural policy.

"I imagine farmers in Western Australia would go up the wall if we suddenly stopped taking their lupins."

This year's crop, to be harvested in September, is likely to be small, if only because demand for seed has outstripped supply.

Source: Sunday Times, 14th April, 1985

Figure 5.12

quantity produced merely encourages the most efficient farmers to grow more and more and to receive more and more subsidies. The situation is made worse as the support prices are invariably set at a high level in order to help the poorer more inefficient farmer but in effect this encourages the wealthy, large, low-cost farmer to overproduce and benefit more. Thus as production soared so inevitably did the cost of this policy (accounting for 95 per cent of the CAP budget). Belatedly the EC has acted to cut back this extravagant policy and in 1984 quotas were introduced to cut back production, notably in the dairy sector but also sugar beet and some cereals, and new plantings of vines were banned until 1990.

The main force behind these reforms was the attempt to get rid of the structural surpluses of foodstuffs built up as a result of the guaranteed price support system. Guaranteed prices simply encourage farmers to grow more and more in order to boost their incomes irrespective of whether they can sell the product. As a consequence the European Commision is obliged to buy up the surplus production, if it is perishable it has to be destroyed, some can be diverted to other uses (cereals for animal feed, wine for industrial alcohol), and the rest goes into store occasionally to be exported at prices much lower than it cost to produce (and those paid by European consumers). To many it seems, at best, to be a clumsy and extraordinarily expensive way of maintaining food supplies to consumers and helping to stabilise farmers' incomes. Not only does its open-ended nature threaten to bankrupt the EC periodically, but the butter mountains and wine lakes have become well-publicised and emotive issues which have led to widespread dissatisfaction with the whole operation of the CAP, which is characterised as an extravagant and wasteful programme which gives agriculture a degree of support unrivalled by any other sector of the economy.

In terms of social justice, the CAP tends to intensify inequalities that already exist. An unwritten objective of the CAP has always been the support of marginal farming areas, mainly the peasant farmers in the Mediterranean countries and Southern Germany as well as hill farmers. But in practice by paying the same level of subsidy irrespective of size of holding, the CAP does nothing to discriminate in favour of the poorer farmer, and, in fact, it makes the discrepancies wider. It has been estimated that there are 2 million efficient, modern farms in the EC as against 6 million inefficient smallholdings but the larger farms are able to achieve economies of scale, thereby making them more productive and profitable. It is the agribusinessman with a large, highly-mechanised farm that has been prospering from the CAP, and not the peasant smallholder, yet the latter is highly dependent upon it for his continued existence.

This inequality is compounded by the fact that the CAP subsidises some products more heavily than others, the bulk of expenditure being on milk products (30 per cent), cereals (16 per cent), beef and veal (9.5 per cent), and sugar (9.2 per cent).

As dairy and cereal production are predominantly to be found in the prosperous, lowland areas of northern Europe – milk production in the Netherlands, eastern and north-west France, and south-west England, cereal production in the Paris Basin, Denmark, and East Anglia – these regions benefit disproportionally more than the poorer peripheral regions especially those of the Mediterranean countries whose principal crops (wine, fruit, olives, rice etc.) receive relatively little support. Research has taken place into the major beneficiaries of the CAP. It shows that in relative terms the already prosperous countries of Denmark and the Netherlands have profited most along with Ireland. In contrast, Italy has fared notably badly.

As stated earlier these inequalities are part and parcel of the structural and geographical redistribution of wealth within the EC. At more refined geographical scale the regional impact of the CAP has been calculated and as

Figure 5.13 More wheat for the EEC grain mountain?

Fig. 5.14 illustrates the principal beneficiaries are the lowland agricultural regions of northern Europe, not the marginal farmers of the Mediterranean, arguably those in greatest need. The CAP clearly acts as a sizeable transfer of wealth out of the urban areas of Western Europe not to the rural periphery but principally to the agricultural regions immediately surrounding the urban core – a sort of 'halo effect' – which further emphasises the importance of the Eurocore as the predominant centre of production, consumption and wealth in Western Europe.

The fundamental dilemma concerning the CAP is the intermingling of the primarily economic motives for which it was formulated with the social, political and environmental issues to which it has become attached. This is perhaps best illustrated in the regions of marginal farming where it makes little economic sense to continue to subsidise inefficient farmers on smallholdings in a generally harsh environment when Western European agriculture as a whole is already producing vast surpluses. Nevertheless the consequences of the withdrawal of support could bring about rural decay and depopulation on a massive scale, social hardship, environmental disruption, and political unrest. It has been argued that a switch from price support to more direct aid for poorer farmers would be beneficial but what then of the other objectives of the CAP (greater productivity, maintenance of supplies, reasonable prices for consumers)? Clearly, the CAP involves mutually contradictory aims which are proving impossible to resolve and it is difficult to see any great change in the system as each member state battles to maintain maximum national benefit from the scheme.

Once again this chapter has shown that despite the massive advances made in agricultural development since the Second World War, only certain parts of Western Europe have benefited fully. Many areas – which have already been identified in other chapters – remain as peripheral in terms of agricultural

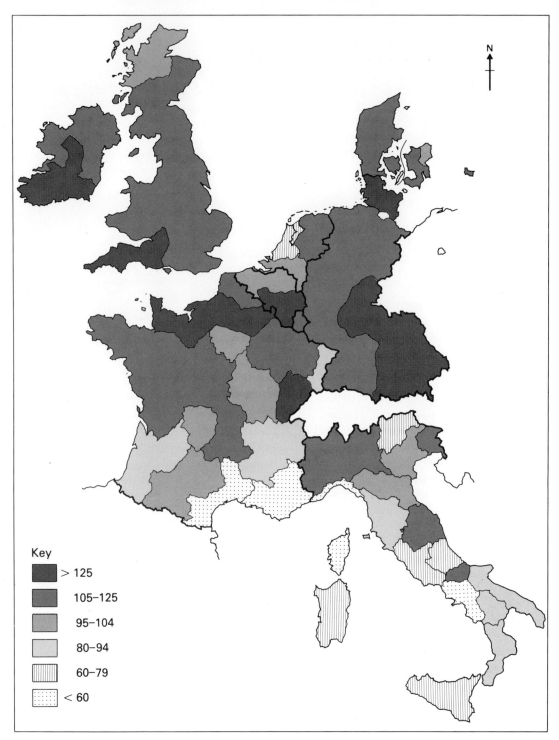

Figure 5.14 The regional impact of the CAP – indicators of support (average European support index = 100)

RURAL WESTERN EUROPE 161

COMMENTARY

The map shows how the northern European regions seem to benefit due to the nature of their production (cereals, sugar beet, milk) than the regions of the south producing fruit, vegetables and wine.

The regions with the strongest indicators specialise either in cereal products (the Paris region) or in milk (south of England, south of Ireland, Franche Comté) or in milk and cereals (German regions and Italian Friuli). In the Italian regions of Basilicata, Molise and Apulia there is also considerable support for durum wheat and olive oil.

The regions with the weakest indicators are those where specialised vegetable output predominates (fruit, vegetables and wine) in areas such as the south of France, western Netherlands and southern Italy.

development as they do in terms of industrial development. Peripherality and inequality do not, however, mean that rural areas are stagnating. We have seen that there is increasing demand for residence and recreation in the countryside which has had a major impact upon *all* rural regions significantly altering the traditional life-style of local society. Furthermore, the results of national and supranational policies, together with a variety of regional incentives has ensured that beneficial changes will continue into the future. It is doubtful, however, whether many rural regions (especially those which are remote) will ever develop as fast as the core regions of Western Europe and although positive changes can be noted the gap between leading (regions of the urban core) and lagging (regions of the rural periphery) parts of the subcontinent continues to increase.

6 Population as an Issue

1 Demographic change as an issue

a) Historical change

The population of Western Europe has been in a constant state of flux for well over a century. Both in sheer size and in geographical distribution, the nature of population has changed drastically during the modern age. The present chapter will centre upon both of these key features of population change. Firstly, in terms of the *growth in numbers*, which gained momentum during the last century but which has gradually slackened over the past four or five decades. Since the 1960s rapid population growth has ceased to occur in most of the countries covered by our study. Even so, it is far from a thing of the past as a subject for study, since it is past growth which is responsible for the present distinctively high levels of population density. Apart from Monsoon Asia, no other major world region approaches Western Europe's density of human occupation. Secondly, *migration*, the movement of people from one nation, region or local area to another, is another process which began in earnest during the Industrial Revolution and which, far from diminishing, has continued to gather pace up to and during our own era.

b) Public anxiety and population crisis

Throughout modern times national political leaders have constantly tended to regard population change as a problem in its own right. During times of rapid population expansion, fears are inevitably expressed concerning the capacity of available resources to cope with the extra mouths to be fed and bodies to be clothed and sheltered. Currently, however, the West European nations are becoming increasingly concerned about their declining birth rates. For almost a decade now the birth rates of most of the EC countries and of Scandinavia, Switzerland and Austria have been gently but persistently falling, to the point where there are now more deaths than births every year. In 1984, a conference of EC ministers expressed anxiety that most of their member nations were now no longer reproducing themselves from one generation to the next. To reproduce the next generation at a constant size there must be a ratio of 2.1 children per woman; yet in the EC the ratio has fallen from almost 3.0 in 1965 to about 1.6 today. If this trend proves to be a long term one, then Western Europe is undoubtedly entering a period of prolonged population decline.

Anxieties about a falling population seem somewhat out of place in a world which is suffering from acute population pressure and diminishing finite resources. Set against the fearful problems of the world's underdeveloped regions, Europeans appear fortunate indeed, protected as they are by sophisticated medical technology and health care systems and enjoying the freedom to plan their families according to individual financial means and emotional needs. Yet we do not need to hark back to the distant past to find a Europe whose population trends were not dissimilar to those of the present day Third World. Up to the first few decades of the present century high fertility was still the rule throughout most of Western Europe, a society in which childlessness and small families were regarded as abnormal and where artificial contraception was chiefly confined to the upper classes. Consequently birth rates remained high even though for many decades death rates had been waning steadily. As a result, the nineteenth century was a time of accelerating population growth, our own European population explosion.

The late nineteenth century was a time of population explosion in Western Europe. In the half century or so preceding the outbreak of the First World War, population increased by 134 million or 80 per cent. Even so this does not compare with the growth rates of some present day Third World nations, which, at present rates, are due to *double* their populations in less than thirty years. Yet, the late nineteenth century saw population growing at twice the percentage rate for the previous half century and almost five times the subsequent rate. The period from 1910 to the present day has seen only a percentage gain of 17.6 per cent.

Within Western Europe there was also a distinct time lag between the population growth phase of north-western Europe and the four countries bordering the Mediterranean, with the latter growing relatively slowly and losing their share of the West European population up until the First World War. Afterwards, they have grown very rapidly in relation to the rest of the continent.

Unlike the present-day Third World, however, the region on the whole suffered few of the dire consequences of overpopulation. Economic development together with emigration ensured that in the long run most countries of Western Europe were able to cope with the increase in their population numbers. In the rapidly industrialising nations like Germany, Britain and Belgium, extra population was largely absorbed by new manufacturing industries which in turn provided exports to be exchanged for imports of food. In this way, the lack of agricultural space was largely overcome. For these nations population growth was a positive boon in that it provided the expanding labour force necessary for industrial development. There was also emigration to the New World and Australasia, which siphoned off large numbers of unemployed and landless workers: this safety valve was especially valuable for areas such as Iberia, Italy and Ireland, where industrial expansion was grossly insufficient for the needs of an expanding population.

The question of balance between population and resources is quite clearly central to the geographer's interest. It is also a controversial one which can be interpreted in several ways. Historians of the last century are able to present much evidence which questions the assumptions made in the previous paragraph. Famines in France, Germany and above all Ireland, the exodus of millions of near destitute migrants to North America and countless other evidence of poverty might suggest a state of crisis due to too many people and insufficient land. Yet people are producers as well as consumers, with hands and brains to create wealth. It is thus possible to view the European population explosion in a positive light, as a great stimulus to economic development and other forms of progress.

What will the future historian make of our own population trends and our attitudes towards them? To say that European population trends have changed over the past century would be a gross understatement: they have been transformed out of all recognition. Both fertility (birth rate) and mortality (death rate) are now only a fraction of their former level and, happily, are far more under human control and less at the mercy of natural hazards. In most West European countries such has been the improvement in health standards that life expectancy is now over 70 years for members of both sexes, almost double that of two centuries ago. Only one hundred years ago, the infant mortality rate (number of babies dying in the first year of life) ranged from 96 per 1,000 live births in Norway to 235 in Germany. Now this rate has dropped below 20 in all countries of our study area except Portugal. In Sweden, where most progress has

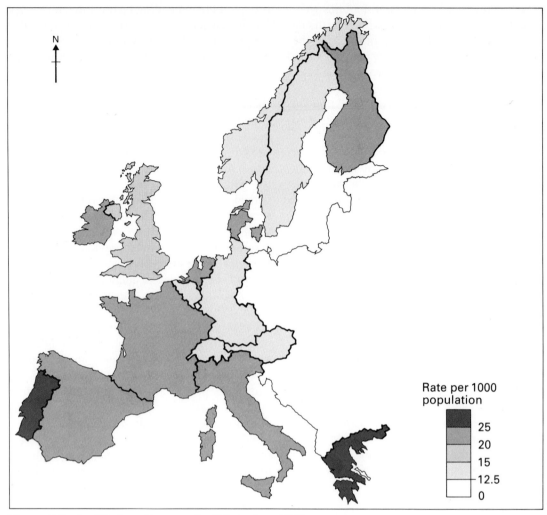

Figure 6.1 a) Western Europe: fertility rates, 1948

been made in post-natal care, the rate is now 7.7 which is less than one-fourteenth of its 1880 level. On the side of births, the crude fertility rate has fallen from over 30 per 1,000 (common throughout Western Europe until 1900) to below 15 per 1,000 today. The only exceptions are the remoter rural regions of the far south and west (see Figure 6.1.).

There has been a general decline in fertility over the whole of Western Europe during the post-war period, the only exception being Ireland whose crude fertility rate has remained around 22 throughout the period. International contrasts were much more pronounced in 1948, when it was almost possible to speak of Western Europe as consisting of two demographic regions: a zone of relatively low birth rates centred on the United Kingdom, West Germany, Norway, Sweden and the Alpine countries; and a peripheral southern and western arc with relatively high birth rates. Only Denmark and the Netherlands interrupted this arrangement. By the 1980s, however, many of the major differences have been ironed out, as the southern nations have experienced a much faster drop in their fertility.

These statistics tell a tale of human progress

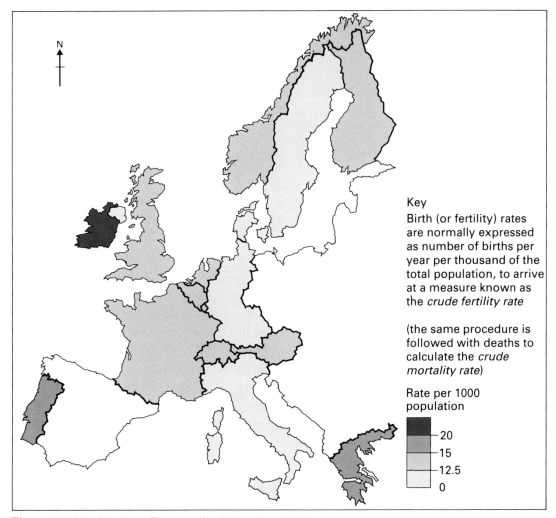

Figure 6.1 b) Western Europe: fertility rates, 1981

and liberation, the coming of an age in which the biblical standard of three score years and ten is now a realistic lifespan; in which families are no longer haunted by the ever-present fear of infant death; and in which women have been released from endless child bearing and rearing.

Yet this new individual freedom is not regarded as an entirely satisfactory state of affairs. In the first place it has led to a moral dilemma. Traditional religious teaching – and notably that of the Catholic Church which is still a dominant force in Ireland and the Western Mediterranean countries – regards artificial contraception and abortion as offences against the sanctity of human life. Not surprisingly, fertility is a major area of political conflict. Even in France, a nation regarded (by outsiders at least) as the acme of cultural sophistication, the sale of contraceptives was prohibited by a law of 1920 which was not repealed until 1967 in response to the weight of public opinion. Abortion law reform followed in 1975 bringing France into line with most of the nations of north-western Europe. It does not follow, however, that there is popular agreement on the subject even in those nations where liberal laws are in operation. Nor does it follow that everyone enjoys equal access to information and contraceptive aids. In France it took over a decade after the passing of the 1967 legislation for contraception to become widespread: now over two-thirds of women in the child-bearing age groups practice some form of contraception but there are still important sub-groups – the young unmarried, the poorly educated and immigrants – who do not enjoy this freedom of choice. This is well illustrated by the fact that in Paris the poorer, heavily immigrant quarters of the east of the city record a fertility rate almost twice that of the richest, most prestigious neighbourhoods.

Considerable variation may occur in social behaviour even at the local neighbourhood level. Like most other large cities in Western Europe, Paris is now falling very far short of reproducing its population. This is most marked in the richer quarters of the city, where the proportion of top professionals and managers is greatest. It is least marked in the east of the city, which is made up predominantly of working-class neighbourhoods (*quartiers populaires*), with a very high proportion of recent immigrants notably from Muslim North Africa. The general lesson to be drawn from this is that small family size is directly related to levels of affluence and education: the more affluent sections of the population tend to have better access to information and a stronger incentive to limit their families as a means of pursuing careers and other non-family activities. This tendency is not confined to Paris, or to France as a whole, but is fairly universal in application.

The second cause for concern is a demographic one. Over the past decade or so, modern Europeans have been using advanced methods of contraception (notably the 'pill') with such enthusiasm that total population has now entered a gentle but definite downward curve. This seems to be more a matter of official governmental concern rather than one which brings sleepless nights to the average citizen. Even so there is no denying that a falling birthrate will mean falling numbers of workers for the future and a possible under use of many expensively provided public facilities such as schools, health services and recreational amenities. Furthermore, when fertility falls below mortality, as is the case in France, West Germany, Scandinavia, Benelux and the Alpine countries at present, the inevitable result is an ageing of the population (see Box 6.1). A dwindling labour force will be obliged to support an increasing proportion of elderly dependants.

Case study: French population crisis

In July 1983, a French weekly magazine *Le Point* ran a feature on declining French fertility. It noted that since the previous October there had been a marked downturn in the number of births and announced, rather melodramatically, that, were this slide to continue, '1983 will be the blackest year for France since 1940'.

To outsiders such a reaction seems unnecessarily pessimistic. It is however entirely understandable in view of the strange events of French demographic history. Throughout modern times French population behaviour has tended to deviate significantly from that of neighbouring countries. Figure 6.2 shows that unlike most of her neighbouring countries, France's fertility fell more or less at the same rate as mortality. As a result national population growth occurred at a snail's pace in the midst of a continent whose population was generally booming. Government and public opinion were particularly worried by the military threat posed by Germany, whose growth of 23 million from 1850 to 1900 outstripped that of France by almost twelve to one. Lack of military manpower was even blamed for the disaster of the Franco-Prussian War.

Even today many French writers are preoccupied by the thought that their nation is falling back in the world population league. In October 1980 the journal *Science et Vie* calculated that the French share of global

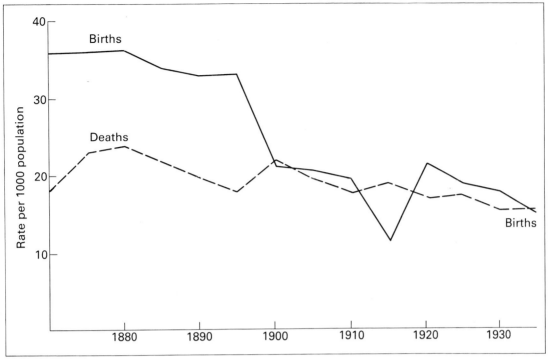

Figure 6.2 France vital rates, 1810–1935

population has fallen from 5 per cent in 1750 to 1.2 per cent today. Taking into account projected future world growth, France would have to increase its population to 370 millions by 2,050 if it were to regain its 5 per cent share!

Between 1900 and 1945, French population grew by only a little over one million. The First World War casualties took their toll both directly and indirectly through the loss of young men who were also potential parents. The close of the Second World War appeared to herald a new demographic dawn, however. In common with the rest of the continent, the return to peace inspired a 'baby boom', which was cause for national rejoicing. Typical of this period was the great feast of media rejoicing in 1967 when population eventually crossed the 50 million threshold. Fertility remained at or above 18 per 1,000 until 1965, a rate which is well below that of traditional or transitional societies but, with mortality below 13 per 1,000, sufficient to sustain a steady growth of around 0.5 per cent per year – generally considered a healthy and manageable expansion. Yet since 1965 it has diminished steadily and in 1975 fell below the level necessary to maintain population at a constant size from one generation to the next. Today, in France, the figure has fallen to 1.75 children per woman, which is significantly below the level required to replenish the next generation (2.1 children per woman).

To place these events into perspective, it is necessary to consider two final points. Firstly, France is no longer demographically unique but part of a continent-wide trend. A recent conference of EC ministers expressed concern at the fact that the majority of member states had fallen below the magic 2.1 figure. Secondly, demographic trends are subject to all manner of short-term oscillations. As yet there can be no definitive conclusions as to whether we are entering an inexorable population decline with critical repercussions for the economy and society of the future; or

> **Box 6.1 The problems of an ageing population**
>
> In 1950, people over 65 made up 5–12 per cent of the national population of Western Europe. By 1980, this proportion had risen to within the range 10.8–15.4 in the EC. If the two main causes of premature death – cancer and cardiovascular disease – were somehow eliminated, then the average life expectancy would rise to 80 years for men and 89 for women. Eventually, one in four of the population would be over 65.
>
> The argument that an ageing population is an economic burden is not clear cut and needs to be set against the following points.
>
> - Even though the retired population is increasing, the child population is declining: hence the ratio of dependants to active workers is not necessarily rising.
> - Even if ageing were to lead to an eventual reduction in the labour force, this might be welcomed as a counter to unemployment. In an era when increasing production can be achieved with diminishing numbers of workers, it seems inappropriate to bemoan any trend which promises to take people off the labour market.
> - Alternatively, any fall in the existing labour force could be used to create openings for women and indeed for any section of the community whose economic participation is at present restricted.
>
> While fears about labour shortage may be exaggerated, there is no doubt that the rising proportion of the elderly poses a social problem in its own right. For obvious biological reasons the elderly are a particularly vulnerable group, with less personal mobility than the young and with greater needs for health care, comfort and security. Unhappily, they are also the least able to provide for these special needs.
>
> - *Poverty.* In the majority of cases, dependence upon state and company pensions means relatively low incomes. Unlike active workers who can organise in trade unions, the retired have no 'industrial muscle' to protect their standard of living. Even if they were to organise as a pressure group, it is difficult to see what effective forms of direct action are open to them.
> - *Social isolation.* In the modern society, fewer and fewer elderly people live under the same roof as their children. Indeed, very many of them (especially women) now live entirely alone. Because married men are on average older than their spouses and because male life expectancy is in any case lower than female, widowhood is the inevitable lot of very many elderly women. In the EC, women 65 and over now outnumber their male counterparts by over 8 millions.

whether this is simply another of the short waves which characterise demographic history.

2 The transformation of Western Europe's population

a) The demographic transition

Most of the nations of Western Europe are experiencing zero population growth but incessant and very rapid migration, both within and across their borders. At this point we should emphasise again that these demographic conditions are very recent in origin and do not constitute an eternal or even a long term feature of the European way of life. On the contrary, they are specific to a particular stage of modern history, standing in complete contrast to past traditions.

It is now generally recognised that, with few exceptions, nations which experience modern economic development also undergo a demographic transformation. We noted in the first two chapters that the Industrial Revolution brought about a critical change in technology, economy and social life. In turn, these changes have exerted a direct influence on population trends: death rates have been reduced by scientific advance and improvements in living standards; birth rates have also been affected by the break up of traditional family and community life and the emergence of new forms. At the same time, the geographical mobility of the population has been vastly improved by developments in transport, the rise of new job opportunities in new locations and the more rapid dissemination of information.

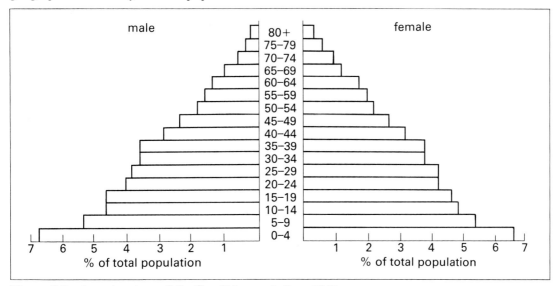

Figure 6.3 Age structure of the Swedish population, 1860

Age structure of the Swedish population, 1860 and 1981 (in thousands)

	1860				1981			
ages	male	%	female	%	male	%	female	%
0–4	259	6.7	255	6.6	246	3.0	235	2.8
5–9	255	5.3	203	5.3	280	3.4	267	3.2
10–14	177	4.6	186	4.8	293	3.5	279	3.4
15–19	177	4.6	177	4.6	301	3.6	287	3.4
20–24	154	4.0	159	4.1	281	3.4	270	3.2
25–29	150	3.9	156	4.1	296	3.6	282	3.4
30–34	138	3.6	146	3.8	332	4.0	317	3.8
35–39	138	3.6	145	3.8	329	4.0	309	3.7
40–44	111	2.9	119	3.1	248	3.0	237	2.8
45–49	90	2.3	101	2.6	218	2.6	216	2.6
50–54	69	1.8	80	2.1	224	2.7	228	2.7
55–59	62	1.6	74	1.9	241	2.9	249	3.0
60–64	52	1.4	64	1.7	235	2.8	248	3.0
65–69	40	1.0	43	1.1	209	2.5	232	2.8
70–74	23	0.6	34	0.9	174	2.1	211	2.5
75–79	12	0.3	19	0.5	115	1.4	162	1.9
80+	7	0.2	13	0.3	96	1.2	171	2.1
Total population			3,838,000				8,320,000	

(*Source: United Nations Demographic Yearbook, 1982*)

170 GEOGRAPHICAL ISSUES IN WESTERN EUROPE

Self-assessment exercise

The population profile in Fig. 6.3 has been constructed from the data for Sweden in 1860. Reconstruct this pyramid on a sheet of graph paper and draw a similar profile for the 1981 population data on a sheet of tracing paper. Then, take the tracing of the 1981 data and overlay it upon the 1860 pyramid and note the differences in population structure.

SOME KEY FINDINGS

By superimposing the two population pyramids you should get a visual appreciation of the contrast between modern Sweden and the same country at an earlier stage of population development. The 1860 pyramid is typical of a population experiencing high birth rates and still relatively high death rates. There is a continuous tapering off from bottom to top of the pyramid, indicating a rather high rate of death at all ages. The 1981 pattern is not a pyramid in the true sense, since there is no tapering off until the age of 40 is reached. In modern Sweden (as elsewhere in Western Europe) a low birth rate means comparatively few people in the youngest age groups, while a high survival rate means that there is no significant death loss until the 40 plus age groups.

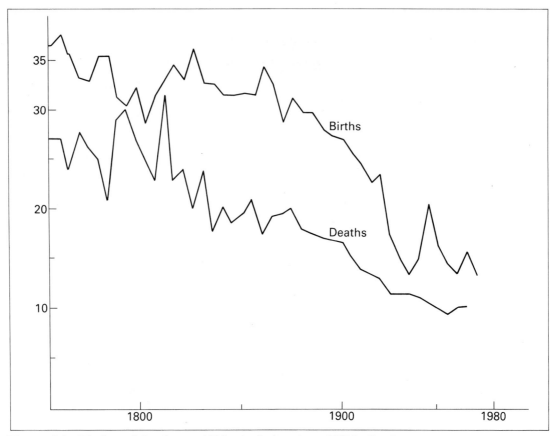

Figure 6.4   Births and deaths per 1000 population since 1750 in Sweden

Demographers recognise three basic stages in the population evolution of developing nations. In the *traditional* or *pre-modern* stage (which would apply to the whole of world history before approximately the late eighteenth century) population exists in a state of near-equilibrium, with high birth rates (fertility) almost cancelled out by high death rates (mortality). With the arrival of the Industrial Revolution, a *transitional* stage is entered, with death rates declining but birth rates continuing for some time at a high level. This dis-equilibrium produces rapid population growth. Finally, society enters the *modern* phase when birth rates also decline and population returns to a state close to equilibrium.

Figures 6.3 and 6.4 and the associated self assessment exercise monitor long term trends in Swedish vital rates. The diagrams show that both fertility and mortality now appear to have stabilised at little more than one-third of their 1800 level, and that Sweden has experienced a protracted interval when fertility lagged behind mortality. This sequence of events has been repeated in outline by most nations in our study area, although at different times and speeds (see Figs. 6.5 and 6.6). In consequence, there has been a

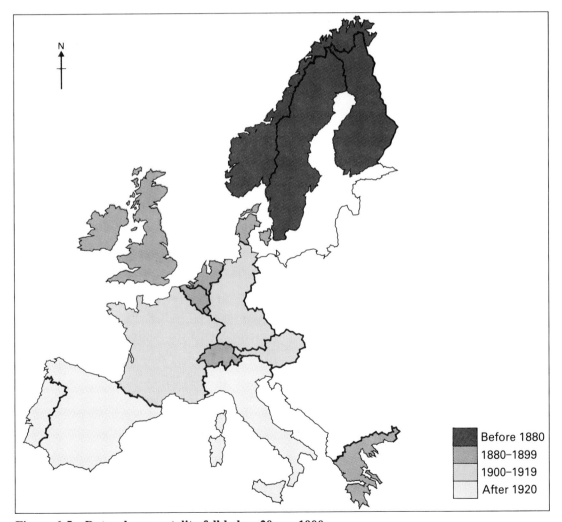

Figure 6.5 Date when mortality fell below 20 per 1000

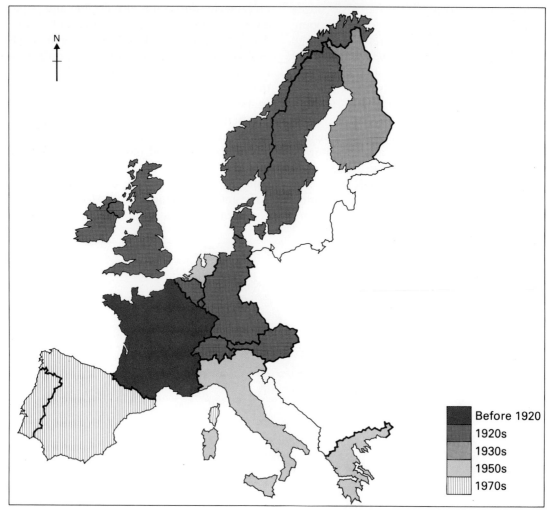

Figure 6.6 Date when fertility fell below 20 per 1000

period of very rapid population growth succeeded by the present era of slow growth or in some cases stationary population. During the period 1860 to 1910, the population of Western Europe increased by over 130 million, but since then, it has grown by only a further 53 million as a result of the shrinking birth rate.

b) Explanations for changing vital rates

I) MORTALITY

Explaining falling death rate is a relatively straightforward matter. We can identify two beneficial advances which taken together have literally altered the entire meaning of life and death.

Scientific medical advance. An eighteenth-century French writer observed that 'more people perish at the hands of doctors than are cured by them', and it is certainly true that the people of pre-industrial Europe were exposed to death from all manner of diseases which are now curable. Box 6.2 outlines some of the crucial breakthroughs which have occurred during and since the Industrial Revolution. It is useful to remember that these advances were part of a general increase in human

> **Box 6.2 The impact of disease in pre-industrial times**
>
> The two major killer diseases of pre-industrial times were bubonic plague and smallpox. It seems that medical advance played little or no part in the elimination of the former which disappeared during the eighteenth century as a result of what Clark calls 'a strictly exogenous zoological event': the disappearance from Western Europe of the plague-carrying black rat. In other cases, however, we can attribute the reduced impact of disease to a long painstaking succession of medical advances. The conquest of smallpox comes first historically, having been set in motion by Jenner's introduction of a successful method of vaccination in 1795. Other noteworthy landmarks are:
>
> - the discovery of anaesthetics (1840s);
> - antiseptics (1850s);
> - antibiotics (1940s).
>
> Even though there was usually a considerable time lag before new methods were accepted and applied, their eventual effect has been the successful control of a wide range of traditionally fatal diseases and to raise the general level of life expectancy over the past century and more.

mastery over the natural environment achieved through progress in science and technology.

Improvements in living standards. A further keynote of the industrial age has, of course, been vast and continuous increase in the production of wealth. As this has gradually become more evenly distributed over the mass of the population, so there has been a general release from the scarcity and want which bedevilled traditional European society.

Thus over the past century or so the peoples of Western Europe have benefited from a combination of improved medical knowledge, expanded health care services and increasing personal incomes giving access to improved diet, housing and other vital resources. Not surprisingly physical health standards have risen also, with the inevitable result of a decline in mortality.

While it is true that Western Europe as a whole has benefited from a prolonged decline in mortality, it is also true that the process has operated in a spatially uneven manner. Figure 6.5 shows that 'death control' appears to have been initiated in the northern areas – Britain, Scandinavia, Benelux, Germany and France – and to have crossed the Alps and Pyrenees at a later date. The pattern represents a form of contagious diffusion, with a space-time lag between the early-developing nations of the core and the laggards of the periphery. Yet more recently this initial gap has been largely closed, as death rates have fallen very rapidly in Mediterranean Europe and there is now no nation which deviates very far from the general norm of very low death rates.

II) FERTILITY

Whereas a diminishing death rate became evident in many Western European countries by the second half of the last century, it was not until the present century that fertility also began to turn downwards (Fig. 6.6). The basic reason for this delay is not hard to find. While the desire to cheat death is shared by almost all humans, the question of birth control is a far more controversial one which has surprisingly little to do with animal instinct, but much more to do with cultural values. From time immemorial and until well into the present century, European society set a very high value upon reproductiveness. Custom, tradition and religious belief demanded large families and was generally opposed to artificial restriction of births. Throughout the long ages of plague, epidemic and famine, this cultural bias towards high fertility was vitally necessary to ensure that the human species would reproduce itself in face of very high (and very unpredictable) death rates.

With the eventual decline in mortality, high fertility was no longer biologically necessary. Cultural attitudes do not change overnight, however, and several decades were to elapse before family planning became socially acceptable and widely used. Why has family planning now come to be the norm in a society which three or four generations ago would have been outraged at the idea? Among the most useful of the explanations commonly offered are the following: (NB These are *not* ranked in order of precedence.)

- *The changing family unit*. The traditional society was dominated by the *extended* family unit – i.e. three or four generations living under one roof. Modern society in north-western Europe has largely replaced this form with the *nuclear* family, consisting simply of parents and children, a smaller unit clearly less geared to the raising of large numbers of children. Since the 1960s this factor has been strengthened by rising divorce rates and an increase in the numbers of single parent households. Elsewhere in Western Europe society still tends to cling to the vestiges of the extended family unit.
- *The changing role of children*. Whereas in the old rural European society children were expected to work from an early age, the rise of mass education has transformed the child into an economic dependant.
- *The changing female role*. The gradual opening up of careers for women has given them an alternative role outside the confines of childbearing. This trend has been reinforced by the more recent rise of feminist political thought with its insistence on women's right to exist as people as well as wives and mothers.
- *Secularisation*. There has been a weakening in the influence of traditional religious teachings on sexuality, marriage, birth-control and related topics.
- *Child survival*. With the decline in infant mortality, it is no longer necessary for any mother to experience multiple births to ensure that some at least of her children will survive until adulthood.

Taken in combination these modernising changes have created a completely new social environment calling for fresh attitudes and behaviour. This new Europe is one in which small families confer more benefits than large. They make fewer demands on income and at the same time give parents greater freedom to pursue career, leisure and other opportunities for personal development. All of this is encouraged by a more relaxed social climate which no longer condemns childlessness and small families.

An intriguing point here for the geographer is that once again there are significant spatial variations, notably between urban and rural, north and south, core and periphery. This does not show up markedly at the international level, where only the Republic of Ireland stands out as lagging behind a continent which has now completed the demographic transition to very low birth rates. It is at the sub-national level that we find clearest evidence of spatial variation, with certain remote peripheral regions proving highly resistant to the diffusion of modernising influences from Western Europe's core. Italy provides a supreme example of the contrast between leading core regions and lagging periphery.

Case study: Italy

In no European country is the core-periphery split more sharply drawn than in Italy. While the northern regions have developed as part of the mainstream economy of Western Europe, the Mezzogiorno remains even now a kind of backwater, partially insulated from the currents of modernisation (see Chapter 3, pp. 90–94). Given the connection between economic development and population trends, it is no surprise to learn that southern Italy retains many demographic traits typical of the old Europe.

Taking the nation as a whole, Italy's birth rate has now fallen into line with the rest of Europe: at around 12 per 1,000 it conforms almost precisely to the EC average. Despite the powerful influence of the Catholic Church with its strictures against artificial contraception and the related fact that abortion is illegal (barring exceptional medical circumstances), the average Italian family is little

POPULATION AS AN ISSUE 175

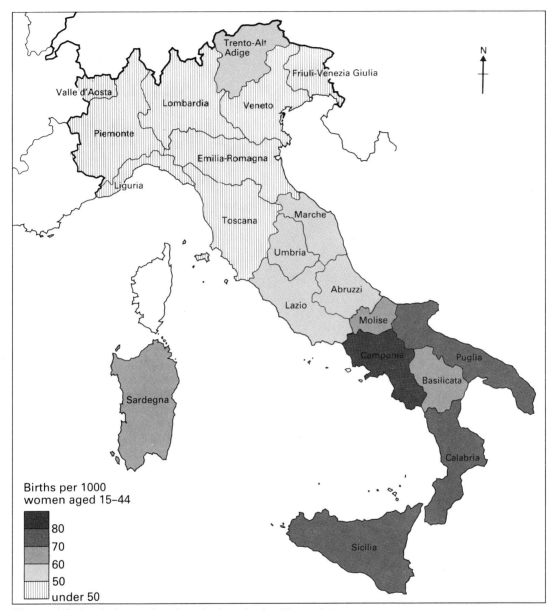

Figure 6.7 a) Italy: regional variations in fertility, 1979

larger than that of the Scandinavian and Benelux countries where official policy is much more liberal. Moreover, it is smaller than in Ireland, which is a similarly Catholic state. Yet as Fig. 6.7(a) illustrates these remarks do not apply to the five southernmost mainland regions, nor to the two islands of Sardinia and Sicily. These seven regions form a cluster, where specific fertility rates are up to twice those of the most industrialised regions of the Po valley.

All this is completely in line with the predictions of *spatial diffusion theory*. In effect, the relatively recent innovation of family planning has been readily accepted in those parts of Italy which are closest

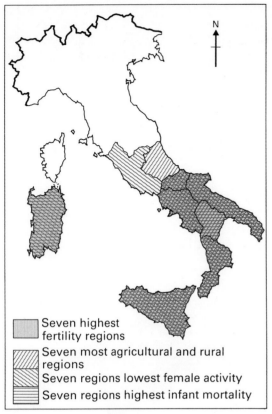

Figure 6.7 b)   Influences on fertility, 1979

opment has ensured the survival of its pre-modern agricultural economy, an economy of small family farms with few urban career openings or jobs for women; a society where the biblical exhortation 'be fruitful and multiply' is still held in high esteem. Liberal ideas about contraception and female emancipation are alien to a male-dominated rural society, where large families are still a mark of social status and where family labour is still a vital part of the local economy.

All this is not to claim that change is not already occurring. The efficiency of modern communication is such that even the most distant regions are constantly bombarded by stimuli from the outside world. Television ownership is little lower in southern Italy than elsewhere in the EC, literacy is now virtually universal and there is a constant return flow of migrant labourers who have worked in the urban core regions of the EC. In the past decade, crude fertility has fallen by over 40 per cent, from its former very high level of 16.5 per 1,000. Southern Italy is certainly undergoing the later stages of the demographic transition but it will be some time before the process is complete.

geographically, economically and culturally the core zones of Western Europe. Here the innovation has made headway even in face of the disapproval of the Church, the supreme arbiter of public morality. South of Rome, however, liberalisation has met with serious resistance from a society still in many ways rooted in the past. The evidence presented in Fig. 6.7(b) suggests that persistently high birth rates are correlated with: 1) geographical inaccessibility; 2) predominance of agriculture; 3) low levels of urbanisation; and 4) low female activity rates.

Essentially, of course, these four traits are interrelated. The key geographical variable is inaccessibility; the Mezzogiorno is remote from all the modernising influences which elsewhere in Western Europe have changed people's attitudes to birth control. In particular, its locational unsuitability for industrial devel-

3   Migration

a) The mobile society

Although in sheer numbers the population of much of Western Europe is virtually stationary, the very reverse is true of its *spatial* mobility. As time has progressed, so the European population has become ever more migratory. If we take the single case of France as an example of current migration, we find that between 1968 and 1975 the number of interregional migrants – individuals shifting residence across regional boundaries – was running at 654,000 per year. Taking into account migration within regions, we find that no less than half the entire French population (26 million) had moved house during the same seven years. Short distance migration is, as it always has been, the commonest form.

b) Migration and modernisation

Just as industrialisation, urbanisation and all the other characteristic processes of the modern age have caused humans to change their reproductive behaviour, so also it has influenced their migratory habits. It is undoubtedly the case that since the Industrial Revolution more people move more frequently over longer distances. This great increase in mobility represents yet another decisive break with the traditions of the past.

Over time, European migration has also changed in direction and purpose. During the first phase of industrialisation, the dominant direction of movement was from rural to urban areas, as rural population increase caused agriculture to become overmanned and as new jobs beckoned in the booming industrial towns. This trend still prevails in some of the less advanced nations of the periphery but in the more advanced economies migration has taken on a more complex appearance during the past few decades. In West Germany for instance the great inward movement from countryside to town which began in the 1870s has now slackened considerably and is outweighed by an outflow in the reverse direction. Hamburg typifies these trends. In 1977 the city experienced a net loss of 10,300 people, gaining 50,900 new arrivals but losing 61,200 former residents, most of them moving out to more spacious housing conditions beyond the city boundaries.

Another new trend of the post-war era has been the great increase in *immigration* from North Africa, the Near East and the peripheral countries of southern Europe. By the 1970s in West Germany the annual influx of workers from North Africa, Iberia, Greece, Turkey and Yugoslavia had reached almost half a million a year; and in France the equivalent figure was in excess of a quarter of a million. Rural-urban migration was still continuing but it had taken on an international dimension by the 1970s, with workers from largely rural agricultural countries migrating long distances across frontiers to take up jobs in the large cities of industrial Europe.

A third dominant trend in the more advanced countries of our study area is the high rate of inter-urban migration – the exchange of population between larger cities. Unlike rural-urban movement which brings about a long term change in the geographical distribution of population, inter-urban migration is more a matter of equal exchange, with each city experiencing a very great movement of people both inwards and outwards but with the two flows virtually cancelling one another out.

Case study: Migration and modernisation in Spain

Internal migration in Spain from the sixteenth century until the present day has been dominated by a movement of many thousands of people from the interior regions to the coastal provinces and major cities – particularly Madrid and Barcelona which today accommodate over 25 per cent of the Spanish population. For many decades this internal migration marked only the first stage of a more distant journey to other Spanish-speaking parts of the world such as Latin America and the Philippines. More recently, especially since the 1960s, Spanish migrants have looked northwards towards the heavily industrialised nations of Western Europe and their abundant employment opportunities. Although emigration has been a pronounced feature since 1960, the rapid growth of the Spanish economy had transformed the nation from a predominantly agrarian society to a modernised industrial country by 1975. Consequently, many migrants from the impoverished interior have had to look no further than the major areas of growth and expansion within their own country. The availability of work in both the industrialised northern provinces and the tourist areas of the

Mediterranean coastal strip provided an attractive alternative to employment elsewhere in the world. Not surprisingly, population growth in these areas between 1960 and 1975 continued unabated. Similar growth occurred in major towns and cities with Madrid increasing in size from 2.5 million inhabitants to well over 4 million by 1975.

Since 1975, however, internal migration trends have changed once again. The results of the international oil crisis, the ensuing economic recession and the political upheavals in the wake of the peaceful transfer of power from dictatorship to democracy with the death of General Franco, abruptly terminated economic growth in Spain. As elsewhere in Western Europe, rising unemployment, industrial restructuring and a check upon investment reduced the once burgeoning opportunities. With little chance of gainful employment anywhere else in Western Europe, or for that matter anywhere else in the world, the era of mass migration ceased. Likewise, internal migration trends were re-oriented and in many areas of Spain, including some of the central provinces of the interior, levels of depopulation have been reduced as a response to a lack of opportunities elsewhere. Additionally, the return of many migrants from other West European countries (see Chapter 5) has boosted the population of provinces such as Burgos, Toledo and Albacete which, for many years previously had witnessed a constant departure of migrants.

Although the main tourist regions have retained their vitality and have continued to attract migrants, other areas of the country which had experienced such rapid growth between 1960 and 1975 have not been so fortunate. In particular the northern provinces have been severely hit. In the coal mining and iron and steel areas of Asturias falling production, plant closures and high unemployment offer little opportunity for would-be migrants whilst further eastwards along the coast the Basque provinces of Guipuzcoa and Vizcaya – the two richest provinces in Spain in 1975 – have received a considerable setback due to the decline of heavy industries such as shipbuilding, iron and steel and heavy engineering. The problem here has been intensified by major political issues in the region namely the inability to retain and attract new investment in the light of Basque terrorism. As a result these two provinces had, by 1981, slipped to ninth and fifteenth richest from their leading position in 1975 (see Fig. 6.8).

4 Historical stages in migration

Once again it is helpful to explain present migration patterns by reference to their past origins. Several writers have noted the way in which migration characteristics – volume, distance, direction, purpose – change over time in response to economic and social developments. Each phase of recent European history seems to have been associated with a distinct set of migration characteristics.

a) Traditional society

On several occasions we have noted the static nature of society before the Industrial Revolution, especially the absence of long term economic and population growth. In the peasant society of old Europe, our rural ancestors were virtually tied to their native village by all manner of legal, social and psychological fetters: obligations to feudal lords, lack of any real sense of opportunity for self-betterment, ignorance of the outside world. Essentially, of course, the pre-industrial economy did not require mass movement. Human migration is first and foremost a response to economic change, a means by which the geography of settlement is reshaped in tune with changing economic geography. Since almost by definition settlement and economic activity are interdependent – people need work, production depends upon workers and consumers – it is virtually axiomatic that a change in the distribution of

POPULATION AS AN ISSUE   179

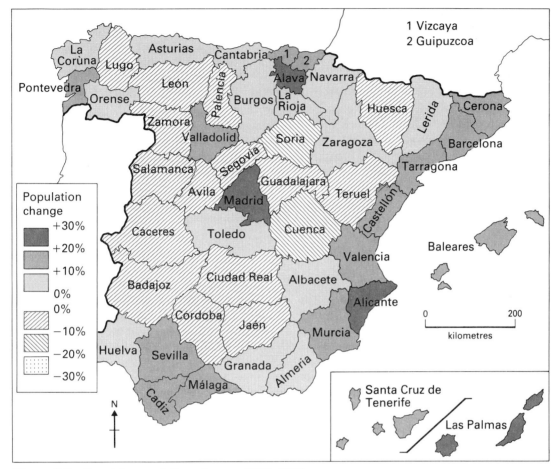

Figure 6.8 Spanish population change, 1970–1981

one will bring about corresponding change in the other. It was not until the Industrial Revolution that profound shifts in the geographical pattern of activity began to require the relocation of large numbers of workers.

b) The transitional stage – rural-urban migration

The predominant trend in economic geography during industrialisation is *centralisation* (see Chapter 2). During its conversion from an agricultural to an industrial economy in the nineteenth and early twentieth centuries, Western Europe also underwent a transformation in its settlement geography. The old pattern of dispersed rural settlement was replaced by one dominated by large urban centres, with concentrations of population growing up around the new localised industrial complexes. By the middle decades of the present century, the great majority of people in the leading industrial countries were urban dwellers, which was a complete contrast to the century before.

Migration has been responsible for most of this shift in the rural-urban balance. For decades there was a continuous flow (interrupted only by war) of individuals moving from country to town, representing a colossal upheaval involving literally millions of individual migrants. Much of this migration took the form of local journeys, from surrounding countryside into the nearest city. Indeed studies of migration have always stressed the importance of short moves and the principle of least effort: i.e. since travel is costly most people move no further than is absolutely necessary to achieve their objectives. Yet, one of the novelties of the Industrial Revolution was that it made mass long distance movement a feasible proposition for the first time. Throughout the period of industrialisation there has always been an important long distance element, with many rural and small town migrants moving to large industrial cities outside their home region. In Italy there have been vast shifts from southern rural regions to the industrialising north; in France from the peripheral west and south to Paris and the north-eastern industrial zones; in the Netherlands westwards to Holland proper; in Norway, Sweden and Finland southwards towards the capital cities. All this is completely in tune with the emerging core-periphery structure of the period. Human labour was literally being transferred, firstly, from periphery to core and, secondly, from rural settlements to urban centres within the core.

In purely functional terms it is clear that migration acts to redistribute population according to the needs of production: to move labour out of areas where it is surplus to requirements and into locations where it is in short supply. Classical economists have seen migration in this way as a process making for equilibrium between firms and workers. Growing firms in labour deficit areas will pay extra wages to attract workers from labour surplus regions. Historians of migration have also tended to account for human movement in this way, emphasising that people are induced to move by a combination of unsatisfactory economic circumstances in their present place of residence (push factors) and the prospect of gain in another area (pull factors).

I) PUSH FACTORS
During Western Europe's transitional phase, the principal push factor in rural-urban migration was population pressure. It is significant that the peak period of rural-urban migration coincided with the middle phase of the demographic transition. Declining rural death rates produced population increases which were surplus to the requirements of agriculture and which could not be supported by the land. Small peasant holdings became even smaller as the land was divided up between large numbers of heirs and agricultural wages fell as the growing army of landless labourers competed for work on the larger estates. The situation was aggravated by two further trends whose combined effect was to reduce rural employment opportunities still further.

The first trend was *deindustrialisation*. Small scattered rural industries using handicraft methods of production were unable to compete with large-scale mechanised enterprises. From the late nineteenth century onwards the traditional industries of the countryside began gradually to diminish in numbers as they became victims of the inexorable trend towards centralisation of manufacturing in large towns and cities.

Secondly, *agricultural mechanisation* was a major push-factor. Towards the turn of the century the larger wealthier farmers had begun to introduce mechanised methods of tillage, harvesting and planting. As the present century has progressed, so agricultural labour has continued to be replaced by machinery and has been subject to a devastating fall in numbers (see Chapter 5).

All over Western Europe, then, the trend in rural areas has been one of high rates of natural increase set against a long-term decline of staple industries and an inability to attract new forms of replacement employment. The only response for rural dwellers has been to migrate elsewhere in search of new opportunities and so rural areas have been, without exception, areas of *depopulation*. Both relatively and in absolute numbers the rural population has consistently declined during the present century. In truth, the term 'push factors' is rather too bland to describe some of the extreme circumstances of destitution and even mass starvation which, as in mid-nineteenth century Ireland, reduced people to the status of refugees rather than migrants of their own free will.

II) PULL FACTORS

The pull effects were those associated with the process of industrialisation. The distribution of population began to shift decisively towards the urban sector from the late nineteenth century onwards, as the great flow of migrants from the countryside arrived to swell the labour force of the burgeoning new urban industries like iron and steel, textiles and engineering. Metropolitan centres, such as Paris and Amsterdam, and ports, such as Hamburg and Rotterdam, also acted as major magnets for migration, as their trading, financial and service activities boomed in response to the industrial growth on the coalfields.

In certain cases, migration produced truly spectacular growth transforming villages and small towns into gigantic urban centres. In Germany, the town of Barmen (now part of Wuppertal, a major textile centre located immediately to the south of the Ruhr coalfield) increased its population from 36,000 in 1850 to almost 170,000 in 1910. This period represents the first phase of German industrialisation, during which Barmen's textile industry was transformed from handicraft to mechanised mass-production. Migration from the countryside had been very largely instrumental in this great demographic upsurge. It has been estimated that if there had been no migration the town's population would have grown to only 61,000 by natural increase (births over deaths). In 1910 four out of every ten inhabitants had been born outside the town and many of those born locally were themselves the children of former migrants. One-third of the migrants were from distant parts of Germany beyond the Rhinelands, which illustrates the considerable pulling power of the industrial town during this transitional period.

Although this vivid example is taken from almost a century ago, we should bear in mind that the general trends it illustrates have continued much closer to our own time. In France the industrialisation/rural-urban migration process was still operating in the early post-war era, with the heavy industrial regions of the Nord, Alsace and Lorraine, the booming tourist centres of Provence and, above all, the metropolitan centre of Paris continuing to attract large numbers of workers away from their rural origins.

c) The transitional stage – emigration

Emigration represents an important secondary migration flow during the transition. Rapidly though the urban economy has grown, it has rarely provided a sufficient quantity and variety of opportunities to satisfy the needs and aspirations of every potential migrant.

Germany and Britain, for example, have experienced some of the greatest emigration losses of all in absolute terms, despite high levels of economic development which they had already attained by the late nineteenth century.

Yet without doubt the greatest impact of migration has been experienced by nations where at the height of the demographic transition there was little compensating growth in urban employment. Here emigration has acted as a vital safety valve. The extreme case is Ireland, whose rural conditions were so catastrophic and whose urban growth so feeble that emigration during the century after 1840 was on such a scale as to reduce the national population to a fraction of its former size. At 4.9 million, the present population of all-Ireland is around three-fifths of its peak in 1841. The 38 million Americans of Irish descent now outnumber the inhabitants of their ancestral island by almost eight to one.

d) The modern phase

In a late industrial society we would expect migration to be less of a primitive affair than formerly: less a matter of rural refugees rushing to the city to escape destitution and more a case of an increasingly affluent population exercising a sophisticated range of choices. In the most advanced nations, the great rural labour reservoir has now become so depleted by generations of depopulation that it supplies only an insignificant dribble of migrants to the urban areas. Rural exodus is now dwarfed by other streams of migrants impelled by entirely different motives.

Current migration patterns in West Germany exemplify the new movements which have replaced the old. Fig. 6.9, which focuses specifically upon Hamburg as a centre of migration, highlights the three forms of flow which most typify the modern phase. These are firstly, *urban decentralisation*, the relatively short distance movement of urban workers out to the suburbs and dormitory satellites; then, *inter-urban exchange*, the flow of migrants between cities and large towns, often over considerable distances; and finally, *immigration*, the international movement of workers from the peripheral regions of southern Europe, the eastern Mediterranean and North Africa. We now examine each of these movements in turn.

I) DEURBANISATION

For two centuries or more the city of Hamburg has been one of Europe's leading poles of attraction, pulling in a huge volume of newcomers and growing from 130,000 in 1800 to a peak of around 2 million in 1960. Since then, however, the number within its boundaries has slumped gradually to 1.6 million, with newcomers increasingly outnumbered by those moving out (a net loss). In the year 1979 alone there was a net loss of 3,500 residents.

When Hamburg's migrants are classified by distance and direction travelled, we find that total migration loss is more than accounted for by *local* movements. More than 38,000 people left to take up residence in the two neighbouring *Länder* of Schleswig-Holstein and Lower Saxony but less than 26,000 people

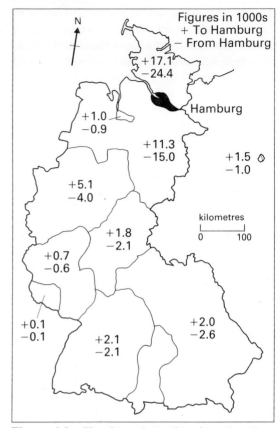

Figure 6.9 Hamburg inter-*Länder* migration, 1978

Self-assessment exercise

Illustrate the data provided on the map by the most appropriate cartographic method. Remember, you need to show both direction and volume of migration.

---

made the move in the opposite direction. Net loss of 12,000 people to the adjacent areas therefore was far greater than total net loss (Fig. 6.9). The motive for this urban exodus is entirely different from that which propels rural migrants towards the city. Most of those leaving Hamburg are not fleeing from unemployment and poverty, since the city actually achieved a slight expansion in employment in the late 1970s. In the majority of cases they are leaving for better housing in the more spacious conditions of new suburbs beyond the *Land* boundary but within commuting distance of the central city. To a growing extent, of course, firms are decentralising in a similar quest for more space (see Chapter 2) but the great bulk of employment is still concentrated within the city itself (hence outward migration is not generally a search for employment).

In all this, Hamburg is entirely representative of large urban centres in general. Of the 20 most urbanised regions of the EC (including Ile de France, Brabant, north and south Holland, Düsseldorf, Copenhagen and Antwerp), 15 are now net 'exporters' of population. Coming after decades or even centuries of inward migration from villages and towns to the great metropolitan centres, this marks a critical reversal of long established trends. This reversal became widely apparent in the 1950s and 1960s which was the end of the transitional period for the most advanced nations. It is now undoubtedly the case that this short distance outward migration accounts for a very high proportion of the total moves made and that much migration of this type takes place for non-employment reasons (that is, reasons connected with home, neighbourhood, physical environment and the need to escape from the negative aspects of urban living).

These motives for migration are exactly those which we would expect of members of an affluent highly developed society. Sights are now set far beyond bare survival and fixed upon the quality of life in the wider sense.

II) INTER-URBAN EXCHANGE

This is the second of the primary migration flows of the modern era. Reference to Fig. 6.9 will show that there was a very high level of long distance movement between Hamburg and non-local *Länder*. Significantly, this is very much a two-way exchange, involving large numbers of migrants both to and from the city but with comparatively little net change (Hamburg in fact made a net gain of less than 1,000 people, even though the total volume of movement in each direction involved over 50,000 migrants).

Evidently, internal migration is no longer mainly a means of redistributing workers from declining to growing areas. This constant circulation of migrants travelling between large cities is largely a middle class affair – managerial, professional and other salaried workers moving in response to the demands of their occupation. Many belong to the category which has been labelled 'spiralists' i.e. employees of large private companies and of government whose career prospects depend on shifting to a new office or a new organisation in a different location. Another not dissimilar category is that of students seeking higher education outside their home area. Indeed, higher education is often the first point on the spiral.

In common with the exodus from the cities, inter-urban migration clearly reflects the higher aspirations of the more mobile members of society and is quite distinct from those older forms of migration where people were driven by economic hardship. It would, of course, be true in a limited sense that even the spiralists of the late twentieth century are not expressing an entirely free choice since it is the nature of their occupation which to a larger extent dictates that they should be nomadic. Even so, the experiences of the aspiring bank manager or university professor are in no way comparable with those of the displaced peasant or destitute farm worker, compelled to move by population pressure and by the dictates of a newly industrialising economy. Nor should we fall into the trap of portraying the spiralists as typical of present day society. Personal mobility varies greatly from one section of the population to another, according to income, occupation, age, sex and individual awareness. The middle-aged, redundant steel worker in Lorraine, for example, has nowhere to run to, since steel is everywhere a contracting industry and his skills are suited only to that or a narrow range of closely related industries; no means to run anywhere, since long distance migration is a costly affair; little desire to escape such life-long ties as the family, the local community, his home (in many cases rented from the public sector); and frequently no reliable information about prospects elsewhere. In reality, the majority of Western Europeans (if not all humans) are still highly immobile, driven to migrate only under duress or by prospects of gain.

III) IMMIGRATION

Few of these remarks about exercising options for personal development apply to the third main migrational stream – the immigrant workers from the peripheral Mediterranean nations. It has been estimated that by 1973 there were over six million foreign workers in the industrialised countries of Western Europe (the EC plus Norway, Sweden, Switzerland and Austria but excluding the UK). Since this 1973 peak, numbers have declined but immigrant workers continue to make a significant contribution to the economies of these countries (see Fig. 6.10). They also continue to be defined as a 'problem'. As a severely disadvantaged group – generally working in some of the least desirable jobs for the lowest pay;

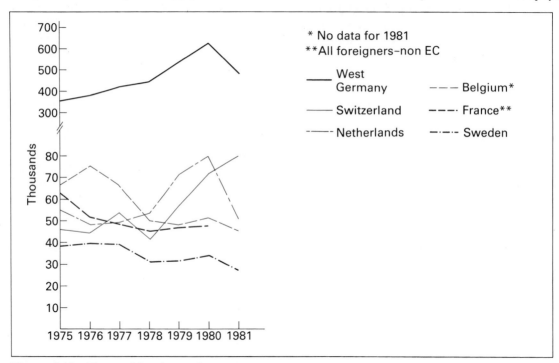

Figure 6.10 Entry of foreigners into selected countries of Western Europe, 1975–1981 (thousands)

	1975	1976	1977	1978	1979	1980	1981
Belgium	62.3	51.9	48.3	44.9	46.7	46.8	n.a.
France	67.6	74.6	66.6	50.1	48.5	51.4	45.2
Netherlands	55.2	48.9	49.9	55.6	72.2	80.1	50.6
Sweden	38.0	39.8	38.7	31.7	32.4	34.4	27.4
Switzerland	46.3	44.9	55.4	43.3	56.8	70.5	80.5
West Germany	366.1	387.3	422.8	456.1	545.2	631.4	501.1

Self-assessment exercise

Table 6.1 Stocks of foreign workers in immigration countries, 1981 (in thousands)

To	Austria	Belgium	France	West Germany	Netherlands	Sweden	Switzerland
From							
Algeria		3.2	382.1		0.2		
Austria				87.0			19.5
Finland				3.0		104.4	
Greece		10.7		132.2	2.3	7.3	4.9
Italy	2.1	90.5	157.6	316.1	10.5		234.9
Morocco		37.3	171.9		40.2		
Portugal		6.2	434.6	57.1	5.6		9.3
Spain	0.2	32.0	128.9	85.9	12.7		63.4
Tunisia		4.7	73.2		1.5		
Turkey	30.4	23.0		637.1	59.5		22.7
Yugoslavia	113.9	3.1		357.7	8.4	23.0	34.3
Others	29.7	121.5	243.6	405.8	97.6	98.8	126.1
Total	176.3	332.2	1591.1	2081.9	238.5	233.5	515.1

From the data in Table 6.1 construct a map to show the flow of immigrants to either France, West Germany or Switzerland. Make sure that the flow lines *accurately* represent the volume of flow.

Compare your map with someone who has chosen to illustrate a different flow from yours. Comment on the findings.

Suggest other ways in which you could map this data.

---

denied many of the citizenship rights of their fellow workers; often subject to the hostility of the indigenous population – their presence constitutes a major political embarrassment for West European governments and a moral dilemma for all Western European citizens.

How do we explain this upsurge in international migration to Western Europe? We must first recognise that the period from 1950 to 1973 was one of very rapid growth in the West European economy, with immigration playing a vital part in supplying the extra workers necessary for the expansion of production. By the late 1950s and early 1960s, most of the industrial nations had exhausted the stock of rural migrants who had formerly supplied the labour for industrial growth. The shortage of labour was further aggravated by falling rates of national population growth. It therefore became necessary for employers to recruit beyond their own borders. Rural-urban migration began to take on an increasingly long-distance international character.

The new sources of labour were principally Iberia, southern Italy, Greece, Yugoslavia and Turkey along Europe's Mediterranean fringe, together with the North African countries of Algeria, Morocco and Tunisia (see Fig. 6.10) which suffer now from a labour surplus similar to that which had pushed Scandinavian or German rural dwellers into the growing cities. Backward agricultural regions with high population growth still function as labour reserves for the industrial growth of the Eurocore: the main difference is that they are now located further away across international frontiers. Table 6.1 helps to explain the logic of migration between poor underdeveloped nations and rich developed ones.

At first sight, international labour migration would seem to offer a remarkably harmonious arrangement, beneficial to all parties. *Employers* benefit from an assured supply of workers who are generally much cheaper than local labour. Because of their inexperience and insecurity immigrants have been generally willing to tolerate lower wage rates than indigenous workers and, through lack of trade union organisation, industrial skills and knowledge of the local language they have been powerless to effect an improvement in their conditions of employment. *The State* benefits because migrant workers can be 'repatriated' when their labour is no longer required, as for example during the present recession. *Indigenous workers* benefit because immigrants have had no choice but to undertake many of the most unpleasant, least rewarded tasks in the modern economy. Thus, for example, the Ford motor works at Cologne employed 15,000 migrants, all but 200 of whom were engaged in manual jobs. The managerial, research, marketing, clerical and other non-manual activities were almost exclusively a German preserve. This concentration of foreigners at the bottom of the employment hierarchy is no isolated instance. Generally, we can conclude that the presence of migrants has allowed native workers to move upwards into the kind of prestigious employment more in tune with the expectations of a modern educated and increasingly well qualified workforce. It might be argued that *immigrants* themselves benefit from regular employment and from earnings which are unquestionably higher than those available in the home country. Even performing some of the worst rewarded jobs in the Swiss or German economies, the Mediterranean migrant can earn many times the average wage of his country of origin.

Case study: The *Gastarbeiter* (guestworkers) question in West Germany

A close examination of West Germany would suggest that the labour migrant system is far less benevolent than it appears on the surface. Up until the oil crisis of 1973, the breathtaking expansion of the German economy led to a demand for labour which could only be satisfied by the importation of foreign workers on a very large scale. Between 1954 and 1973 the foreign labour force expanded from 100,000 to over 3 million. Table 6.2 gives the basic reasons why workers from southern Europe should have been so eager to enter the Federal Republic: it also demonstrated the economic logic of an exchange of labour between core and peripheral nations. There can be no doubt that the vast majority of immigrants make considerable material gains from the West German economy, exchanging a life of unemployment and under-employment in a low wage economy for substantially higher earnings in an advanced industrial economy. Against this, however, is to be set a long list of social and psychological costs.

DISPLACEMENT

As anyone who has ever moved house will testify, any change of residence is an occasion of personal trauma. When, as in the case of 'guestworkers', this involves movement across international frontiers and language barriers into a completely alien culture, then the sense of displacement is hugely magnified. In the case of Muslim Turks, who are now the largest group of foreigners in the Federal Republic, migration has also meant moving into a new religious and cultural environment. Immigrants usually attempt to find security by congregating together in tight-knit family or work groups, a natural form of behaviour which is often seen as provocative ('clannish', 'aloof', 'stand-offish') by the native population.

CONTROL

A report by the Minority Rights Group notes that 'the German foreign labour force is the most tightly controlled, organised and supervised of all European countries'. All 'guestworkers' are recruited by a Federal Government Agency which rigorously vets all applicants, weeding out those deemed to be physically or medically unsuitable and directing successful job seekers to their appointed employer. Critics of this system suggest that it smacks more of a cattle market than an exchange of human labour.

POPULATION AS AN ISSUE 187

Table 6.2 West Germany: its main sources of immigrant labour

	Per capita income	Agricultural proportion	Rural proportion	Population growth	Crude birthrate
West Germany	13,450	4	15	0.0	10
Turkey (1,598)*	1,540	55	53	2.0	33
Yugoslavia (611)	2,790	29	57	0.6	17
Italy (610)	n.a.	11	30	0.6	12
Greece (280)	4,420	37	37	0.9	15
Spain (188)	5,640	14	25	1.1	14
Portugal (94)	2,520	24	69	0.6	16
Definition	dollars 1981	per cent of labour force 1981	per cent of population 1981	per cent annual increase 1970–81	Births per thousand population 1981

* Figures in brackets indicate number of people from each nationality who were resident in West Germany in 1982 (figures in thousands).

COMMENTARY
Despite its international character, migration from the southern periphery of Europe to the Federal Republic of West Germany has all the classic hallmarks of rural-urban migration – a flow of surplus workers from relatively poor agricultural areas with high rates of population growth to a rich core region suffering from labour deficit.

INFERIOR STATUS

'Guestworkers' exist primarily to carry out the thankless tasks which are still essential to the functioning of economic life but which few people in a modern advanced society would voluntarily perform. Hence West Germany's leading employers of foreign labour are metal foundries (arduous, dirty, dangerous), hotels and catering (low pay, often menial), textiles (monotonous, low pay) and plastics (noxious, unhealthy). As late as 1977 only one per cent of immigrants had graduated to non-manual jobs. Coupled with inferior working conditions is inferior housing. Until recently the standard practice has been to house single male foreign workers in factory owned barrack camps for which rent is deducted from wages. Whether or not this is an improvement on the *bidonvilles* (shanty towns) which house much of France's immigrant population is a questionable point. An advert placed in an Ulm newspaper is an interesting comment on the exceptionally low esteem in which foreign workers are held in Germany:

Barn in large wooded meadow . . . excellent opportunity for stabling horses or for use as accommodation for *gastarbeiter*.

Evidently those who do the donkey work must also expect to live like draught animals.

In theory, the past decade should have seen the end of the *gastarbeiter* 'problem'. With economic recession and rising unemployment, a Federal ban on foreign recruitment was imposed in 1973 and workers returning home on the expiry of their contracts have not been replaced. Perversely, however, the foreign population has actually risen since 1973 from 4 million to 4.7 million, as increasing numbers of southern Europeans have contrived to settle permanently and to bring their families to join them. Many immigrants have now attained a much more secure position than formerly. In the case of the Greeks, for example, recent EC membership gives them virtually unrestricted entry to West Germany, a status now also enjoyed by Spanish and Portuguese. Even Turks and Yugoslavs, the

188 GEOGRAPHICAL ISSUES IN WESTERN EUROPE

Figure 6.11 Turkish gastarbeiter in West Germany

two largest groups (whose nations are not EC members) have continued to increase in numbers since 1973. It is the Turks who, because of their relatively large numbers and their most obviously 'foreign' appearance and behaviour, bear much of the brunt of the widespread German resentment against foreigners. In the eyes of the most racially prejudiced of the Germans, it is the Turks who are the chief 'stealers' of jobs and houses; and Turkish children who create educational 'problems' with their language difficulties. All this provides inflammatory fuel for racist, anti-immigrant propaganda, which makes much play of the simple minded equation between 2 million foreign workers and 2 million unemployed West Germans. There is, of course, no evidence that if the Gastarbeiter were to disappear the German unemployed would eagerly take their place manning iron furnaces, shifting dustbins and emptying bedpans. A study commissioned by the city of Düsseldorf found that the disappearance of its legion of foreign workers would be followed by the virtual disappearance of its public transport and cleansing services and the severe disruption of its hospital and educational system.

Case study: Migrant labour in Switzerland

It will perhaps be surprising to many that Switzerland has the highest proportion of resident foreigners of any major West European country, yet these workers fulfil a vital role in supporting the nation's high level of prosperity. There are some 930,000 foreign residents in Switzerland within a total population of 6.4 million. The Swiss authorities rigorously control numbers through quotas for both economic sectors and geographical districts (*cantons*) and they define various categories of foreigners:

- *Frontier workers* live in neighbouring countries and commute into Switzerland each day.
- *Seasonal workers* employed mainly in the

WORLD NEWS

The miracle workers that Germany no longer wants

● **By this weekend, a new West German law will have persuaded about 350,000 Turkish workers to leave the country. GITTA SERNEY reports on its inequities, and on the troubles facing the young outcasts.**

AT 4.45 one morning, on a street corner in the ill-famed Reperbahn district of Hamburg, Semra Bilir, a 24-year-old Turkish girl, poured petrol over her body and lit it, in protest against the "Auslanderfeindlichtkeit" — hostility against foreigners — of the Germans. She died, leaving behind her "guestworker" parents, three sisters, and 361 poems.

Her last poem, written the morning of her death, on May 28, 1982, describes what it is like to live, despised and taunted, in the Turkish ghettoes of Germany. It ends with the words "...the Germans say, if you don't like it, get out: go back to Turkey." ["Turken Raus" — "Turks Out" — a disturbing reminder of the Nazi party cry of the 1930s "Juden Raus", is scrawled on the walls all over West German cities.]

Semra's younger sister, 15-year-old Reysa, recalls: "She rang up NDR (North German Network) a few hours before she did it. They laughed at her, they didn't believe her. She died alone.

"They took her seriously enough afterwards," Reysa said bitterly. "German television made a film of the funeral. We buried her in Turkey, in our town on the Black Sea. Soldiers carried her coffin and 2,000 people walked behind it." She shrugged wearily. "But of course, that was two years ago. It's all forgotten now."

Resya is dark-haired, slim, and beautiful. Exceptionally bright, she is in an academic school — rare for children of Turkish guestworkers (a euphemism coined in the 1950s for those foreigners invited in to help run West Germany's "economic miracle").

There are 1.5m Turks in West Germany — 560,000 workers and their families — making them the largest single national group among the 4.53m foreigners. In the 1950s and 1960s, when single men began streaming in from the poor countries of southern Europe to fill the unskilled and "dirty" jobs that German workers no longer needed to accept, Turks were an essential part of Germany's flourishing economy.

Now, however, there are 2.5m unemployed, and of the 274,000 foreigners who are jobless, 40% are Turks. They cannot be pushed into easy-to-isolate dormitories, for they have their families with them now and so take up housing space, though almost invariably in outrageously overpriced slums.

Dark moustachioed men with women in the traditional headscarf are conspicuous in the shops, buses and trains of many German cities. In schools in some districts, Turkish pupils — who speak virtually no German — outnumber blond little Germans.

As the "miracle" has soured and work has become scarce, Turks have faced increasing competition from desperate Germans for their "dirty" jobs. The most numerous, most conspicuous and least adaptable community among the foreign population has come to be seen as a burden on the economy.

Under a new federal government law, a payment of DM10,500 (£2,840), plus DM1,500 per child, has been on offer to any foreign worker who, unemployed through redundancy, or over the past six months continuously in part-time work, had registered before yesterday his willingness to leave the country permanently and depart within 30 days of registration. Everybody knows the law is aimed at the Turks alone: EEC nationals are free to work in any EEC country, and workers from Spain, Yugoslavia, Italy and Portugal have in general adapted and been accepted by the Germans.

A second part of the law tries to entice Turks to leave by enabling them to collect a partial refund on their pension contributions.

One Turkish father told me that after agonising over what to do, he had decided his family would leave. His two children, 16-year-old Ayda and 18-year-old Cetin, want to stay, at least until they finish their education.

Cetin has a year of his apprenticeship in electro-mechanics to go. He earns DM300 (£80) a month. "In Turkey", he said, "they call us Germans and envy us, while despising and, I think, fearing our 'foreign' ideas. Here in Germany, they call us Kanaken [bumpkins]and say we smell."

Reysa's parents, too, are leaving, and want to take her with them. She left their home in Kiel, a small city in north Germany, a year ago to join her older sisters, Zuhal and Ismet, in Munich.

Reysa's father, unlike many of his compatriots, has a secure job. An architectural draughtsman by training, he has worked as a machine-fitter for the past 23 years in Kiel. None the less, the Bilirs have decided to return to Turkey.

"They have no energy left", said Reysa. "They are giving up. But I'm not going."

Walter Weiterschan, who directs the foreign section of Bavaria's labour welfare bureau, claims that the so-called "prämie" (bonus) on offer to Turkish workers is but a fraction of the unemployment benefits *and* pension contributions due to them. If they remain in Germany, unemployed, they, like any other workers, would receive about 70% of their earnings for a year, and after that social security at about 60% of their earnings — in addition to health insurance.

Payment of even this "bonus" — which barely covers the move to Turkey — proceeds only when the entire family has crossed the border, at which point they must post back a "border-crossing certificate". They must then wait nearly four months for the cash.

Many MPs have protested against the hypocrisy and miserliness of the "returnee" law, but to no effect. Also, the German government has tried for years to make arrangements with Turkey to facilitate the return of these millions of temporary emigrants. But the Turks can't set up the necessary machinery.

Figure 6.12 *The Sunday Times*, 1 July 1984

construction and tourist industries. They are allowed a residence permit for a maximum of nine months, have no right to bring in dependants and are not counted as residents.
- *Annual permits* can be granted to seasonal workers who have completed four seasons but they are renewable and depend on labour market conditions. Limited social security rights are available to holders who can also bring in dependants after fifteen months.
- *Established status* (renewed every three years) can be applied for after ten years as an annual permit holder. It carries full social security rights and twelve years residence gives eligibility for Swiss citizenship.

This strict categorisation and quota system has been used by the Swiss authorities to regulate the number of foreigners in response to changing social and economic conditions. A peak of 655,000 foreign annual permit holders and 317,000 established migrants was reached in 1969 but the impact of the 1973 oil crisis and subsequent recession was to produce a public reaction against foreign workers in Switzerland as economic conditions worsened. There was a drastic reduction in the number of seasonal workers from 148,000 in 1973 to 60,000 in 1984, whilst the number of frontier workers has stabilised at just over 100,000 during the same period. Similarly, the number of annual permits has been cut back to almost one-third of its 1973 level of 328,000. Conversely, the total number of those with established status has grown from 267,000 in 1973 to 401,000 ten years later as migrants qualified under the ten year rule. The net effect of these changes was that Switzerland experienced a net return of over 150,000 foreigners back to their countries of origin between 1974 and 1978 and a further net loss in 1983 so that the proportion of foreign residents in the total Swiss population fell from 16.9 per cent in 1974 to 14.5 per cent in 1984. Moreover, those gaining established status are perceived to be less of a threat by the host population in terms of fertility as an increasingly larger proportion of them reach the age of forty.

Table 6.3 Nationality of foreign workers in Switzerland, 1984 (in thousands)

Nationality	Total	% established
Italian	404.8	91.0
Spanish	104.2	79.3
German	83.5	77.6
Yugoslav	58.9	51.2
Turkish	48.5	64.5
French	46.8	81.5
Austrian	30.1	88.0
Portuguese	19.7	31.0
Others	129.1	52.8
Total	925.6	77.4

(*Source: La Vie Economique*, June 1984)

The origin and status of Switzerland's foreign residents are shown in Table 6.3. These people are primarily concentrated in the urban areas, notably Geneva (where they comprise 29 per cent of total population), Vaud (20 per cent), Basle (18 per cent) and Zürich (17 per cent). Foreign residents are much less common in rural areas (with the exception of Ticino, see Chapter 5) but seasonal workers are more important here where they work extensively in the tourist industry. Interestingly, the spatial distribution of migrant labour in Switzerland parallels international divisions with Spanish and Portuguese workers concentrated in French-speaking cantons and Turks and Yugoslavs in German-speaking areas.

The role of migrant labour in Switzerland has been a means of regulating labour supply in line with fluctuations in economic performance. In effect, foreign labour is 'sucked-in' during times of prosperity but the strict system of control enables the Swiss authorities to react quickly, to re-export surplus labour when it is no longer required, or when public resentment grows when foreign workers are seen to be taking 'their' jobs – a feature all too common in many

Western European countries in the late 1970s and 1980s. With a peak unemployment rate of 1.1 per cent in February 1976 and May 1984, the effectiveness of the system is not open to doubt. Nevertheless, Switzerland remains extremely dependent on foreign labour and will continue to be so for many years to come.

Clearly immigrant workers – in Switzerland and West Germany – are vital to the functioning of the West European economy. They happen also to be human beings, an inconvenient fact which Western European governments failed sufficiently to acknowledge at the outset of mass immigration in the 1950s.

5 Conclusion

One of the key themes emerging from this discussion has been the increase in human freedom enjoyed by the citizens of a modernising society: freedom from disease and premature death; freedom to plan one's family; freedom of movement in seeking a career and a place of residence. These are the human benefits concealed behind such statistics as declining vital rates and increasing migration.

As always, however, there is a reverse side to the coin. Freedom of choice in reproductive behaviour seems likely to lead to a contracting labour force, which may or may not have serious economic repercussions according to the rate at which machines and robots replace human beings in industry. Increasing freedom of movement is achieved at the cost of the stresses and strains of migrating into a new environment. As always benefits and costs are not equally distributed as the plight of migrant workers in Western Europe would testify.

7 Transport in Western Europe

Western Europe is characterised by an extensive and largely efficient system of transport and communications resulting from the great commercial activity of the region and the high standard of living enjoyed by the majority of the inhabitants. Because Europe is an advanced trading society an extremely complex network of road, rail, air and waterways, together with an expanding network of pipelines, has evolved over the past 200 years as a response to market forces. Yet despite the high level of development the European transport system has grown up without any real long-term strategic planning and, as a consequence, a number of serious problems exists today which affects all these different modes of transport. The range and scale of these problems is exhaustive; they are both economic and technological, social, spatial and environmental; many problems are local, others are international – at best they can all hinder the smooth flow of operations whilst, at worst, they can defeat the uninterrupted movement of goods and people. Inevitably, in a part of the world so densely populated for many centuries some of today's transport problems stem directly from the past. Others, however, are more recent and have often occurred as a response to the late twentieth century demand for an ever improving lifestyle. This chapter will explore the most important issues, their origins and review the attempts which have been made to solve the problems created.

1 The historical legacy

Although important trade routes were established throughout Europe in Roman times and were further extended and reinforced during the Middle Ages, the first major impact of transport upon the region's landscape was imprinted during the period of industrial revolutions. As the world's first great industrial region there was a need to import increasing quantities of raw materials and fuels to keep the wheels of industry turning and foodstuffs to keep the rapidly increasing population fed. In turn the industrial output from the world's 'workshop' created a massive trade in the export of manufactured goods of all shapes and sizes. This period of rapid growth marks not only an industrial revolution but, more specifically, a transport revolution which was necessary to meet the ever increasing demands of the new industrial society. Within the short space of 200 years the transportation of goods and people underwent a series of far reaching changes as new modes of transport were developed. In turn these developments gave rise to a massive transport industry based upon the manufacture of vehicles and equipment for use at home or abroad, providing employment for hundreds of thousands of workers throughout the emerging industrial regions of Western Europe. It was a period of frantic investment, development and construction as attempts were constantly made to try and reduce the problems of distance. Despite the pace of events little effort was made to integrate transport networks throughout the continent. It was a time of great international rivalry and competition both within the continent and further afield. The expansion of national economies and colonial adventure was reinforced by political fragmentation which characterised the map of Europe. As a result the nations of Europe developed their own unique transport systems constructing networks tailored to meet the specific demands of their own economies. Consequently transport facilities were developed in a vacuum in which emphasis was placed upon particular transport modes, routes were organised in different ways and systems were established with varying degrees of governmental influence and control reflecting individual policies and needs. Little relationship existed between the transport systems and networks of neighbouring countries, a situation which was to last well into the decade following the end of the Second World War.

With the cessation of hostilities came a political realignment of Europe with the partition of Germany and the imposition of the 'Iron Curtain' effectively sealing trading links with Eastern Europe for a number of years. Western Europe, like its Eastern neighbour, had been devastated by the war and industrial production had been brought almost to a standstill. Years of armament production followed by the havoc of total warfare had ruptured the economies of all but a few neutral nations throughout the sub-continent. It soon became clear that the only answer to recovery was through international cooperation and economic integration. Consequently, one of the principal post-war aims heralded by all Western European governments was to restore and stimulate national and international trade.

Trade, however, requires smooth and efficient transport and communications and although such conditions were largely adequate *within* most countries of Western Europe, the transport infrastructure *between* countries – even neighbours – was poorly developed as a result of the historical legacy. Only Germany had made any real attempt before the war to develop transport links with neighbouring states and this was more a reflection of military planning in the 1930s than a desire for European economic integration. After the war the division of Germany severed the predominant east-west links which had been established under the Third Reich to connect Berlin with the principal regions of Hitler's Germany. As a consequence many of the important trading links between west and east Germany, Czechoslovakia and Central Europe were broken overnight. Future trading partners for West Germany now lay in different directions, westwards and, more importantly, southwards where a poor and often inadequate transport infrastructure existed.

Not surprisingly the advantages of transport integration in Western Europe have been championed far and wide as the key to future growth and development, yet the historical legacy left by successive generations of transport development and the emphasis upon the development of national systems and networks has proved extremely difficult to overcome. Consequently, integration remains the single most important problem in the development of an efficient transport system in Western Europe and it is to this area that we shall return, in more detail, later. At this stage we need to familiarise ourselves with the pattern of transport development throughout Western Europe and to identify the particular issues associated with the various modes of transport.

a) Inland waterways

Any chronological discussion of transport development over the past two hundred years starts, of course, with the development and utilisation of inland waterways. Since the first phases of European industrialisation in the eighteenth and nineteenth centuries waterborne traffic has played an important role in the whole development process. Above all the impact of the River Rhine and its tributaries, navigable between the Netherlands and Switzerland, flowing through or past five nations has always been and still remains Europe's principal commercial waterway, providing the key to the continental inland waterway system (Fig. 7.1). Furthermore, it is a truly international river and, as such, transport integration between 'user' countries has been less

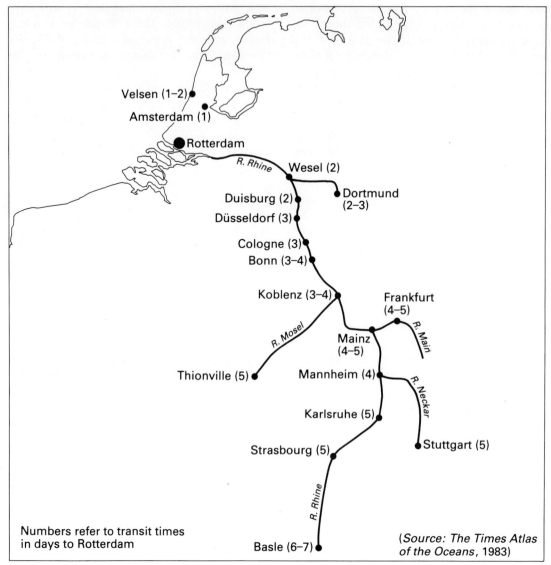

Figure 7.1 The Rhine: its major tributaries and inland ports

Freight transport on the Rhine, 1970–1982 (in 1,000 tonnes)

	1970	1982
Total goods carried	274,424	257,335
Between Netherlands and West Germany	110,339	122,005
Transit through West Germany	12,217	11,675
Between Netherlands, Belgium and France via the Rhine	17,580	24,924
Mixed Rhine and maritime traffic	2,050	2,869
Within the Netherlands via the Rhine	63,500	47,799
Within West Germany via the Rhine	62,733	45,069
Within France and Switzerland via the Rhine	2,260	2,822

Figure 7.2 Rhine barges

of a problem, especially when compared to the difficulties in establishing an integrated highway or railway network. The importance of the river and the accessibility that it provides is a major reason for the concentration of industry and population alongside its lower reaches and tributaries. Additionally, the Rhine is at the hub of continental Europe's inland waterway system from which canals have been constructed – often to link other river systems such as the Rhine-Rhone and the soon to be completed Main-Danube – forming a dense network of water navigation responsible for the carriage of a massive volume of traffic especially in north-eastern France, the Benelux countries and West Germany. Today, despite the decline of inland waterways in other European countries – most notably Great Britain and Sweden – and the increase in competition from other modes of transport – initially railways, then highways and, more recently, pipelines – the waterway network is still an integral part of the West European transport system. Inevitably some under utilised stretches have disappeared but the core of the system has been consolidated and the overall network has been rationalised and modernised. Furthermore the new key routes linking some of Europe's principal rivers represent significant investments and will go some way to improve the internal accessibility of the whole European network allowing water traffic to compete more effectively with other means of transport.

The principal use of inland waterways is for the carriage of bulk cargoes such as iron ore, coal, grain, scrap metal, chemicals, oil and petroleum products, where significant savings can be made over other forms of transport (see Fig. 7.3). Nevertheless, despite these advantages inland waterways have met with stiff competition throughout Europe as other modes of transport have substantially eroded

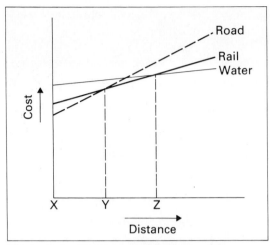

Figure 7.3 Cost curves for three major modes of freight transport. We can see that road transport is cheapest for the short haul (XY), whilst water transport is cheapest for long hauls (beyond XZ).

Self-assessment exercise

i) What factors might influence the alteration of these distances?
ii) What about pipelines? Where would their cost curve appear on the diagram (first see Figure 7.2)?
iii) What about passenger transport – do cost curves act in the same way, tapering off as distance increases?
iv) Find out the cost of passenger coach, rail and air travel from London to other major cities in the United Kingdom and find the distance between cities, and plot a graph showing cost against distance. Do the same for rail fares on the commuter lines from any one London station. Compare the results of the two graphs – do they differ, and if so, why?

Box 7.1 The cost of transport

Transport costs have an important bearing upon the scale and type of transportation system which operates in any particular region. These include *terminal costs* (infrastructural costs such as roads, marshalling yards and lock systems) and *line-haul costs* (vehicle, fuel and staffing costs for example), which tend to taper off as distance increases. However, the degree of tapering varies greatly from one transport mode to another and we can see from the diagram that the cost of sending freight by road, rail and water transport each has a *cost advantage* over each other for certain distances. Although it is difficult to state with any accuracy at which point rail costs become cheaper than road transport, and water transport becomes cheaper than rail, the principle is still valid. It has been suggested that in Western Europe, road freight is cheaper up to a distance of 80 kilometres, at which point rail freight becomes more cost effective up to a distance of 560 kilometres, after which distance water transport comes into its own. In other parts of the world these arbitrary distances differ. In North America, for example, it has been calculated that road transport is cost effective up to 180 kilometres, and rail freight is economical up to 824 kilometres after which water transport becomes the most cost effective. Nevertheless, the principle that different modes of transport are more competitive over certain distances holds true. As a result, the principle helps us to explain why particular emphasis has been placed upon specific modes of transport throughout Western Europe.

the commercial viability of many stretches of water. The transport by pipeline of chemicals, oil and petroleum products from the refineries and chemical works adjacent to the port of Rotterdam to the industrial heartland of West Germany, Belgium and elsewhere in the Netherlands is a particular example. Furthermore, many inland ports in heavy industrial regions have suffered as a direct result of the structural changes that have taken place in West European industry during the past 25 years. Ports such as Duisburg and Dortmund in the Ruhr region of West Germany have been seriously affected as the major steel industries of the region – with their, previously, constant demand for coal, iron ore and scrap metal – have fallen into decline.

With the loss of traffic to other forms of

transport it may seem surprising that there is still considerable interest in the development of canals and water navigation. In both France and West Germany as we have seen substantial investment is taking place in the construction of 'missing links' in a network that will eventually connect the North Sea with the Mediterranean and the Black Sea. Although the reasons for completing these missing links is fundamental for improving the accessibility of the whole European network the scheme is a controversial issue in both France and West Germany. There remains much scepticism as to whether or not any commercial or economic advantage will be gained from them. There is a feeling in both countries, nevertheless, that these links have to be completed in order to provide a balanced waterway system in Western Europe, opening the Eurocore to important growth areas within the continent. In France the Rhine-Rhone link is seen as a means to further stimulate growth in eastern France by providing a counter attraction to the magnetism of the Paris Basin – it is hoped that the new axis will become one of the principal zones of international goods traffic. The West Germans, likewise, see the Rhine-Danube link bringing many new jobs to middle and upper Bavaria. Already 2,500 people have been employed in the construction of the new canal harbour at Nuremburg which is an area which has traditionally suffered from long-term structural unemployment. Additionally, the Austrians see the new link as fundamental to the promotion of industrial growth in their country assisting the national economy with a direct outlet to the North Sea as well as to the major trading nations of the Eurocore.

Despite these optimistic predictions about regional growth being stimulated by the new canals there is no guarantee that development will necessarily take place, or that the volume of traffic forecast will even meet the most pessimistic of predictions. Apart from giving the European waterway system the visual appearance of overall connectivity there is a distinct possibility that both schemes will become little more than expensive 'white elephants'. Despite technological advances which have occurred in canal transport during the past few decades, competition from other forms of transport is as fierce as ever. The Rhine-Rhone link faces severe competition from other north-south transport links; the electrified railway line between the port of Fos, on the Mediterranean coast, and Strasbourg, the motorway between Marseilles and Dijon, the south European pipeline from Marseilles to Karlsruhe and Mannheim and the Rhone-Saone canal which, when widened, will allow large capacity barges to travel from the Mediterranean to Lyons. Furthermore there is a genuine fear that the new links with Eastern Europe will allow the passage of 'cutthroat' competitors from eastern bloc shippers throughout the Rhine basin and associated waterways which will prove as threatening as the Comecon challenge in road haulage and maritime shipping has been during the past two decades. Furthermore, there are still question marks over the cost of the project which, per kilometre, costs nearly three times as much as the equivalent motorway construction (see Box 7.2).

b) Railways

Despite the initial success of inland waterways their total dominance over the European transport system lasted only as long as it took the railways to develop a sufficient grip upon traffic flows. Although water-borne transport even today retains much of its pre-eminence in the carriage of long haul, bulk cargoes across international frontiers, railways – towards the end of the nineteenth century – offered a quicker, more flexible, cheaper service over short distances. Although competition in continental Europe (especially in the Eurocore) was met with greater resistance than in Great Britain, for example, the volume of traffic carried by canal and river has gradually fallen, being maintained only along the most accessible parts of the waterway system influenced by the River Rhine.

> **Box 7.2 The COMECON challenge**
>
> The Council for Mutual Economic Assistance (COMECON) was established in Moscow in 1949 to promote and develop trade between socialist states, primarily in Eastern Europe. Today most East European states – Bulgaria, Czechoslovakia, East Germany, Poland, Rumania and the Soviet Union – are members together with Cuba, Mongolia and Vietnam.
>
> In recent years many of COMECON'S East European members, particularly the Soviet Union, East Germany and Poland, have increased the size of their maritime shipping fleets (at a time when West European fleets have been rationalised) and have made substantial inroads into international sea transport, profiting considerably from the fact that shipping is still largely unregulated and accessible to any trading nation. As a consequence they have captured a large share of bilateral traffic within Western Europe. The EC has estimated that the Soviet Union now handles 60 per cent of shipping to and from the United Kingdom, 75 per cent between the Soviet Union and West Germany and 80 per cent to and from Belgium. Estimates also suggest the COMECON countries are responsible for handling 95 per cent of all seatrade between themselves and the Netherlands. By operating at lower freight rates, creating such a price differential, COMECON shippers have created a situation in which West European shippers cannot compete effectively.
>
> Road haulage and inland shipping are now beginning to face similar problems in Western Europe as Eastern bloc traders have started to control certain routes (Hungary and Bulgaria dominate road haulage between Western Europe and the Middle East, for example). Inland waterway operators, especially in Austria and West Germany, are being faced with increasing competition, especially along the Danube which, with the completion of the Main-Danube link at the end of this decade, will open up the markets of the Eurocore to Eastern bloc operators via the many arteries of the River Rhine.

Railway development in Western Europe was, from the outset, strongly influenced by national policies – the state taking a great interest in railway development to ensure that expansion served the best commercial interests of the nation. By 1900 the basic rail network of Western Europe was well established and a common gauge (with the exception of Spain) had been adopted throughout the sub-continent. The era of rapid railway expansion had given rise to extremely dense networks in Germany, the Low Countries, Great Britain, Austria and Switzerland. The alignment of these networks was, however, intra-national serving the particular needs of individual countries rather than allowing the ease of movement between countries. Nevertheless, despite this failing, which would become more noticeable in later years, the railways largely dominated the transport systems of the majority of Western European nations until well after the Second World War. Since then, in the short space of 40 years the railways of Western Europe and their future have become one of the major transport issues of the century. From being 'the pride of state' a few decades ago, railways throughout the sub-continent have been relegated to a position of inferiority by the spectacular developments which have taken place in other forms of tranport – most notably road transport.

The declining fortunes of the railways have been almost as rapid as their staggering rise to prominence in the latter half of the nineteenth century. Today, governments in Western Europe are faced with an unenviable task of trying to revive an extensive and largely uncompetitive railway network which is unwieldy in both its extent and organisation and consumes vast and increasing sums of public money. Yet the majority of governments, as well as supra-national bodies such as the EC, recognise that railways, like inland waterways, have a part to play in the overall transport system of Western Europe and are

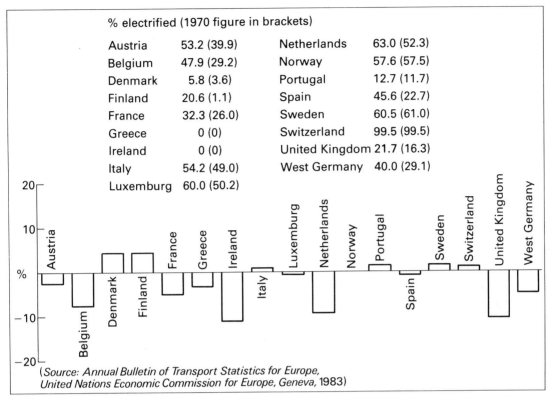

Figure 7.4 Changing length of rail network (including sidings) in Western Europe, 1970–1983

therefore prepared to back their judgement with considerable new investment as well as propping up unremunerative services. Just why this should be the case and how it is being tackled will now be examined in more detail.

Firstly we need to understand why railways have been reduced to such a difficult position. The easy answer is that they have not been able to adapt swiftly enough to the changing social and economic conditions which have swept Western Europe since 1945 – inevitably, however, the real circumstances are rather more complex. To start with, their inflexibility and high capital costs have been exposed to other forms of transport which have the ability to offer door to door services of both passengers and freight at considerably cheaper rates. Furthermore it can be argued that railways, being state owned, differ from their two main competitors – road and inland waterway – and that unfair competition benefits small operators who are more responsive to changes in demand, technology and markets. State control, on the other hand, has proven unresponsive to change – only a few countries, for example, the Netherlands, the Irish Republic and the United Kingdom have made any attempt to arrest the decline by a process of severe rationalisation (Fig. 7.4). Other countries (West Germany and Austria for example) have only recently put forward new overall planning proposals and still, today, signs of their practical implementation are not yet clearly visible.

Apart from the various structural problems, railways have faced increasing competition from road transport. The rise in car ownership, most rapid since the 1950s, parallels the decline in rail passengers. In comparison the total flexibility bestowed upon road users sealed the fate of many urban and rural railway lines throughout Western Europe. For

Table 7.1 Changing length of oil pipelines in Western Europe, 1970–1983 (in kilometres)

	1970	1983
Austria	604	777
Belgium	52	261
Denmark	0	68
France	3,533	5,386
Italy	1,939	3,266
Netherlands	323	391
Norway	0	350
Spain	267	2,017
Sweden	222	239
United Kingdom	1,634	3,281
West Germany	2,058	1,715

(*Source: Annual Bulletin of Transport Statistics for Europe*, 1982)

longer journeys increased competition from both road and air transport has also clouded the picture. The expansion of motorway networks and the introduction of short-haul air services have started to severely test the viability of railways throughout the sub-continent.

A similar crisis has befallen rail freight which, like passenger traffic, has been badly hit by the large scale highway developments which have taken place during the past twenty five years. Motorways offer both flexibility and constant speed allowing a reduction to be made in 'turnaround' times, thus raising productivity. Additionally, the footloose nature of much of today's industry has favoured location adjacent to the new, expanding motorway networks. Perhaps more seriously, though, are the changes which have taken place in the structure of demand, especially reduced requirements for the transport of coal and base metals such as iron ore as well as the transfer of many chemicals, oil and petroleum products to pipelines, robbing rail freight of its traditional bulk commodity traffic. Such a situation is little more than another reflection of the changing structure of industry in Western Europe in which the transport of materials for heavy industry is being replaced by the transport of high value/low weight goods in small consignments

> **Box 7.3 The impact of pipelines**
>
> Despite the sensitive political issue surrounding the future transportation of natural gas from Siberia to West Germany, pipelines have been absorbed very quietly into the transport infrastructure of Western Europe. Nevertheless, they have had a major impact upon the viability of other forms of surface transport – in particular, railways and waterways.
>
> Pipelines represent the final stage in the development of surface transport in Western Europe, having emerged since 1945. The formation of trading blocs in the 1950s and the desire for an integrated transport system created an ideal opportunity for pipeline development which, despite their inflexibility and fixed carrying capacity, offered a low unit cost method of transporting suitable products great distances and across difficult physical barriers. As a result where there was a constant demand, pipelines were constructed having considerable cost advantages over other modes such as waterways and railways. For the landlocked countries of Western Europe – Austria and Switzerland – as well as for the more distant inland centres in France and West Germany, pipelines, most commonly transporting crude oil, have become vital.
>
> The orientation of pipelines, especially for the transport of oil, have been from the major oil terminal ports – Rotterdam, Marseilles, Genoa, Trieste, Wilhelmshaven – to inland refineries where refined products are pumped by an internal system of pipelines to industrial consumers.

– which are eminently more suitable for carriage by other modes of transport (see Box 7.3 and Table 7.1).

Clearly then, the railway systems of Western Europe have largely failed to adapt to the changing circumstances of the new market and they have now reached a situation whereby major decisions have to be taken about their future. Faced with this dilemma a number of positive steps have been taken to try and remedy the situation. Throughout Western Europe there has been a greater emphasis upon speeding up passenger services

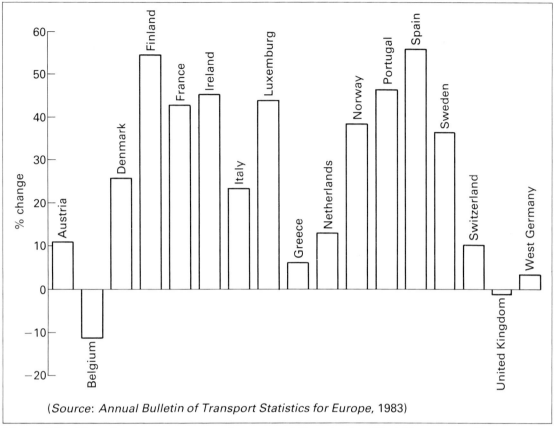

Figure 7.5 Changing passenger usage (passenger-kilometres) – West European railways, 1970–1983

and this has been achieved by the development of fast inter-city services. In West Germany, for example, inter-city services were introduced in 1971 linking 33 major cities and covering a distance of 3,115 kilometres. Services initially run at two hourly intervals have now been increased to hourly intervals over a network comprising four main routes, interlinked at Dortmund, Cologne, Mannheim and Wurzburg. The inter-city system has meant a 20 per cent reduction in travelling time and ensures that all main stations served by the system can be reached with a maximum of just two train changes. As a consequence passenger usage has increased as the frequency and timing of trains has progressively improved. On long distance routes average speeds of between 120 and 140 kilometres per hour are now regularly achieved. The success of the inter-city services in West Germany (as well as in France, Italy and Great Britain) has helped to stabilise the fall in passenger traffic as these new and faster services compete more effectively with other forms of transport (see Fig. 7.5).

In France, as well as the introduction of inter-city services, there has been considerable investment (to the tune of approximately 10 billion francs by 1983) in the development of the high speed passenger train (*train à grande vitesse* – TGV) which commenced operations on a new purpose built track between Paris and Lyons in 1981. Such a development of upgrading and rationalising the existing system marked a departure from normal rail policy. The TGV, capable of a top speed of 270 kilometres per hour, represents not only a major technological breakthrough

Figure 7.6 The TGV train

but also a major commitment on behalf of French Railways (SNCF) and national government to invest heavily in rail transport. Government sees such investment as a stimulus to regional growth in and around the Lyons region as well as along the eastern corridor adjacent to the new route. For the SNCF the development of the TGV is seen as a major boost to the image of the French railway system. With the network having now been extended from Lyons via St Etienne to the Mediterranean coast at Marseilles, and from Paris to Geneva via Dijon and Besançon, the reduction in journey time between the capital and the urban centres of southern and eastern France is impressive (Fig. 7.8). Future plans to extend the TGV to parts of western and south-western France as well as to the north of the country (to link up with the Channel Tunnel?) are well under way. More recently, one response to the crisis in Lorraine (see Chapter 3) was, in 1984, to lay the plans for a TGV route through the troubled region. In all these cases government regional planning priorities are being met and, despite the heavy investment costs, France has decided that railways – for the transport of passengers at least – have a long term future.

The solutions aimed at resolving the problems faced by rail freight are, in many respects, rather more serious. Although limited growth has occurred with the introduction of inter-modal traffic (the carriage of containers which, because of their standard size, can be carried by road, rail, air and sea freight) in the form of container and piggyback trains, the changing nature of the market and the continually improving accessibility afforded by the European motorway network have generally retarded recovery. On certain routes and within certain countries, however, limited success has been achieved in this area of operations. In West Germany, for example, container and piggyback traffic has increased more than threefold between 1970 and 1979 and 60 inter-modal terminals have been opened throughout the country to handle the special segregated services which have been introduced to meet the demand. In Sweden, too, the establishment of 37 road-rail terminals by Swedish Railways has been completed to handle similar traffic linking

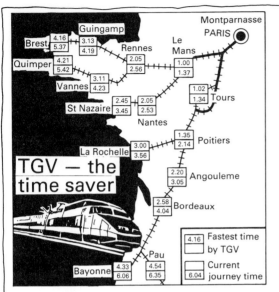

France speeds up rail success

THE FRENCH are having a love affair with their railways. While "we're getting there" on a shoestring, SNCF, with the biggest railway network in Europe built around its successful 168mph TGV service, can claim to have proved that throwing money at a problem does pay.

On current performance, the £600 million loan and interest charges for the existing line from Paris to Lyon will be paid off in about 15 years which, even at preferential rates, is some achievement.

Now, 18 months after completion of that project, work has already started on the next section of high-speed line, the TGV Atlantique. It will clip an hour or more off journey times between Paris and all the major destinations in the western half of the country. It will cost more than £800 million for the infrastructure alone, with the state contributing 30 per cent. The construction will create about 5,000 jobs. The rate of return for SNCF is reckoned to be more than 10 per cent, but for the country as a whole it may be more than 20 per cent.

Approval for the preparation of the project was given less than four years ago and, in November 1982, the planning inquiry procedures were started for a section of high-speed line to go from Paris Montparnasse in a Y-shape to both Le Mans and Tours, a total of 175 miles, before picking up the existing lines to the Atlantic seaboard.

Some existing rail land could be used for part of the route but because the whole point of the high-speed line is to provide tracks as straight as possible, much of the land would have to be compulsory purchased.

There was even the problem of driving a route through some of the best wine-growing districts in France such as Vouvray and many equally-famous names in the Loire Valley.

In Britain we have not built a railway on this scale for more than a century and we are not likely to in the foreseeable future. But for a road project, the average time taken from the publication of plans to the start of construction is now 11½ years.

The delays make nonsense of any suggestions that Britain could rapidly reduce unemployment by gearing up transport construction.

The French system employs many of the changes proposed in a report by the National Economic Development Council last December which suggested that the British system was too adversarial, too much based on the intellectual games of the civil servants involved, and did not devote enough cash or effort to solving the problems of the people threatened by a project.

The last Stansted airport enquiry, for example, is reckoned to have cost more in legal fees than the entire cost of building the mini-airport planned for East London.

In the French example, the cash we waste on protracted legal hearings is used to boost the price offered to the farmers whose land is being taken away. If, for instance, their land is split they are given an access tunnel under the railway track, and perhaps a new access road or free land drainage as a sweetener.

The planning appeals are kept short because once the project has the seal of official approval it is up to the inspector, skilled in judging land values as well as the technicalities, to make the system work efficiency. In Vouvray the problem of desecrating the vineyards will be solved by tunnelling underneath them — the first time that TGV will have operated at its maximum speed underground.

The half-mile-long tunnel will no doubt become one of the sensations of the new line, as trains with a closing speed of up to 330 miles an hour pass only inches apart.

The trains will be more streamlined and longer (because the Atlantique route avoids the steep inclines of the existing line). The one main criticism aimed at the first set of trains will be answered by making big improvements in the catering service for second class passengers.

Otherwise, SNCF will rely on the recipe that has carried 34 million passengers on the Paris-Lyon route, filled 70 per cent of the seats, achieved 97.5 per cent absolute punctuality (not within five minutes of schedule), and led to widespread cuts in air services and car journeys along the route.

Work on the Atlantique will be in full swing by the middle of next year, with the entire system becoming operational in the autumn of 1990 — less than 10 years after the first tentative plans were laid and a few months before British Rail completes its east coast main-line electrification. If we're getting thre we are coming a very poor second.

The Guardian, 23 April 1985   Geoff Andrews

Figure 7.7

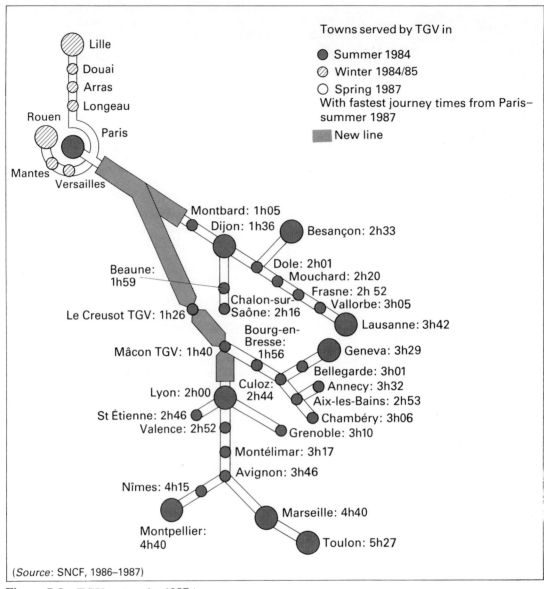

Figure 7.8 TGV network, 1987

national inter-modal traffic with 470 other terminals throughout Europe (see Table 7.2).

Despite the obvious success of inter-city services, the introduction of high speed trains and the limited expansion of inter-modal traffic throughout Western Europe the central issue – that of recovery – still remains. Although there is, undeniably, investment in railways it would be incorrect to think that all parts of the system are on the road to recovery for it is clear that only the more remunerative routes and services are benefiting. State railway systems, to compete effectively with other forms of transport, have to rely upon the few operating advantages that they still retain – the carriage of businessmen on fast inter-city services is a case in point. Elsewhere, the picture is rather more bleak with

Table 7.2 Inland freight transport for selected countries in Western Europe, 1983

	Rail	Road	Water	Pipeline
Austria	53.4	16.5	6.6	23.5
Belgium*	28.5	46.5	24.0	1.0
France	34.2	46.5	5.5	13.8
Italy	10.8	83.0	0.2	6.0
Netherlands**	7.0	24.4	60.6	8.0
Norway	19.2	43.7	0.0	37.1
Spain	8.9	88.4	0.0	2.7
Sweden	41.5	58.5	0.0	0.0
Switzerland	47.1	44.4	0.4	8.1
United Kingdom	13.5	78.9	0.3	7.3
West Germany	28.3	42.7	24.6	4.4

* 1970 data
** 1982 data

(*Source: Annual Bulletin of Transport Statistics for Europe*, 1983)

considerable pressure by governments to urge their railway organisations to concentrate upon what they can do best and to relinquish unprofitable services. Such strong advice has not gone unheeded as the route mileage of the West European network has continued to contract. As a social service it appears as though the railways have only a minor part to play with their high cost making them unattractive compared to other forms of transport which can operate more efficiently and economically outside the main sphere of interurban influence. In West Germany between 1960 and 1979 the total length of the Federal railway network was reduced by 26 per cent from 36,000 to 28,600 kilometres, the majority of which occurred in the remoter parts of the country. Inevitably such drastic closures have met with fierce public opposition not only in West Germany but elsewhere in the sub-continent especially where alternative forms of public transport have not materialised.

c) Road transport

The declining importance of the railways for the carriage of passengers and freight is in marked contrast to the massive growth which has taken place in road transport since the 1950s. In turn such growth has given rise to a new mobility which has had a most profound effect upon the location of housing and economic activity throughout Western Europe. In particular this massive rise in car ownership since the Second World War has given people throughout the sub-continent flexibility and freedom to choose their place of residence relaxing, once and for all, the spatial constraints previously dictated by public transport (see Fig. 7.9).

The benefits of car ownership are patently obvious – its flexibility and convenience have given rise to a changing urbanised society in which residence in car based commuter suburbs, shopping at out-of-town centres and hypermarkets and travelling to some distant workplace has become the norm. By 1982 there were nearly 110 million private cars on the roads of Western Europe and nearly 90 million of these were registered in the Common Market countries – a total which has more than doubled since 1965. To give some indication of this phenomenal growth in car ownership the West German situation is relevant. Although levels of ownership remained low in the Federal Republic until the late 1950s (under 600,000 cars in 1950, 2 million by 1956) rising personal affluence and increasing purchasing power was reflected in a quadrupling of the number of private cars on West German roads between 1956 and 1966. During the next ten years this figure almost doubled again from 10.3 million to 18.9 million vehicles until 1982 when there were over 24 million private cars representing a figure of 391 cars per 1,000 people, exceeding the rate of car ownership for all European countries with the exception of Luxemburg.

To meet the growing demand from this new mobility extensive road building plans were inaugurated and the task of motorway construction was put into motion. Although the concept of motorways was not a particularly new idea – introduced in the United States of America and, as mentioned earlier,

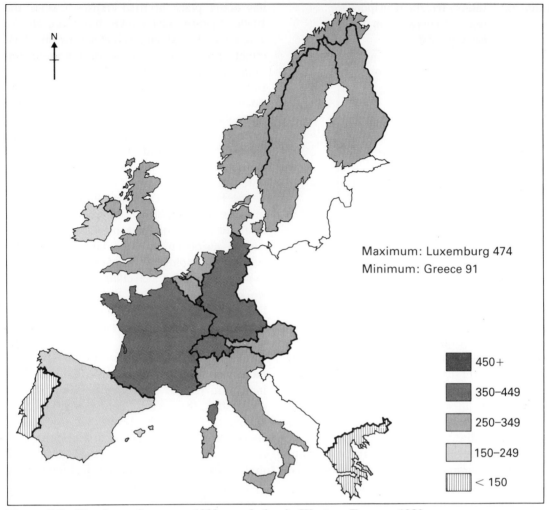

Figure 7.9 Numbers of cars per 1000 population in Western Europe, 1980

in Germany during the inter-war years – their widespread diffusion throughout the landscape of Western Europe did not take place until ten years after the Second World War. By that time the road transport revolution was well under way and there was a pressing need for improved road systems as ever-increasing volumes of traffic, goods vehicles as well as cars, began to snarl up the main traffic arteries. As a consequence, motorway development was seen as an ideal solution and the first links of what has become a well integrated, international system throughout much of Western Europe, was laid. Initially much of the construction was to meet national needs but since the 1970s emphasis has been directed towards the improvement of international routes to meet the requirements of present day trading patterns.

The new mobility together with the fundamental improvement in road communications has also been instrumental in scattering new industries far and wide. Although some firms are still tied to mineral reserves or markets, the structural changes which have taken place since 1945 have created a new industrial structure which is essentially 'footloose' giving individual companies a much greater flexibility when it comes to locational decisions. As a consequence factors other than the availability of raw materials or the market come into play and, increasingly, the desire to be close to the motorway network and to be situated in a pleasant environment have become uppermost in the minds of those taking such important decisions. Not surprisingly areas which meet the new 'accessibility' and 'environmental' requirements have become favoured zones for the location of new industries and rapid growth. The Frankfurt-Mannheim-Stuttgart corridor, the Maastricht-Aachen-Liège triangle and the often quoted *Autostrada del Sole*, stretching 750 kilometres from Milan to Naples along which 600 new factories were established between 1968 and 1973, have all become prime examples of highway related growth. Such highways have become, in the words of Hugh Clout, 'the industrial boulevards of Western Europe'.

The new mobility has, however, not been achieved without a considerable cost not only in economic but also in social and environmental terms, and it is these issues which have dampened much of the original enthusiasm lavished upon road transport. Although the potential for unhindered movement exists throughout Western Europe, actual movement is often greatly interrupted. Traffic congestion on intra-and inter-urban highways and at international frontiers has become commonplace. Elsewhere roads, not built to motorway standards, are being severely tested as both the volume of traffic and size of vehicles increase. Motorways themselves have come under increasing pressure from heavy vehicles as a rising volume of freight traffic is transferred from other modes of transport. Some of the busiest motorway stretches in the more densely populated parts of Western Europe now require constant repair and maintenance to enable them to handle the ever increasing volume of traffic – traffic which, incidentally, was not anticipated when they were first constructed. It has been suggested that road congestion in Western Europe has occurred because the automobile and the heavy lorry were adopted without fully accepting the burden of creating a first class

Figure 7.10 Traffic congestion in Rome

highway system. There is certainly plenty of evidence throughout the sub-continent to support such a notion.

In most urban areas of Western Europe, the issue of traffic congestion has been paramount for many years and it has been described as 'the thrombosis of our cities' transport arteries'. There is still little consensus about how to remedy the situation because no two cities are alike. Instead solutions to the problem of congestion have focused on either *accommodating* traffic or else attempting to *restrain* it. Some of the earliest post-war attempts to tackle the problem involved accommodation policies. In Paris, for example, the *Boulevard Périphérique* encircling the city 5 kilometres from the centre was constructed, with an outer ring road, still to be completed, (the ARISO) at 10 kilometres from the city centre linking the suburbs of Versailles, Orly and St Denis. These two ring roads together with the flow of traffic through the city centre by means of an extensive one-way system are seen as a potential solution to traffic congestion. Additionally, eight radial routes linking Paris to the provincial cities of France were all upgraded to motorway standard. Although similar solutions have been applied in London and other smaller cities such as Turin and Munich, in recent years this 'accommodating' approach has been largely rejected on financial, social and environmental grounds. In Paris, the *Boulevard Périphérique* has itself become congested, especially at peak times of the day, week and year and has proved to be both dangerous as well as inadequate. It is now commonly appreciated that in densely populated areas such as Paris even the most ambitious road-building programmes are incapable of accommodating the large flows of traffic which invade long established urban centres. Since the 1970s attempts to alleviate the problem of traffic congestion have focused mainly upon a variety of restraint policies aimed at reducing the volume of traffic entering city centres. In particular this policy has centred upon pedestrianisation which has become commonplace throughout many cities in Western Europe. Within this broad framework a number of approaches have been adopted; some do not allow cars and public transport (Bonn, Dortmund, Essen), some permit public transport (Amsterdam, Hannover, Bremen) and others allow taxis and delivery vehicles or permit vehicles at certain times of the day (Innsbruck and most West German cities) or in certain pedestrianised zones (Göttingen). Such schemes offer only a partial solution, however, and although extensive zones do exist in some cities (Copenhagen, Munich) their spatial extent is often restricted by the constraints placed upon existing businesses and residents. In addition to policies which, in the main, exclude motor vehicles, the wholesale introduction of other restraints such as parking meters and the preferential treatment given to public transport are used widely and go a considerable way in controlling the extent of traffic congestion in many of the cities of Western Europe.

Whilst making life more difficult for the private motorist many cities have reinvested heavily in public transport to complement the policies mentioned previously. In many cases the modernisation of public transport systems has received much popular support as the cost of building urban motorways has soared. Taxpayers, residents housed in the planned route of new road schemes and conservationists concerned about the built environment and the high energy costs associated with private transport have amalgamated to form influential transport lobbies supporting the renewed interest in public transport. In many cases the re-evaluation of public transport represents the first serious investment since systems were established during the nineteenth and early twentieth centuries.

Although buses have traditionally carried the greatest number of public transport passengers throughout the towns and cities of Western Europe their declining patronage over the past 25 years has largely hampered attempts which have been made to revitalise the system. Instead it is rail transport which has responded most readily to the new challenge, with many cities constructing or modernising urban rail systems at or below street level. To date, over sixteen cities in West Germany have developed the *U-Bahn* as well as public transport authorities in Brussels and Antwerp. Some cities, such as Lyon, Brussels, Milan and Amsterdam, have built (often at extremely high cost) new high-capacity metro networks under the city centre; other cities, such as Paris, have added new links to complement the existing underground network, and in so doing have improved overall accessibility. Above ground too, urban railway lines have been modernised as in Hannover and elsewhere in West Germany with the development of five *S-Bahn* networks. New urban rail links connect the Hague with its overspill town, Zoetermeer, providing a purpose-built commuter railway. In Paris also, the construction of the RER (*Réseau Express Régional*) has provided a mass-transit commuter network spearheaded by twin cross-city routes linking the outer suburbs and the SNCF network with the metro lines of the inner city. Elsewhere, principally in West Germany, existing surface tramway systems have been redirected underground with the construction of shallow tunnels immediately below city centre street level, forming the first part of a truly integrated urban transport system connecting up with bus and rail networks at major intermodal locations as, for example in Cologne, Bonn and Nuremberg.

In all these cases attempts to revitalise the public transport system can be viewed as a remedy to the interminable problem of urban traffic congestion. All these policies, together with the adequate provision of cheap and plentiful parking space outside the inner-city combined with the establishment of park and ride facilities, will go a considerable way in alleviating some of the worst problems. However, there is little room for complacency as the volume of private vehicles continues to grow and, as a consequence, the search for new solutions is an on-going process.

Figure 7.11 An accident on an autobahn

Although traffic congestion is a particularly harmful bi-product of the new mobility – interrupting the smooth flow of operations – there are a number of more frightening and sinister effects which are, once more, directly related to the rapid growth in the number of road vehicles. Of these perhaps the most worrying is the massive number of fatal and serious road accidents which, apart from the distress and suffering which they directly cause, are also a severe burden upon the economies of all car-owning societies. Although the number of road casualties in Western Europe has declined since 1970 (when 72,000 people were killed) over 56,000 people were killed in 1982 with nearly 2 million seriously injured. Despite this reduction, the 1982 figure is still needlessly high and, furthermore, it is evident that not all countries in Western Europe are improving their road safety standards. In some countries, however, the introduction of successful road safety measures – the strict imposition of speed limits, the compulsory wearing of seat belts, the use of

210 GEOGRAPHICAL ISSUES IN WESTERN EUROPE

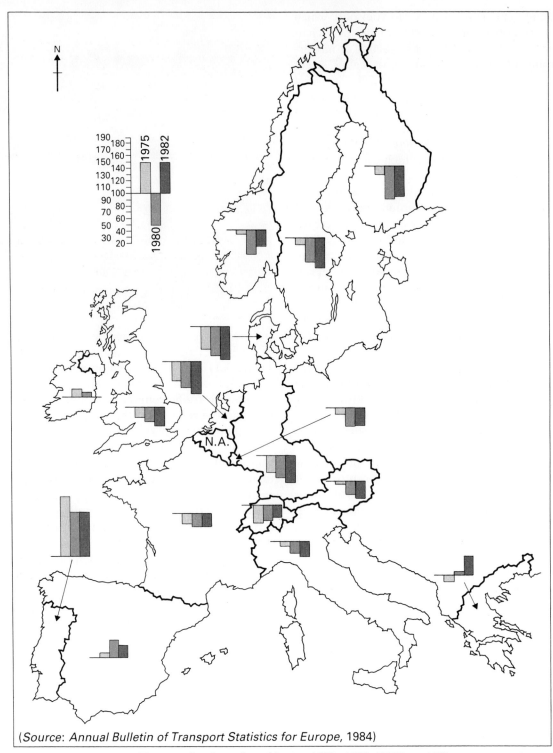

(Source: Annual Bulletin of Transport Statistics for Europe, 1984)

Figure 7.12 Western Europe: fatal road accidents, 1970–1982 (1970 = 100)

COMMENTARY

Fig. 7.12 shows the incidence of fatal road accidents for countries in Western Europe between 1970 and 1982. If 1970 is taken as a base year (= 100) then bars above or below the base line represent an increase or decrease in the number of fatal road accidents.

Self-assessment exercise

Considering Fig. 7.12 write an account in not more than 250 words of the major trends illustrated by the map. Secondly, suggest reasons why some countries have significantly improved their road-safety figures over the 12 year period.

---

constant and hazard lighting on cars and the tight control of alcohol and driving – have significantly reduced the number of casualties (Denmark, Sweden and Finland, for example). Elsewhere, there is still considerable room for improvement with some countries – Spain, Portugal and Greece – recording accident figures significantly higher in 1982 than those published in 1970 (see Fig. 7.12 and self assessment exercise).

It is unlikely that the central issue is ever going to be totally solved but safety measures can, and do, save lives. The introduction of safety legislation should therefore be a matter of upmost priority in Western Europe, for a range of countries which are so advanced in many other respects.

Apart from the direct assault inflicted upon the public by road accidents, there is a wide range of harmful effects associated with traffic related air and noise pollution. Layers of smog which have begun to settle over many West European cities are maintained by the noxious gases emitted in the exhaust fumes of automobile engines. In particular vehicles with petrol engines are the worst offenders pumping harmful oxides of carbon and nitrogen into the urban atmosphere in increasing quantities. Apart from the harmful effects upon the surrounding natural environment (which will be discussed in more detail in Chapter 8) these substances, together with the emission of lead, are extremely hazardous in concentration. In particular, oxides of nitrogen and carbon are especially harmful to people with heart and lung complaints and can cause headaches, nausea, loss of vision and abdominal pains to normal healthy individuals. Lead, on the other hand, is added to petrol to improve engine performance but in the process a certain amount is emitted in exhaust fumes. It is a particularly harmful substance and acts as a cellular poison, often difficult to diagnose, but known to effect the neuro-muscular system, circulatory system, the brain and the gastro-intestinal tract.

Inevitably, where traffic congestion is most severe the levels of atmospheric pollution recorded are significantly higher. In any crowded town or city street throughout Western Europe people are at risk. In particular, those who spend their working day outside, at or below street level are subjected to the greatest danger. Despite available technology for removing these harmful substances few countries in Western Europe have adopted such measures as a legal requirement. Similarly, until they are told to do so, the environmental conscience of motor manufacturers and petrol companies is not strong enough for unilateral action. Not surprisingly, these multi-national companies are not prepared to introduce this new technology because of the fear that the additional cost (however slight) to the product may affect their overall market share.

d) Air transport

Air transport, like the advent of mass car

ownership, is very much a product of the late twentieth century (Fig. 7.13). Although scheduled air services have been operated in Western Europe for well over 60 years it is only during the past few decades that air transport has established itself as a principal mode of passenger transportation competing effectively with all other modes of transport, especially on medium and long-haul international routes. Despite the steady growth which has been recorded annually since 1945, air transport in Western Europe has not dominated passenger transport as it has in North America where intense competition since the Second World War has resulted in the virtual demise of passenger rail services as a viable form of transportation. Just why air transport has not been so influential in Western Europe can be explained by a variety of human factors. Firstly, the bulk of Western Europe's population is clustered within and immediately adjacent to the Eurocore, contrasting vividly with the population distribution of North America which tends to be peripheral, within 100 kilometres of the coastline. This is further compounded by the fragmented nature of the nations of Western Europe, with the division into relatively small political units creating a situation whereby domestic travelling distances are considerably shorter than they are in the United States or Canada. Consequently, domestic air services, as well as many routes within the Eurocore, have not been developed to the same extent as other, established forms of transport, which can operate more economically over such short distances.

Secondly, the present situation has been exacerbated by the high cost of air travel in Western Europe which remains significantly more expensive per passenger kilometre than in North America. Additionally, the cost of air travel in Western Europe is also excessive when compared to surface transportation, and this has allowed competing forms of transport (such as the railways) to retain a major share of the market. Why the cost of air travel should be so much more expensive in Western Europe than North America remains a central issue, which we shall now look at in more detail.

Basically, it has a lot to do with the structure of Western Europe's major airlines, the majority of which (Alitalia, Air France, KLM, Lufthansa, SAS, Swissair, for example) tend to be state owned organisations having, often, a virtual monopoly upon domestic scheduled services. Furthermore, national authorities responsible for civil aviation decide upon the licensing of new airlines and negotiate operating rights with other countries, which means that civil aviation throughout Western Europe is almost totally influenced by government decisions. As a consequence scheduled international air services are regulated by bilateral agreements between the two countries wishing to offer a service and, in almost all cases, is operated by the state airlines of the countries involved. Such a system has impeded the opening up of new routes and has encouraged existing services to be concentrated between major airports. Additionally, the airlines of the 'cartel' have focused their attention upon the business end of the travel market attracting customers for whom the high fares are outweighed by the advantages which can be gained from saving time. Inevitably, other sectors of the travelling public for whom cost is more important than time-saving, remain largely excluded from scheduled services.

Scheduled traffic, however, only forms part – albeit an important part – of total passenger traffic. A considerable number of flights are offered by a multitude of charter airlines operating throughout Western Europe primarily for the transport of holidaymakers to the major tourist resorts and regions of the Mediterranean, West and East Africa, the Far East and North America. As the demand for international tourism has grown since the 1960s these airlines have cashed in on a lucrative and expanding market. Today, airlines such as Sterling of Denmark, LTU of West Germany, Martinair of the Netherlands, Aviaco of Spain and Brittania of the United Kingdom are larger, in all respects, than some

TRANSPORT IN WESTERN EUROPE   213

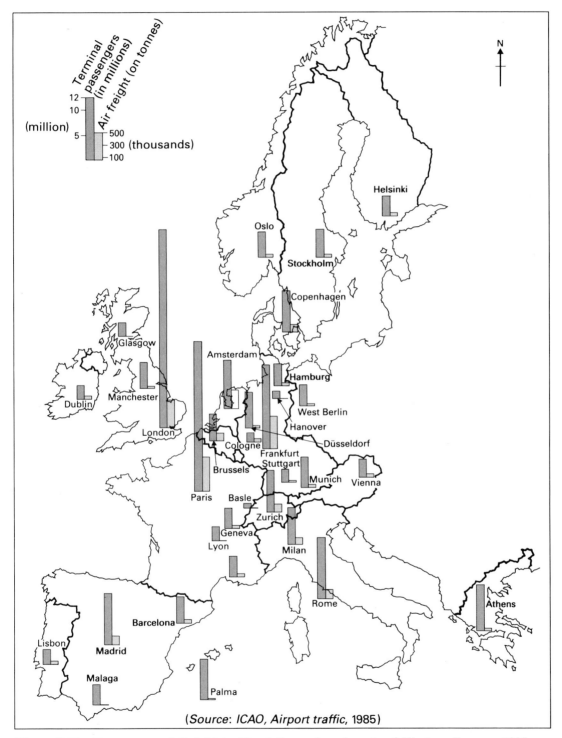

Figure 7.13 Passenger and freight traffic at the major airports of Western Europe, 1985

of Western Europe's state owned carriers. More recently, the booming charter market has encouraged state owned airlines to enter the business and a number of them have established wholly owned subsidiary companies for a share of the profits Air Charter of France, and Condor of West Germany deal exclusively in charter operations and are all subsidiaries of their state owned airlines). Charter traffic, being relatively free from the strict regulations applied to scheduled services, has led to intense price competition between not only themselves, but also between charter and scheduled operators (particularly over the North Atlantic routes) which has meant cheap travel for the flying public.

Such a situation has had little effect upon fare reductions for scheduled services although a number of operators in Western Europe were proposing some unilateral or bilateral action on fares in 1985. In recent years, the Scandinavian airline SAS has had great success in charging an 'economy fare' for all seats; whilst British Airways and the Dutch national airline, KLM, are involved in serious talks about the possibility of introducing cut-price scheduled fares. If these, and other discussions between the continent's major airlines bear fruit in the near future and if economic pressure leads to deregulation (as it has done in the United States of America), the cost of domestic and international air fares may well fall considerably.

For air travellers this will be good news. For many state owned airlines, however, it will not be greeted with such enthusiasm. Many major state operators have, in recent years been faced with increasing financial difficulties. Airlines such as British Airways found it increasingly difficult to maintain an operating profit to the extent that the government in the United Kingdom in 1987 sold off its assets in the state airline to the public, opening up the airline's vast route network to the numerous small, independent operators eager to offer a cheaper and more reliable performance. The events in the United Kingdom will undoubtedly be studied very carefully by other West European governments and we might just be witnessing the start of the break up of one of the subcontinent's greatest cartels.

Despite these issues, airport traffic – both scheduled and non-scheduled, passenger and freight – has continued to grow throughout Western Europe and the region is now covered by a fairly dense network of routes concentrated particularly in the Eurocore. By the end of 1984, 114 airports had at least one direct, international scheduled service within Western Europe, although the scene was dominated by six major airports – London (Heathrow and Gatwick), Paris (Charles de Gaulle and Orly), Amsterdam and Frankfurt – which between them offered nearly 250 direct international scheduled services throughout the sub-continent. Apart from this all four cities were at the hub of a multitude of domestic, charter and inter-continental services, illustrating once more the dominance of the Eurocore over the surrounding regions (see Table 7.3).

In recent years perhaps the greatest issue concerning the future development of air transport has revolved around the siting of new airports. Inevitably, as passenger traffic has grown since 1945, airport capacity has been progressively outstripped and new airports have had to be built to cope with the increase in demand. In Paris, for example, the city's original airport at Le Bourget, 12 kilometres north-east of the city centre, was supplemented by Orly Airport in 1948, 15 kilometres to the south of the city. In 1974 a third airport was opened on a greenfield site at Roissy, 12 kilometres north-east of Le Bourget (which closed to scheduled traffic in 1977). A similar chain of events has occurred since the Second World War in other capital cities – Madrid, Stockholm, London, Rome – as airports close to the city have become constricted as a result of urban encroachment. In all these cases new airports have been constructed on the 'new' periphery of the built-up area.

Table 7.3 Air transport indicators for the capital cities of Western Europe*, 1985.

	a	b	c	d	e
Amsterdam	16	59	9.7	139.7	370.4
Athens	14	23	9.3	103.8	75.5
Brussels	15	39	5.0	77.8	122.2
Copenhagen	15	44	8.3	142.5	143.6
Dublin	13	27	2.5	50.3	40.4
Frankfurt	16	47	17.0	204.4	628.3
Helsinki	15	20	3.8	56.6	33.1
Lisbon	13	18	2.9	35.6	48.9
London*	16	78	39.1	402.4	575.4
Luxemburg	8	12	0.6	14.0	62.2
Madrid	14	28	10.2	113.4	153.9
Oslo	10	19	4.9	74.4	31.0
Paris*	16	63	29.6	283.4	667.2
Rome	15	28	12.1	144.0	120.8
Stockholm	14	24	5.7	88.6	45.4
Vienna	12	23	3.1	55.4	34.9
Zurich	16	44	8.3	120.3	179.8

Key
a = Number of capital city connections (regular scheduled services, direct)
b = Number of direct international scheduled services
c = Terminal passengers (millions)
d = Aircraft movements (thousands)
e = Air freight ('000 tonnes)

* Frankfurt and Zurich are assumed to have a capital city function in terms of air transport.
* London figures for Heathrow and Gatwick
* Paris figures for Charles de Gaulle and Orly

(*Source*: ABC World Airways Guide)

Self-assessment exercise

With regard to the air transport indicators left, columns (a) and (b) represent the accessibility of the 17 capital cities in Western Europe. In column (a) you will note that five cities – Amsterdam, Frankfurt, London, Paris and Zurich – are connected to all other capital cities by regular scheduled services. In column (b) the same cities exhibit a high degree of connection with many other cities in Western Europe. The final three columns give some indication of the relative importance of air traffic in each of the cities and can be used in conjunction with Fig. 7.13.

For each of the five columns rank the cities according to their relative positions and then derive a composite rank by adding up the rank scores for all five columns. Having done this, then rank the cities in terms of their total scores – those with the lowest being the most influential and vice versa.

Comment upon the final ranking.

For many years airport development was considered yet another trapping of urbanisation and planning, and construction was encouraged and progressed largely uninterrupted. It was appreciated by government and public alike that major airports were not only a source of major revenue, providing much direct employment, but they also acted as 'growth poles' in their own right attracting a wide range of industries to the immediate locality. In the last decade or so, however, the economic benefits of airport development have been overshadowed by environmental considerations. During the 1970s the scale of air transport operations moved up a gear with the introduction of wide-bodied jets such as the Boeing 747, the Lockheed TriStar, the McDonnell-Douglas DC–10, and the European Airbus. Although not requiring any greater length of runway for landing and taking off, these jets, carrying up to 500 passengers, created major problems for airport authorities as passenger handling facilities were completely swamped when two or three of these aircraft arrived shortly after each other. The answer, it seemed, was to build larger airports with greater terminal capacity. The first (and possibly the last!) of this new generation of airports in Western Europe is Charles de Gaulle Airport, Paris, covering an area of over 3,000 hectares of farmland. Fortunately for the authorities this airport escaped the major public demonstrations which have accompanied similar developments elsewhere (Montreal-Mirabel and Tokyo-Narita). Plans to expand Frankfurt's Rhein-Main airport with the construction of an additional runway has met with violent confrontation from local residents and environmental groups united in their opposition to a development which would remove a vast area of woodland, create additional noise, increase air and surface traffic, increase air pollution and bring with it the increased chance of an air disaster. Against these formidable and sustained protagonists the airport authorities have been forced to back down and to reconsider their plans.

The time has passed when airport development was a foregone conclusion. Today it has become a major public issue and one which, in Western Europe at least, has so far largely defeated the bureaucrats.

e) Seaports

No discussion of transport is complete without a mention of seaports, which for many centuries have been the pivot of European trading relationships both within the continent as well as throughout the rest of the World. The present day pattern of port development along the varied coastlines of Western Europe reveals a multitude of seaports, especially concentrated in north-western Europe and the western Mediterranean, with over 70 of considerable size, competing with each other for the trade of the European hinterland. Inevitably, as transport history has always shown, changing physical conditions, patterns of seaborne trade and advances in maritime technology have upset the hierarchy of port development, favouring some locations whilst discriminating against others.

The impact of changing technology upon maritime transport has also had a fundamental effect upon the majority of ports in the region; the increase in vessel size, the greater depth of water required in port approaches and within the vicinity of berths, and the growth of containerisation for the carriage of general cargo have all led to far reaching changes in port systems (see Fig. 7.14). Invariably some seaports have often been unable to cope with the changing requirements of maritime operations whilst others have adapted to the new circumstances successfully. As a result the traditional port hierarchy in Western Europe, which had been established during the late nineteenth century, has been disrupted in the short space of 30 years. The effects of this 'transport revolution' have reverberated throughout the old port cities of the sub-continent and have led to major issues regarding their future prosperities. Some of these issues will now be examined in greater detail.

Many seaports in Western Europe are located on the banks of rivers and estuaries, owing their development to the fact that they were the lowest bridging point of that particular river, or at the head of the large estuary, offering a safe anchorage. In recent years, however, the survival of many of these ports has been put at risk by the changes, mentioned previously, in maritime transport, together with the restrictions upon port expansion imposed by surrounding urban development. As a result many traditional port cities such as Amsterdam, Gothenburg, Hamburg, Marseilles and London, are being faced with the unavailability of land for terminal development in the surrounding city centre and have been forced to build facilities downstream where deeper water is available, narrow navigation channels are avoided and greater areas of land are available for the expansion of facilities. Gothenburg is a perfect example of such a process. Being Sweden's main port and major trade outlet on the River Gota it has, over the past three and a half centuries, been forced to move progressively downstream to deeper water and available land reclaimed from the Kattegat. As a result the focus of port operations is now centred upon the Skandia and Alvsborg harbours where a new container berth and car ferry terminal have been constructed. The old river quays and harbour adjacent to the commercial centre of the city have been progressively closed or redeveloped to meet the needs of new traffic (such as recreation) often only indirectly associated with seaborne trade.

TRANSPORT IN WESTERN EUROPE 217

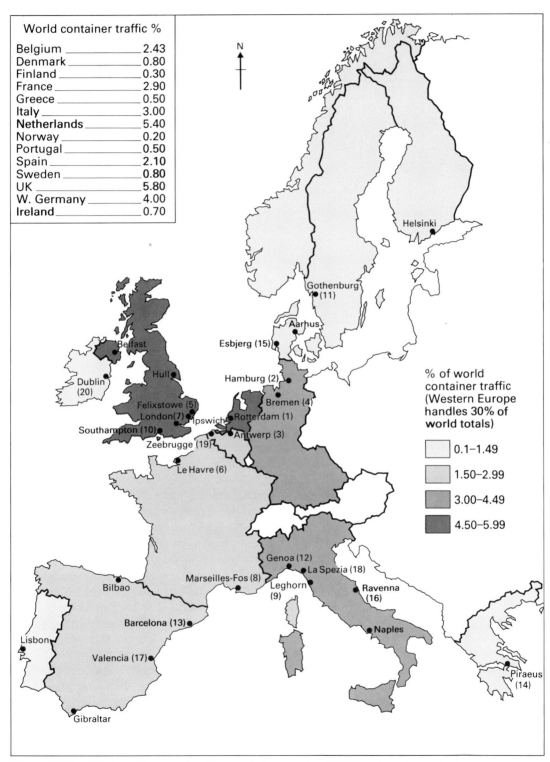

Figure 7.14 Western Europe: leading container ports, 1984 (top twenty ranked)

Figure 7.15 Europoort containers

Ironically, the commercial decline of city centre port installations has presented many port cities in Western Europe with an ideal opportunity for urban redevelopment, transforming such areas into housing, industrial and recreational uses, whilst often maintaining the maritime flavour of the surroundings (often by the incorporation of a maritime museum charting the city's development).

Although there are many sizeable ports in Western Europe expansion in trade since the 1950s has witnessed the growth of what has been termed 'Euroports' which handle a larger tonnage and serve a much wider hinterland than other ports. Such international gateways are exemplified by the ports of Rotterdam which handles more cargo (mainly oil and related products) than any other port in the world – Marseilles, Antwerp, London, Genoa, Le Havre, Hamburg and Amsterdam. Between them these eight ports dominate the pattern of total port traffic in Western Europe. Rotterdam, however, remains the undisputed leader handling the seaborne trade of many countries and having an international hinterland encompassing most of the highly industrialised areas of Western Europe, in which 250 million inhabitants live. The port has become the natural focus of the transport and communications network in north-western Europe (see Fig. 7.16).

Rotterdam's growth is due, not only to trade in its narrowest sense, but also to the growth of major industries of which oil and petrochemicals are by far the greatest. Inevitably traffic increased originally as the port became the oil refining centre of the world with its five refinery complexes stretching for many kilometres along the south bank of the estuary. Additionally the port has handled increasing amounts of other bulk cargoes – such as iron ore, grain and coal – and has rapidly become the major container port in Europe. Despite this phenomenal growth, recent years have witnessed a decline in the volume of cargo as the volume of oil imports has constantly fallen since the world oil crisis of 1973. Inevitably, a port so highly dependent upon one commodity is susceptible to market fluctuations and, despite expanding other sectors of goods traffic, Rotterdam has not yet fully recovered. Nevertheless, the port

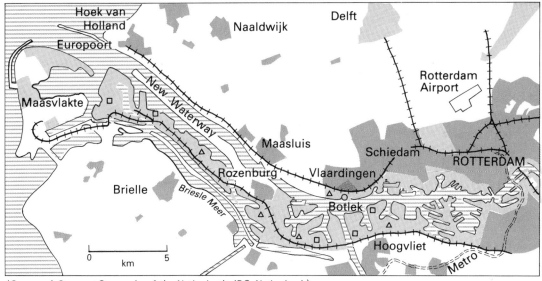

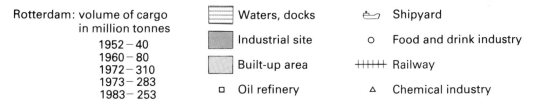

Rotterdam: volume of cargo in million tonnes
1952 – 40
1960 – 80
1972 – 310
1973 – 283
1983 – 253

Figure 7.16 Rotterdam/Europoort

has adapted to changing world circumstances and continues to expand seawards creating facilities for other bulk commodities and container traffic. The most recent extension at Maasvlakte has been built to handle the largest bulk carriers and oil tankers, whilst elsewhere on this reclaimed island a new container berth has been constructed and land to the south has been earmarked for industrial development.

2 Transport and the future development of Western Europe

Having focused upon the specific transport modes and looked at the particular issues relevant to each of them, it is now necessary to turn to some of the wider issues facing the development of transportation in Western Europe and, in particular, the problem of transport integration which is seen by many as the key to the future growth and development of the whole region. Although considerable debate still exists as to the role of transport in the development process, it can be said that the spatial distribution of population in Western Europe is largely organised by the location of employment opportunities which, in turn, are heavily influenced by *accessibility* to markets, sources of supply and the capital city. As a result the populous regions of the Eurocore owe their phenomenal growth partly to their central location on their national transport systems as well as the fact that they have always been major centres of attraction, overshadowing all other urban areas. As a result transport has had a major impact upon reinforcing the traditional core-periphery relationships mentioned in other chapters. Although we may suspect a link between this striking pattern of economic development and accessibility and consider, as a result, that greater transport integration and the completion of 'missing links' will lead to greater economic integration and the wider dispersal of 'footloose' industries, available evidence suggests that despite the knowledge that economic progress is dependent upon an adequate transport infrastructure, rarely does transport have direct consequences upon the location of economic activity. Thus, rather than core-periphery relationships being relaxed, research has shown clearly that in Western Europe the trend is towards *greater* industrial concentration in the Eurocore at the expense of the periphery. Although new sub-cores have, and will probably continue to emerge (such as the Lyons–Grenoble region, Frankfurt and the M4 corridor in England) they are but part and parcel of the amorphous whole, reinforcing the traditional relationships.

Despite the limited effect of transport development in stimulating regional growth there is still a vital need to improve the transport system in this age of high mobility. In particular, the advantages to be gained from transport integration have been championed since the Second World War. Such a process is seen as a fundamental measure enabling nations in Western Europe to compete effectively with other major trading regions of the world, by allowing the smooth and efficient flow of goods, materials and people within a common market. Yet as we have already seen such a process has been severely hampered by an historical legacy which focused upon the development of national transport systems and networks. Consequently, the most pressing task of recent times has been the various attempts which have been made to superimpose a truly integrated transport system upon the landscape of Western Europe. Exactly what has been achieved in the years since the idea was first mooted will now be examined.

3 Transport integration – the way ahead?

The desire for transport integration was reflected soon after the War, in 1953, when the majority of West European nations (except Finland and Iceland) jointly set up the *European Conference of Ministers of Transport* (ECMT), a permanent body whose principal aims were:

- To take whatever measures may be necessary to achieve, at general or regional level, the maximum use and most rational development of European inland transport of international importance.
- To co-ordinate and promote the activities of international organisations concerned with European inland transport, taking into account the work of the supranational authorities in this field.

Despite ambitious and far-sighted objectives, the ECMT has, unfortunately, neither the power nor the resources to implement agreed areas of policy and, as a result, its main function today remains as a research forum investigating contemporary transport problems and issues throughout Western Europe.

As for the major trading blocs, they too made little headway in the direction of an integrated transport policy until the 1960s. In the meantime, however, a number of international bodies took the initiative upon themselves. The first of these was the International Union of Railways who, in 1957, inaugurated a system of Trans European Expresses (TEE) throughout the original six member states of the EC, Austria, Switzerland and the United Kingdom. With the principal aim of providing fast and regular rail services to the main centres of commerce and industry, the network was designed to meet the specific needs of the international businessman around whose requirements the services were scheduled. The network was further extended in 1964 to incorporate a number of long distance domestic services such as that from Paris to Nice. Similar developments were also being promoted by the United Nations Economic Commission for Europe who, from their base in Geneva, identified and designated a system of international 'E' roads throughout Europe. More recently, international air transport has benefited from the development of 'Eurocontrol' established on behalf of a number of West European countries to co-ordinate and control air traffic movements in the region.

The need for an integrated transport system was recognised by the EC from the outset in order to supplement existing national transport systems throughout the region. Furthermore the undeniable importance of the transport sector – employing some six million workers, attracting 11 per cent of total private investment, 40 per cent of total public investment and accounting for 6 per cent of the Community's gross domestic product – reinforced the desire for a smooth and efficient international transport system. To achieve these ends a Common Transport Policy (CTP) was proposed and published in 1961. The necessity for such a policy soon became clear as traffic between member states continued to grow (since 1960 this growth has been far greater than the expansion of domestic traffic and has doubled between 1969 and 1979). Despite the available evidence, many governments were still reluctant to improve transport links with their neighbours. In the Netherlands, for example, there was little incentive to build railways to connect up with rail systems in France and Luxemburg when transport priorities had been traditionally directed towards inland waterways. Additionally governments throughout Western Europe, and especially within the EC, had been allocating a disproportionate amount of funds to the development of roads and motorways, often at the expense of other potentially viable forms of transport such as railways which, as we have already seen, declined significantly in their overall importance during this period.

However, with the passage of time, changing attitudes and the development of new markets have shown the relevance of the Common Transport Policy and it has begun to exert an increasing influence upon the transport policy of all member states. From its inception the Policy has championed two long-term objectives; firstly, to identify the major transport links in the community network at present, as well as in the future, and to evaluate particular areas where investment will be necessary to improve the overall system; secondly, to provide a sound basis for research into areas of future investment. In addition to these on-going objectives, a number of short term objectives were drawn up including:

- *The identification of bottlenecks* likely to hinder the smooth flow of traffic between member states. In particular the improvement of frontier crossings such as the Alps. Within countries too, the eradication of both regional and local bottlenecks is a prime objective in order to improve internal as well as international accessibility. Improvements to the transport infrastructure of the Ruhr region in West Germany (particularly the construction of new north-south routes) and the completion of London's outer orbital motorway (the M25) are cases in point. Elsewhere, although bottlenecks have been identified, such as Paris, solutions have still to be found.
- *To improve international links between major cities and regions* such as the upgrading of rail links between Brussels and Cologne, Cologne and Utrecht and between Strasbourg and Amsterdam, Brussels and Luxemburg and the general improvement of road and rail links between France and West Germany which, for historical reasons, are traditionally weak.
- *To improve links with peripheral regions* such as the Iberian peninsula, the Irish Republic and Northern Ireland (in particular, communications between

> The Sunday Times, 6 January 1985
>
> ## Swiss road plan taxes foreign drivers's patience
>
> **by Brian Moynahan**
> **European Editor**
>
> IN one of those acts of extreme civic responsibility that have become as legendary as the cuckoo clock, the Swiss voted in favour of a road tax in a referendum last year. Not much of one, £9.90 a year, but a self-imposed tax nonetheless.
>
> Admiring observers were less impressed when the Swiss went on to reject the advice of the government and the transport ministry and voted that foreign car drivers should pay it too. Since January 1, any foreigner driving a car on a Swiss motorway has had to fork out the same annual fee as a native, however short his stay. If he is in a truck, the cost can run from £178 to £1,000 depending on vehicle weight and frequency of trips.
>
> Until now, Swiss road maintenance has been paid for largely through fuel tax, with 20% coming from the government. The new tax is aimed at cutting the state contribution.
>
> "We were really after foreign trucks", says a spokesman for the federal transport ministry. "You get these big trucks barrelling through the north-west transit from Basle to Chiasso. They don't stop. The drivers don't eat or buy fuel – it's more expensive in Switzerland. They were wearing out the roads, and we got nothing."
>
> Taxing foreign cars was not on the agenda. "We didn't propose it, and neither did the government", says the spokesman. "Parliament put it into the referendum and the people passed it. We are obliged to accept it."
>
> The West Germans are furious and plan to retaliate by charging Swiss drivers a special levy from March or April. The Swiss are threatening to counter this retaliation with a form of "corkage": drivers who filled up in Germany – where diesel is much cheaper – to drive through Switzerland, would be taxed on the fuel in their tanks.
>
> The Belgians are also angry although they are on shaky ground since they considered similar proposals to tax foreign motorists. These were turned down on grounds of administrative difficulties rather than altruism. Some Belgian motorists are vowing to drive for holidays in Italy though France, sidestepping Switzerland. This gesture will cost more in extra petrol and French motorway tolls than the Swiss tax disc.
>
> The Swiss were resigned to protests from West Germany and Belgium, since motorways in these countries are free. The French have also threatened to impose special charges on Swiss trucks, despite the high tolls on French motorways. French truck drivers blocked crossings into Switzerland last month. Although other protests have come from Sweden and Finland, the British, who rate eighth in a Swiss transport league of visitors dominated by the West Germans and French, have been virtually silent.
>
> The new taxes should bring in £20m a year from trucks and £70m from cars. The Italians may reduce the take. Within hours of the imposition Italians were selling well-made forgeries of the tax discs on the roads leading up to the frontier at Chiasso. The asking price – 90p.

Figure 7.17

Dublin, Belfast and Derry and between Dublin, Cork and Galway) the Mezzogiorno and the Italian and Greek islands.
- *To develop links with the accession of new member states* in particular Greece, but also with Spain and Portugal. Greece, unlike the latter two states, remains disconnected from the rest of the Community and in this case the development of maritime links would seem to be the most appropriate course of action.
- *Links to overcome natural obstacles* several of which still remain as a hindrance to the free flow of goods and passengers, in particular the Channel crossing between continental Europe and Great Britain, but also between West Germany and Denmark via the Fehmarn Channel. Elsewhere, additional Alpine crossings are also highly desirable as the volume of traffic continues to grow between the north and south of the Community and also within countries, such as Italy, where the Appenine barrier retards the flow of trade between regions of the country.
- *The completion of missing links of existing networks* such as those which still exist on the inland waterway system (the Rhine-Danube link, for example), as well as the

extension of existing waterways in Belgium and France to enable them to handle the largest type of barge. Additionally, there are still missing links in the motorway network including those between the ports of Holyhead and Fishguard on the Irish Sea coast of Wales, within the Irish Republic and between France and Spain.

Despite agreement over the main objectives of the Common Transport Policy, considerable disagreement still exists among the member states of the EC regarding its implementation. Unlike other areas of community policy (such as agriculture) transport systems between nations still remain fragmented, bound up with an historical legacy and the determination of both national government and private enterprise to pursue a unilateral form of development. Inevitably, the chances of developing a fully integrated transport system in the near future remain very much in the balance.

Such scepticism should not conceal the fact that the next few decades will reveal major transport developments throughout Western Europe, and these advances will continue to speed up the smooth and efficient flow of goods, people and services. Additionally transport systems of the future will be more cost effective, have an added degree of safety and will have a greater regard for environmental standards. We can only look forward to such developments. However, future events do not promise to solve some of the basic issues which have threaded their way through numerous chapters of this book. Transportation, like other parts of the West European economy, is most highly developed within the Eurocore. Major axes of communications identified in the 1960s have, to all intents and purposes, been reinforced today and few areas outside the central core have been able to break the stranglehold. Although new industries have largely been freed from the constraints of locating near raw materials, there are still significant advantages to be gained from siting plants in *accessible* locations. Consequently, where transport and communications are most highly developed industry will be attracted. Elsewhere, even the stick and carrot of regional incentives offered by national and supra-national government will do little to break the 'established' pattern of events.

8 Environmental Concern

1 Environmental concern

From the Arctic reaches of Scandinavia southwards to the warm temperate Mediterranean coast, from the Atlantic fringes of Iberia to the High Alps in the east, Western Europe is characterised by a landscape of infinite variety and a natural environment of great diversity. Not surprisingly, some of the world's most fortunate inhabitants live in this well endowed region where the expectation of long life, freedom from disease, comfortable accommodation, an adequate and often luxurious supply of food and drink as well as the accumulation of a vast array of material possessions allow an exceptionally high standard of living.

However, as Chapter 1 emphasised, there is a price to pay for this seemingly enviable life style. Western society which, on the one hand provides us with so many tangible benefits seems, on the other, to be capable of its own self destruction. The depletion of natural resources, the spread of industrial dereliction, the presence of increasing amounts of air pollution, the treatment of rivers, lakes and seas as open sewers, the destruction of wildlife and landscape and the transfer of agricultural land to urban and industrial uses are commonplace. Even more worrying is that these problems, although spatially concentrated, are not geographically restricted to a few select countries in Western Europe – all share in the overall contribution to environmental degradation. Additionally, all environmental problems are intimately related to the natural systems with which they interact, thus the transportation of water and air-borne pollution generated in one country or region can become a problem which has to be faced by another.

What, then, lies behind this state of affairs? The purpose of this chapter is to investigate some of the reasons behind these problems and issues, to evaluate their effect and to gauge their spatial extent. In so doing, attempts to discover solutions and to put ideas into practice will be examined.

2 Origins of environmental degradation in Western Europe

Surprisingly, many of Western Europe's environmental problems are not new as they originate from an earlier historical stage well before the period of industrial revolutions in the nineteenth century. The smelting of iron ore in areas such as the Ruhr in West Germany and the Nord region of France, and its processing into crude forms of steel had been active for a considerable period of time. Similarly, the discharge of wastes into water courses was a frequent occurrence throughout the sub-continent and was, in places, starting to exert considerable influence upon the carrying capacity of the natural environment. Fortunately, this emerging industrial base was essentially small scale and those problems which did exist were largely hidden by an agrarian landscape interspersed with vast areas of forest.

Nevertheless, it was the period of industrial change in the late eighteenth and nineteenth centuries that unleashed environmental degradation on a grand scale. The industrial revolution which spread across Western Europe was fuelled by the harnessing of steam power, so that the new industrial society was dependent upon an increasing and unending supply of coal and iron ore. Areas that had previously known only localised mining activity erupted with the scars of industrialisation and as settlements grew, coal burning factories and domestic hearths started to belch acid pollutants into the atmosphere. As new industries sprang up old sites were renovated or else fell into dereliction, more wastes were discarded, untreated effluents flowed into rivers and streams and spoil heaps became characteristic features of many previously unblemished landscapes. Northern England, north-western Germany, northern France and Belgium witnessed phenomenal industrial growth. The image of the Ruhr was one of

> pit heads and factories, miners black with coal and panting steel workers, slag heaps and boggy industrial wastelands, smog and thunder, the sky full of the smoke of chimneys in the daytime, and at night the growing red reflections of a thousand soot-belching fires (*Source: Changing for the Future*, Kommunalverband Ruhrgebiet, 1982).

It is an image which has characterised so many of these regions ever since. Furthermore, the massive drift of population from the countryside to the towns and cities of Western Europe also casts an indelible shadow over so many unspoilt landscapes. Rapid urbanisation brought in its wake a multitude of problems on a scale previously unexperienced and unimaginable. Massive quantities of domestic waste requiring disposal, increasing clouds of air pollution, low standards of housing, interminable traffic congestion, the difficulties of obtaining an adequate and pure supply of drinking water and a severe shortage of open space were serious threats to human health and welfare.

Despite this image of filth, overcrowding and despair, for many people this was a time

---

Box 8.1

Étienne now commanded a view of the whole district. It was still very dark, but the old man had peopled the darkness with untold sufferings, which the young one could sense all round him in the limitless space. Could he not hear a cry of famine borne over this bleak country by the March wind? The gale had lashed itself into a fury and seemed to be blowing death to all labour and a great hunger that would finish off men by the hundred. And with his roving eye he tried to peer through the gloom, with a tormenting desire to see and yet a fear of seeing. Everything slid away into the dark unknown, and all he could see was distant furnaces and coke-ovens which, set in batteries of a hundred chimneys arranged obliquely, made sloping lines of crimson flames; whilst further to the left the two blast-furnaces were burning blue in the sky like monstrous torches. It was as depressing to watch as a building on fire: as far as the threatening horizon the only stars which rose were the nocturnal fires of the land of coal and iron.

(Extract taken from *Germinal* by Emile Zola, Penguin Edition (1978) page 23)

Germinal was written by Zola (1840–1902) to draw attention to the misery prevailing among the poor in France during the Second Empire. . . . The novel depicts the grim struggle between capital and labour in a coalfield in Northern France.

(Penguin frontispiece)

---

of great prosperity in which fortunes could be amassed from the cinders, smog and fumes. Throughout Europe it was the heyday of the great industrial families – of Krupp, Siemens and Thyssen in Germany, of Michelin in France – exerting their influence upon economy, society and political life yet having little or no regard for the environment which they were polluting (see Box. 8.1).

The distasteful environmental legacy left behind by the first phase of major industrialisation was hardly improved upon by successive eras of development. Despite the changeover to seemingly cleaner sources of energy – such as oil and hydro-electric power – and the introduction of piecemeal environmental legislation, the growth in economic prosperity in Western Europe since 1945 has also been marked by environmental abuse. There has been, however, the beginning of a much greater awareness of the fragile state of nature and the realisation that a number of vital resources were starting to show serious signs of depletion (see Chapter 9). Such feelings were reinforced by a number of environmental tragedies which occurred during the 1960s and 1970s.

Case study: The Amoco Cadiz disaster

Within the short space of eleven years two major disasters had occurred on the Atlantic fringes of the English Channel, both accidents involving oil tankers and the massive spillage of crude oil. The first of these environmental tragedies happened in the early spring of 1967 when the 118,000 tonne oil tanker, *Torrey Canyon*, sailing between the Persian Gulf and Milford Haven in South Wales, ran into rocks 25 kilometres west of Land's End just east of the Scilly Isles.

In March 1978 a second similar disaster occurred when the Liberian registered oil tanker, *Amoco Cadiz* ran aground into rocks off the coast at Portsall, a small fishing village situated on the north-western tip of Brittany. In this case the stricken vessel discharged 1.5 million barrels of oil, (three times the amount released by the *Torrey Canyon*) which polluted over 500 kilometres of the Brittany coast (see Fig. 8.2).

Both these disasters can be considered as true environmental tragedies; although human life was spared, marine life was seriously affected, seabirds were decimated by ingesting the clots of oil or else suffocating after surfacing under the oil slick, or by the disruption of insulation and buoyancy normally provided by the bird's feathers. Ironically, the detergents which were widely used to disperse

Figure 8.1 Big waves rolling over the wrecked tanker *Amoco Cadiz* **while oil still pours out of the damaged holds**

the oil and the resulting detergent-oil mixture proved to be even more harmful to a wide range of other marine organisms. The massive death of all coastal life-forms – fish, crabs, lobsters, shellfish, starfish and shrimps – occurred in waters as deep as 12 metres up to a distance of 0.25 kilometre from the shore. In the inter-tidal zone all organisms, apart from one or two species of sea anemone, were destroyed.

Apart from the horrific but inevitable effect upon the natural environment, there were also significant costs borne by the local economy in both Brittany and Cornwall. In both cases tourism is a major source of revenue, yet within the space of twenty-four hours the viability of that economy was seriously threatened. In the case of the *Amoco Cadiz* disaster

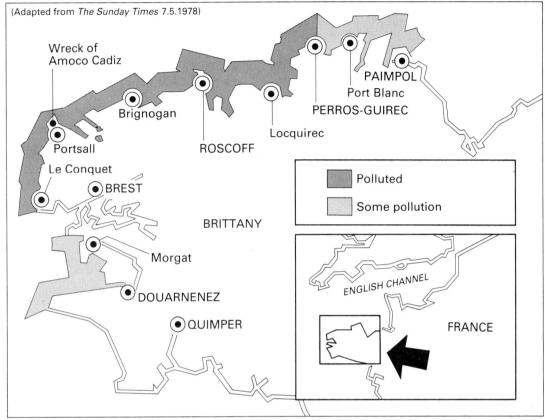

Figure 8.2 Oil pollution on the Brittany coast after the Amoco Cadiz **disaster, 1978**

it was estimated that 1.5 million tourists expected in Brittany during the spring and summer of 1978 stayed away. The subsequent loss of income to hoteliers, fishermen and shopkeepers and to the local economy in general was extremely serious.

In assessing the costs of disasters such as these consideration has to be given to the loss of the ship and its cargo, the clean up costs (which in the case of the *Amoco Cadiz* amounted to more than 84 million dollars), the loss of income to those who make their living directly (fishermen) or indirectly (hoteliers) from the polluted sea, property damage and the damage to natural resources. In addition to these costs there are also subjective evaluations which have to be made to assess the cost of damage to recreational areas, views and the effect upon non-commercial and non-endangered species. Inevitably, the process of determining fair compensation and the task of making the polluter pay the true cost of the damage caused, is an extremely difficult, long drawn-out process.

The cause of both disasters was quickly established and the claims for liability eventually settled. Marine life has returned slowly back to normal and the tourist industries of the affected parts of Cornwall and Brittany have revived. However, there is still no guarantee that similar tragedies involving even larger ships will not occur in the future. One of the penalties that we must pay for living in a highly industrialised society dependent, largely, on external supplies of energy is crowded shipping lanes. The English Channel, southern North Sea and the Mediterranean

Is this your holiday beach?

THE television pictures of the dead birds and the carpet of oil along the beaches may already have faded from the public memory. But it is still impossible to go to the worst-hit parts of Brittany without being appalled at the effect of the 200,000 tons of oil dumped into the sea by the wrecked tanker Amoco Cadiz nearly two months ago.

The problems of fighting this seaborne invasion are immense. Langoustes from Roscoff had to be rescued, from their tanks and driven to Douarnenez, south of Brest; unpolluted sea-water from the same area has had to be shipped up to the little village of L'Aber-Wrach for the local crab tanks, now occupied by imported English crabs.

A truly amazing job is being done of the major task of cleaning up the coast for the summer trade. The French army — which is also doing an amazing job supplying statistics — now has 3,753 men involved in 'Operation Polmar', and, by last week, they had collected 158,498 sacks of polluted sand.

The director of the Brittany tourist office, encouraged by all this activity, said last week that the beaches are clean, that the rocks are being cleaned — and that Brittany "will be clean by June."

That is, I fear, an optimistic assessment. Of course, opinions differ as to the exact meaning of a "clean" beach. When I saw the impressive wall map at Operation Polmar headquarters it showed that the Plage des Blancs Sablons, near Le Conquet, was clean; but the woman who, earlier in the day on that beach, had been trying to remove the oil from her child's clothes certainly did not consider it "clean."

Not all of Brittany is polluted. South of Le Conquet, the coastline is clean except for a few points around Morgat, Reserve du Cap Sizun, and Baie des Tréoassés. So, all the way down the coast from Brest there should be few problems; from the Pointe du Raz there should be none at all.

To the east, the pollution starts to taper off after Port Blanc, and beyond Paimpol the beaches are clean. But between Le Conquet and Port Blanc, it is impossible to claim that the Brittany coastline is clean.

Certainly the odd beach has been cleaned and a very few have not been affected by oil at all. In Locquirec for instance, the mayor — who should have no trouble in getting re-elected — had the sand beside the little port bulldozed up beyond the high-tide mark before the oil arrived.

But the overall picture remains depressing. Although not on the same scale as in that terrible first week, small pebbles of oil are still being washed ashore on some beaches. On almost every beach I visited it was still possible to see thin traces of oil in the marks left by the waves.

The army is persevering, but one of the joys of Brittany, climbing along the rocks to your own secluded beach or pool, must be out this summer. As one fisherman put it; "What the rocks need are some good winter storms."

TRAVEL

In Portsall, probably the worst-hit port of all, the mayor, Jules Legendre, says he expects the visitors this year to be the "curious or the sympathetic." He looks depressed by the catastrophe that has hit his little town of 1,500 people; but he claims that he is simply very tired. He is also worried that nothing has been done to prevent the same thing happening again.

In Roscoff, where the car ferry comes in from Plymouth, they are not only worried about protection, but also about money. The season usually starts there in April, because of the local curative "baths." But so far there has been no season at all, because of the natural reluctance of people to take cures in waters they believe might be oily. At Jean-Paul Chaplain's Brittany Hotel, only two rooms were taken one night last week; usually the place would have been at least half full. He has lost 130,000 francs in cancellations so far, and he says that if the government does not move fast, many of the hotels in the town will have to close.

Baths and beaches, of course, are not Brittany's only attractions. There is still the sea air (with only the odd whiff of oil) and the walks — still unaffected, despite tales of the whole countryside being given over to the army.

Tales about the fish and seafood also seem to have been exaggerated. One Breton fishmonger in Douarnenez told me: "In Paris they taste oil in the fish; we cannot taste it. They are imagining things."

And certainly at Locquirec the local crab, and some local fish, were both excellent. However, a French government survey suggests that, although there is no danger to health, fisherman should not take fish from within five miles of the shore between Roscoff and Le Conquet.

It will take time for the full effect on sea life to become apparent. The guillemots at the bird reserve at Cap Sizun, now having to dive through a "thin tablecloth of oil"; the seaweed harvest, an important part of the seaside economy; the oyster beds in villages like L'Aber-Wrach — all are still at risk.

Amid all this despondency, there had been talk about positive measures to tempt visitors back. The Brittany tourist office says that, to reward the faithful, tourists who stay in the affected areas for more than two days in June or September will get one free night next year. But hoteliers I spoke to either had not heard of the idea, or stressed that it was no more than an idea.

The tourist office is also considering an insurance scheme, under which a tourist who was promised a clean beach but got an oily one would be able to claim his money back. However, the office emphasised that the scheme was only in the planning stage.

Personally I would not rely on either a free night or an insurance scheme. If you are committed to a holiday in the worst areas, I suspect that if you look hard enough you will eventually discover a tolerable cove. If you are not committed you should remember that much of the rest of Brittany does still remain unspoiled by the Amoco Cadiz.

Will Ellsworth-Jones

The Sunday Times, 7 May 1978

Figure 8.3

are congested routeways bounded by hazardous coastlines – there is no doubt that the risk of collisions or groundings is extremely high. Events such as the *Amoco Cadiz* disaster and the forecast of ever increasing growth rates suggested to many concerned individuals that the advantages to be gained from modern technology were, in many cases, outweighed by their harmful environmental effects. As a reaction to this state of affairs a group of the world's leading industrialists and scientists began to question whether the global environment could continue to support such persecution indefinitely. To study these urgent problems the group founded an organisation – the Club of Rome – which in 1972 published the *Limits to Growth* which powerfully suggested that the unbridled gallop towards continued economic growth carried with it its own seeds of destruction.

The global implications of pollution and resource depletion led the United Nations to call a conference at which the major problems associated with the environment could be discussed at an international level. Stockholm was the venue in 1972 for the 'Conference on the Human Environment' which led to the establishment of a separate international agency to deal with such problems (United Nations Environmental Programme). In Western Europe, too, the member countries of the EC organised a summit meeting in Paris in 1972 to discuss similar problems. It was clear that the quality of the environment within the Community had declined in the years following its inception, especially in the Eurocore, and that public opinion was starting to reject not only the values of a consumer society which had been championed during the 1950s and 1960s, but also the economic policies which had maintained this pattern of development. The sentiments which were widely expressed in 1972 were reinforced the following year with the onset of the world energy crisis and the subsequent economic recession which has affected the developed world, and Western Europe in particular, ever since.

3 Environmental pollution

Pollution may be defined as the disruption and contamination of the environment by the creation of waste and its improper disposal by both producers and consumers. There are a number of specific and well documented causes resulting from four major areas – industrialisation, urbanisation, rural change and resource depletion. These will be examined in turn (see Fig. 8.4).

a) Industrialisation

Industrialisation, as we have already seen in Chapter 2, has spread in various ways and at different rates throughout Western Europe during the present century, and more particularly in the past 40 years. All forms of industry produce waste and effluents some of which can be highly dangerous, producing noxious substances (certain chemicals and heavy metals such as cadmium and lead, for example) which can cause widespread damage to the natural environment as well as posing a direct and severe threat to human health. Industries concerned with mining, metallurgy, chemicals, oil refining, pulp and paper milling, textiles and some branches of the food industry are the principal offenders, their effluents being responsible for polluting all natural systems. In the atmosphere certain chemicals, gases and dusts – by-products of many industrial processes – are transported great distances by the normal circulation of wind systems, being washed to the ground at a later date with the onset of precipitation. Of these harmful substances, perhaps the most serious – and easily the most extensive – is sulphur dioxide (SO_2). When combined with water in the atmosphere SO_2 forms 'acid rain', poisoning not only lakes and rivers but also contaminating crops, damaging forest and woodland, corroding metal structures and paintwork and risking the health of the local population. The land, lakes and rivers, from Scandinavia to the Mediterranean, are being subjected to more obvious forms of pollution

230 GEOGRAPHICAL ISSUES IN WESTERN EUROPE

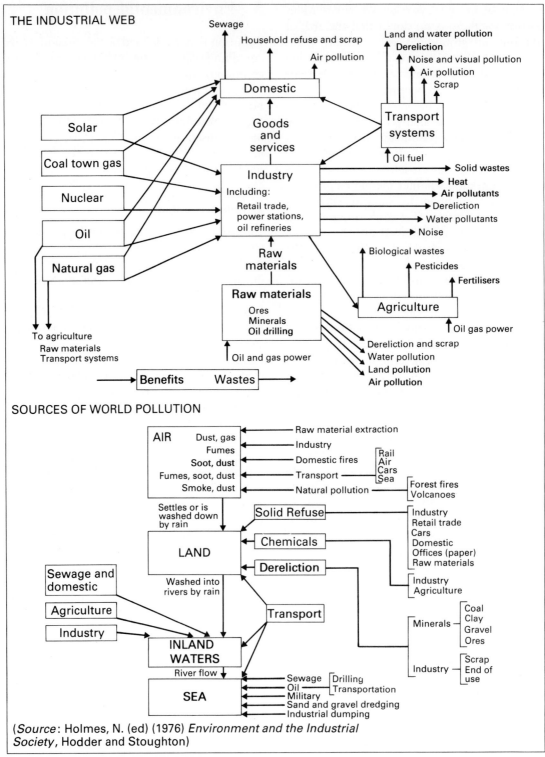

Figure 8.4 a) The industrial web; b) Sources of world pollution

as industrial plants discharge increasing quantities of waste into the nearest available water course. The resulting 'cocktail' causes a fall in the life-supporting oxygen levels, dramatically rendering aquatic systems lifeless or, at best, leaving them to face a slow biological death. Not only is such water harmful to all living organisms contained within it, but it is also potentially dangerous for a wide range of other functions – for domestic consumption, as well as recreational activities – which are so vital to Western Europe's densely packed population. Industrial wastes are also dumped on the land, at official or unofficial tips, and these too can have a disastrous effect upon the environment leading to the destruction of the natural structure of the soil, causing infertility and sterility as pollutants seep through the soil to contaminate aquifers which, ultimately, pass on health risks to all life forms.

Even when industry has gone, the wastelands left behind are often little more than a visual reminder of the past. The structural changes which have shaped West European industry since 1945 have led to widespread dereliction contributing to progressive decline of whole regions. The Ruhr, Sambre-Meuse, Lorraine, the Nord coalfield and Limburg are all regions where environmental degradation caused by a past industrial age continues to dominate the landscape. They represent regions once at the heart of their nation's economy but now bypassed and forgotten in the quest for new economic and industrial horizons.

b) **Urbanisation**

With population densities within the Eurocore not matched to the same areal extent anywhere else in the world, Western Europe is one of the most crowded regions of the globe. To contain this continual and rapid growth of population which has occurred since the end of the Second World War there has been rapid urban development throughout the sub-continent. Despite having managed to control some of the worst excesses of overpopulation and urbanisation – as exhibited by many cities in the Third World – the heavy concentration of people in Western Europe has, nevertheless, exerted considerable pressure upon the environment. The problems which have resulted stem directly from the density of population and their insatiable demands for the maintenance of a high standard of living.

The urban environment is threatened on a number of fronts not only by the socio-economic problems of the 'inner city' (discussed in Chapter 4) but also by increasing difficulties involved with the transfer of goods, materials and people. Furthermore, all modern urban systems function today by utilising high-level technology which is ultimately dependent upon vast supplies of energy secured at considerable environmental cost.

The motor car can, perhaps, be singled out as a major cause of environmental pollution in towns and cities. As ownership has increased (see Chapter 7) so has traffic congestion which, in turn, leads to the accumulation of harmful pollutants such as lead and nitrous oxides, thickening smogs and the over presence of noise. In fact, unwanted sound (a useful definition of noise) continually invades the urban environment – throughout any twenty-four hour period there is seldom total silence, cities are rarely motionless. Instead noise levels reach easily identifiable peaks at certain times of the day, reducing progressively at other times, but seldom disappearing altogether. The environmental threats of noise are felt not only by those individuals who have to live and work in the city, but also by the urban fabric which is constantly bombarded by periods of intense vibration. This has had disastrous effects upon the physical structure of many towns and cities in Western Europe, such as Florence, Rheims and Amsterdam.

Environmental problems are further compounded by the accumulation, collection and disposal of rubbish and other wastes. In our 'waste maker society' the consumption of both domestic and industrial products involves

> **Box 8.2**
>
> The Ruhr in West Germany, being a densely populated and highly industrialised region, produces great quantities of domestic waste and rubbish. It has been estimated that each of the five million inhabitants of the region produces 2,880 kilograms of rubbish annually, representing some 16 million tonnes for disposal. To get rid of such vast quantities each year is a daunting task and one which has received serious attention from the local authorities. Fortunately, the region has many derelict industrial sites and open spaces which can benefit from being infilled by domestic waste, which can then be reclaimed, returning the land to other uses aimed at improving the image of the region.
>
> One such example, developed with the financial backing of the Federal Government and the *Land* of North Rhine-Westphalia, was started in 1968 when the regional planning authority acquired the abandoned site of the former Graf Bismarck coal mine in the Emscherbruch area of Gelsenkirchen. Covering 2.6 million square metres, one part of the derelict mine was recultivated as a recreation area immediately, whilst the remaining area was to be used as a central dump. As infilling has progressed it is estimated that the tip will accommodate about 30 million cubic metres of compressed waste. Controlled by Federal and *Land* legislation the dump provides a model example of what can be achieved by careful planning and reclamation. At present the outer embankments of the individual tips are being grassed over, trees are being planted and the final effect will create a line of low hills blending well into the natural landscape of the Emscherbruch, helping to restore and improve the image of a region blighted by past periods of industrialisation.

the creation of vast quantities of garbage (see Box 8.2). Not only is the handling of this waste extremely expensive but it is often associated with more serious environmental threats as a result of improper disposal methods. Rubbish tips are not only unsightly but can also provide suitable breeding grounds for vermin and other pest species which can have a detrimental effect upon human health.

The production of great quantities of domestic sewage is another feature of all towns and cities in Western Europe and although many local authorities pride themselves on their efficient disposal methods it is often a case of removing it from the immediate vicinity and pumping the pollution problem elsewhere.

Finally, urban sprawl itself has a considerable influence upon the environment not only in terms of visual intrusion and over-hasty development but also by the consumption of vast quantities of open land. In particular the demand for more space in which to live and the creation of 'infra-structural improvements' such as new airports and urban motorways consume hundreds of square kilometres of countryside in Western Europe annually.

c) Rural change

Throughout the rural regions of Western Europe the adoption of contemporary farming practices in collusion with a multitude of national and supra-national farm support schemes, such as the Common Agricultural Policy of the EC (see Chapter 5) have paved the way to increasing specialisation and intensification of production. Farming has become an 'industrial' way of life in Western Europe and 'agri-business' has become 'big-business' in which the emphasis is upon maximising production, often regardless of both demand and the natural environment. As a result of this desire to achieve greater productivity from a finite land resource the agricultural landscape has been radically modified; hedgerows have been uprooted, woodlands cleared and ploughed up, marginal land reclaimed, rivers and streams straightened and permanent pastures have been replaced. Additionally, widespread use has been made of a variety of inorganic chemicals which act as pesticides, fertilisers and herbicides. Although important in promoting high crop yields, these can cause severe damage to biological systems far removed from the areas of application. The chemical compounds used in pesticides are extremely toxic to man and wildlife and, as a consequence, require great care in their use. Fertilisers, too, although less toxic and creating less of a problem to living organisms directly, can have serious implications for water courses if washed through the soil after heavy rainfall.

As a result of these new practices the shape and structure of much agricultural land throughout the sub-continent now takes the form of bleak, open arable landscapes, lacking the traditional variety of wildlife habitats and associated species. All too often the movement towards monoculture creates an unattractive landscape and one which is increasingly difficult for public access.

d) Resource depletion

Modern industrial society is dependent upon a continual supply of materials and resources. Although such a system operates smoothly, it only does so as long as natural resources are available. It has become plainly obvious, however, that present levels of production and consumption are using up many of the world's natural resources at an alarming rate — fossil fuels are burnt, agricultural land is overworked and exhausted, mineral deposits are depleted — so much so that it is estimated by many sources that within the next half century, if rates of consumption continue to grow, the depletion and exhaustion of many vital resources will be so serious that society will be faced with problems the magnitude of which have never before been experienced (see Chapter 9 for a fuller discussion).

Box 8.3 The Seveso disaster

In the summer of 1976, the town of Seveso, an industrial suburb 21 kilometres from Milan, was subjected to a deep, rumbling explosion coming from the vicinity of the sprawling chemical plant situated on the north-eastern edge of the town. Rising from the centre of the explosion a white cloud of smoke drifted towards the town and settled over 283 hectares of public housing. Within hours residents of the district reported becoming nauseous and feeling dizzy, suffering from severe headaches and diarrhoea. Those people who had been outside while these events had been taking place started to develop sores on exposed skin surfaces (chloracne). Two days after the explosion vegetation throughout the contaminated area wilted and domestic animals and pets became ill and died.

Nine days had passed before officials from the Swiss-owned chemical plant (ICMESA – a subsidiary of the Swiss multi-national Hoffmann La Roche) informed the local authorities in Seveso that the white cloud contained the deadly poison *dioxin*. Within hours of this news, government officials had ordered and carried out the evacuation of 739 residents in those areas most seriously affected by the toxic cloud. The accident had occurred, it was later revealed, when a chemical process used in the production of the antibacterial – hexachlorophene – had gone amiss causing a reactor vessel to explode and, in so doing, release the deadly cloud.

The full impact of the disaster upon the health of the local population may not be known for many years, but the seriousness of the event can be appreciated by the fact that the most affected core of the town (zone A) covering 108 hectares still remains uninhabitable and, according to some experts, may remain so forever.

---

4 The impact of environmental pollution

Although pollution has manifested itself in a number of different directions, the immediate threat to human health and welfare is perhaps the most serious issue of environmental deterioration. There is now a wealth of evidence proving conclusively that many pollutants are extremely damaging to mankind in both the short term and also the long term (see Box 8.3). In a densely populated and industrialised region such as Western Europe we are continually subjected to these risks through the food we eat, the water we drink and the air that we breathe. Ironically, this shameful state of affairs exists despite the fact that the harmful effects of many industrial processes have been known and well documented for some considerable time. Although we are now certain of the dangers of lead and mercury pollution we still persist in releasing vast quantities of both lead and mercury into our air and water and on to our land. Similarly, the burning of fossil fuels in domestic hearths has been related to specific diseases following the infamous London smogs of the early 1950s, in which a clear relationship was shown to exist between bronchial and respiratory disease and the volume of air pollution hanging above the metropolis. In southern Europe, warm and sunny city regions, such as Madrid, Marseilles, Naples and Athens, are regularly faced with what is termed photochemical smog resulting from the interaction of sunlight and the exhaust emissions generated from car engines and factory furnaces. Such pollution effects not only man – causing eye irritation and respiratory problems – but also plants and animals, causing injury to both domestic plants and agricultural crops which are grown within the area of exposure. In spite of the available evidence which proves beyond doubt the harm that these substances have upon all living organisms, the legislation which has been introduced to deal with these *man-made hazards* has been intermittent and piecemeal. In Western Europe today, although there is a general awareness of the harmful effects these pollutants are having upon the environment, there is still a reluctance to take the required safety measures so as to eliminate the hazard.

Case study: Air pollution in Athens

In Athens, a sprawling city with over three million inhabitants, the past 40 years have witnessed an alarming growth in air pollution. With high temperatures, low rainfall and skirted by mountains the city's ideal natural setting is also one which is highly susceptible to the creation of photochemical smog, the vital ingredients of which (smoke, dust, carbon monoxide, sulphur dioxide and oxides of nitrogen) have been added with increasing regularity as both industry and population have grown. Today the smog (referred to locally as the '*nefos*') has become a topic of everyday conversation in the city. Athenians have reluctantly grown accustomed to the heavy, noxious chemical soup, accepting regular headaches, nausea and burning eyes as an unfortunate by-product of modern city living. Yet despite the extent of the problem the main culprits are easy to identify: heavy industry across the Bay of Eleusis, the manufacturing belt and residential suburbs between Athens and its contiguous port city of Piraeus, the burning of heavy oil (with its high sulphur content) in domestic heating systems and the interminable and increasing amounts of road traffic.

Apart from the unpleasant human side effects the smoggy conditions also pose a severe threat to many of the ancient monuments, such as the Parthenon, in central Athens, as the ancient marble is dissolved. So serious is the problem that a recent UNESCO report concluded that atmospheric pollution had caused more damage to the city's historical buildings in the past four decades than had taken place in the previous 2,400 years put together.

Although piecemeal environmental legislation has been introduced recently, it remains ineffective against the scale of such a problem. Apart from attempts to control the pollution at its source (it has, for example, been shown that if the burning of domestic heavy oil was banned, the sulphur content of the air would be reduced by 50 per cent) there have also been attempts to reduce the amount of traffic in the congested city centre. One solution considered has been to ban cars from a series of traffic-free zones in central Athens, whilst another remedy has been to ban vehicles with odd and even number plates on alternative days when the smog is at its worst. Unfortunately, both solutions have put undue stress upon the city's public transport system and although plans have been put forward to extend the underground system, the completion of this and other public transport projects is still sometime in the distant future.

In addition to the environmental effects which have a direct impact upon human health, there is also a less tangible consequence associated with aesthetics. These aesthetic effects do not cause damage directly to either natural systems or to the population at large, but are bound up with the subjective notion of the 'quality of the environment'. In many cases aesthetic forms of pollution have little or no effect upon the general public unless they can actually 'taste it', 'see it', or 'smell it'. Once it becomes obvious to the senses then questions begin to be asked and remedies sought, although there are many historical examples throughout Western Europe of regions, such as the Ruhr, where rapid economic growth and the clamour for material prosperity overshadowed the call for environmental protection. Similarly, areas of attractive landscape have, in many cases throughout the sub-continent, been subjected to the unscrupulous motives of politicians and developers alike in their pursuance of an 'ideal urban society'. In this respect, the drainage of wetlands, the flooding of valleys, the removal of woodland and the creation of vast new towns, motorways and international airports serve to illustrate society's profligacy.

5 The costs of pollution

All too often the desire to raise national income is at serious odds with the need to safeguard the environment. Environmental protection has been, and still is, seen by many as a luxury which can only be afforded in times of steady economic growth. Not surprisingly, in Western Europe, environmental protection has often had to play second fiddle to the need to recapture and maintain a desirable level of growth. Financial remedies to offset persistently high inflation, increasing balance of payments deficits, the structural problems of industrial change and the rise in unemployment have left sparse funds available for the care and maintenance of the environment.

It has been calculated by the OECD that member countries spend between 3 per cent and 5 per cent of their Gross Domestic Product each year repairing damage caused by pollution. In France, in 1978 for example, some 110,000 million francs were spent on the repair of pollution related damage and on measures to prevent further destruction, and this figure was broadly equivalent to the amount spent on the 'cost' of unemployment in France that year. Inevitably, the cost of repairs is ultimately met by the taxpayer and can only be reduced if the government concerned is willing to impose severe penalties upon the offending industries. The theory of 'making the polluter pay' is widely accepted throughout Western Europe and has been achieved by protective legislation in the form of environmental quality acts and tighter pollution controls. Unfortunately, during the curent recession the advances made in the 1960s have often not been sustained by governments who have seen their major priority as rekindling industrial prosperity. As a result national administrations have been loath to impose additional financial burdens upon awakening industry in the short term, as opposed to the longer term benefits to society of a healthier environment in the future. Ironically, the cost to industry of introducing pollution control measures are often one-off expenses – filters and other purifying equipment requiring little additional cost after installation – and make sound financial sense to national governments faced with massive annual cleansing bills.

Although many of the costs of pollution can be evaluated fairly accurately – the destruction of crops, the cost of purifying drinking water and the cost of health care as a result of pollution related problems – certain other

Austria

Au, ouch

The Austrian government has suffered a humiliating defeat at the hands of the country's ecologists over its proposal to build a dam on the Danube at Hainburg, near Vienna, in the primeval Au Forest. Less than a month after the official go-ahead for the project, the chancellor, Mr Fred Sinowatz, has been forced to shelve it. Work on clearing the site which should have been resumed this week, after 27 people were injured in battles between police and protestors on December 19th, has been postponed "to allow tempers to cool". Mr Sinowatz has promised to look at alternative sites, and has humbly admitted that he will pay more attention to environmental considerations in future.

This delights Austria's small but vocal ecology lobby. The names of a number of prominent intellectuals figure among the 10,000 or so signatures on a petition demanding a referendum on the Hainburg issue. The protestors claim that the building of a dam in one of Europe's last remaining primeval forests would produce an ecological disaster. They have attracted support abroad, including that of the president of the World Wildlife Fund, the Duke of Edinburgh, who criticised the project in Vienna last February.

The government replies that no such disaster is even remotely likely; but it has failed to publish evidence in support of its argument. It had also given the impression of wanting to push the project through as quickly as possible in order to avoid the inconvenience of a referendum. It would probably win a referendum. The dam is strongly supported by industry and the trade unions, on the ground that it would both lessen Austria's dependence on imported energy and create new jobs for the building industry. Austria's trade unions recently organised a march through Vienna, during which the Hainburg dam's student opponents were attacked as "gentlemen's kids" who live comfortably but prevent the creation of new jobs for less fortunate youngsters.

The Au affair is likely to speed up the growth of Austria's ecological movement, which last autumn managed to win four seats in the parliament of the western province of Vorarlberg. It will also increase the strains between the two parties which make up the coalition government, the chancellor's Socialists and the small Freedom party. Devising a new energy policy will not be easy after the Au debacle. In 1978, before the Freedom party joined the government, the Socialists conducted a similar retreat over the (already built) Zwentendorf nuclear power station. Austria's reputation for consensus politics could be in jeopardy.

The Economist, 5 January 1985

Figure 8.5

costs are more difficult to quantify, requiring the application of particular techniques for evaluation. It is difficult, for example, but often necessary to be able to evaluate hidden costs or intangibles. What is the 'cost', for example, of 'acid rain' upon the lakeland environment of Scandinavia? Apart from the biological costs there is also a loss of revenue associated with the harmful effect upon recreation and leisure as well as commercial fishing. Similarly, the effect of land drainage upon the wetland habitat of the Camargue, in southern France, not only imposes environmental penalties upon the habitat itself, but also upon tourism and tourist expenditure in and around the region. By way of a final example, it is often desirable to evaluate the social impact of living, working and sleeping in the urban environment. A value has to be put upon the subjective phenomena of stress and discomfort associated with such a lifestyle. Quite obviously, any procedure involved in the evaluation of these, and many other examples, relies heavily upon a subjective assessment and all the limitations that this implies.

a) Atmospheric pollution and the effect of acid rain

Although many of the effects of environmental pollution in Western Europe have a global dimension there are, nevertheless, more localised and specific issues. Perhaps the greatest environmental problem facing the countries of Western Europe today is a truly international one – the spread and impact of acid rain. Almost all domestic and industrial activities involve the burning of fossil fuels which, in addition to producing the usual elements of combustion – water and carbon dioxide as well as partially burnt particles of carbon and other hydro-carbons – produce substances such as sulphur dioxide and oxides of nitrogen which have a significantly more harmful effect upon the environment. Despite the presence of clean air legislation, introduced by many countries since the mid

Figure 8.6 Acid rain damage to trees in Norway

1950s to rid the immediate environment of smoke, pollution caused by these compounds has continued largely unchecked because they are mostly invisible to the naked eye.

During the 1960s, however, a group of Swedish scientists became alarmed at the dramatic decline in fish stocks in many lakes throughout Sweden. Studies showed conclusively that these lakes had, over the years, become highly acidic to such an extent that young fish could no longer survive. Research showed, without question, that the acidification was being caused by air pollution emitted from the heavy industrial regions of Great Britain and West Germany, drifting hundreds of kilometres by way of the prevailing wind systems and being washed to the ground again by the normal process of precipitation.

The inevitability of trans-frontier pollution – of which the problem of acid rain is a

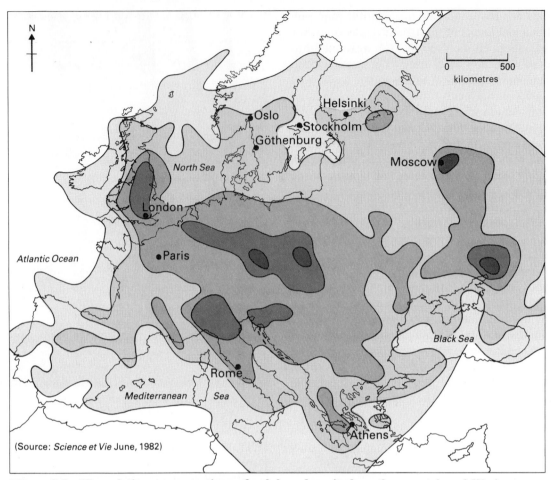

Figure 8.7 The relative concentrations of sulphur deposited on the countries of Western Europe. The darkest zones are evidently those areas where the greatest quantity of sulphur is emitted. The remaining area is where the deposits fall, having drifted by means of the normal circulation of air masses above the continent.

perfect example – can be fully appreciated if one considers the location of Western Europe's major industrial regions, situated adjacent to or within a short distance from an international boundary. The Ruhr, the Antwerp and Rotterdam agglomerations, the Nord coalfield in France, Sambre-Meuse in south-west Belgium, the Midlands and northern England all lie in close proximity to neighbouring states – and not surprisingly export pollution to their immediate or more distant cousins depending upon the strength and direction of the prevailing winds. During the past twenty-five years the problems associated with acid rain have been intensively studied in northern Europe, especially Sweden, and it has been estimated that half of the SO_2 deposited in that country comes from abroad – three-quarters from Great Britain, West Germany and the Benelux countries (see Fig. 8.7). Further research suggests that nearly a quarter of Sweden's total land surface is affected by increased acidification and that 20,000 lakes, almost all of which are in southern Sweden, covering an area of 5,000 square kilometres are in the process of a slow biological death. In Norway, too, it has been found that 1,650 square kilometres of lakes have been poisoned of which 50 per cent contain no fish at all. It is not only Scandinavian countries which are faced with this acid downpour, the rain that falls in West Germany has been contaminated with SO_2 from its neighbours in England, France, Belgium and the Netherlands as well as an indeterminant amount from sources further afield. Ironically, West Germany, with its own acid rain output, exports just about as much as it imports!

What effect does acid rain have? An understanding of this question requires some knowledge of a number of biological and chemical processes. When released into the atmosphere, emissions of SO_2 react with oxygen and moisture to form clouds of weak sulphuric acid (H_2SO_4) which is then dispersed by the normal circulation of air masses. With the onset of precipitation – whether it be rainfall, fog, hail, sleet or snow – the weak acid solution is washed from the atmosphere on to the land, rivers, lakes and streams. The effect of this increased acidification is most severely felt by both aquatic and terrestrial ecosystems, although it is in the former that effects are best documented. Although rain water is slightly acidic (with a pH value approaching 7) acid rain can have a pH value of less than 2 – often as acidic as vinegar or lemon juice (see Fig. 8.8). In southern Sweden, for example, some lakes have pH values of below 5 which is extremely serious if one knows that when values drop below 6 the viability of a wide range of micro-organisms is severely tested, affecting the variety of food for fish and other animals in the food chain. Acidity greater than pH 5.5 greatly reduces the reproductive capacity of many species of fish – such as salmon, trout and roach – and, as a consequence, many lakes have become lifeless or else inhabited by one or two species – such as eels – which are able to tolerate the extremely acidic conditions. As acidification increases the viability of practically all life disappears.

Table 8.1 West Germany: forests affected by acid rain, 1983

Land	Forest area in million hectares	% affected by acid rain
Schleswig-Holstein	137	12
Lower Saxony	977	17
North Rhine-Westphalia	855	35
Hesse	834	14
Rhineland Palatinate	771	23
Badan Würtemburg	1,303	49
Bavaria	2,444	46
Saarland	85	11
Total	7,406	34%

(*Source: Science et Vie*, April, 1984)

Paradoxically, such 'vinegar lakes' turn a striking turquoise blue as a result of the 'cleanliness' of the water in which no life exists.

Apart from aquatic ecosystems, terrestrial ecosystems – in particular forests – are also affected by atmospheric pollution. In the summer of 1982 the Ministry of Agriculture in West Germany released information which showed that over 562,000 hectares of forest – 7.7 per cent of the total area – was contaminated. Initially it was thought that only coniferous species were at risk, the chemicals in the air causing the needles to turn brown and eventually breaking off, progressively weakening the tree until it is uprooted by the wind. Indeed, over 60 per cent of the fir tree *Apies Alba*, covering 10,000 hectares, were in the process of dying or were already dead. Although not of great commercial importance the fir is a species of considerable ecological importance in the German Alps and the Black Forest. The same was happening to the Norwegian Spruce, covering 40 per cent of all forested land in West Germany and forming the basis of the national forestry industry, of which 9 per cent (270,000 hectares) were seriously affected. By the autumn of 1983, a census of tree stocks throughout the Republic revealed that the contamination rate had increased fourfold in the space of only a year. Even more worrying was the realisation that

240 GEOGRAPHICAL ISSUES IN WESTERN EUROPE

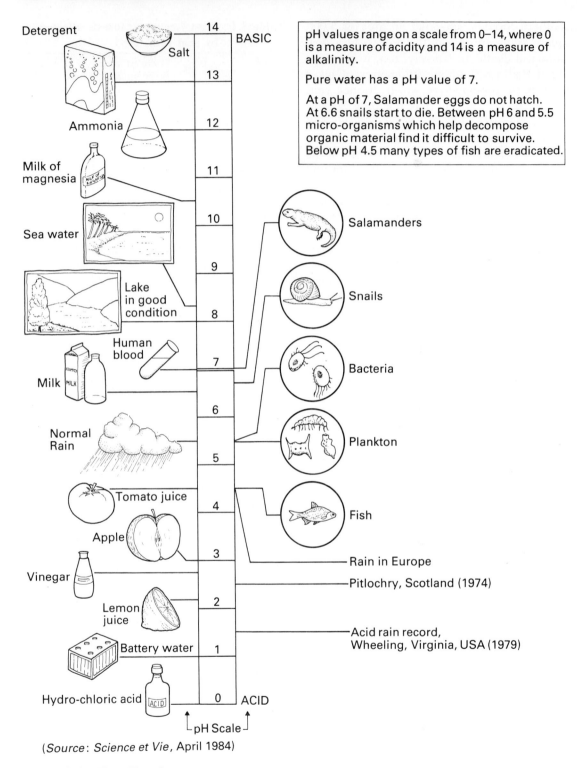

(*Source*: *Science et Vie*, April 1984)

Figure 8.8 The pH scale

deciduous trees, such as beech and oak, were being equally poisoned. The survey concluded that more than a third of the total forest area in West Germany was affected (see Table 8.1).

At first sight acid rain was believed to be the sole cause for this extensive degradation but research has suggested that the causes are rather more complex. Recent evidence from West Germany suggests that oxides of nitrogen, released in towns and cities primarily by the exhaust emissions from motor vehicles, react with the ultra-violet rays of sunlight breaking down into basic chemical constituents of which ozone (O_3) is one. The increased level of ozone reacts with the waxy layers which cover both the needles and leaves of trees allowing the absorption of greater quantities of acid rain.

Apart from damaging natural ecosystems, atmospheric pollution has a destructive effect upon mankind's own structures. Some of Western Europe's most historic monuments – the great cathedrals of Rheims, Cologne, Canterbury and Chartres, the priceless buildings of Venice and Florence, the remnants of Greek and Roman civilizations – have been seriously affected by the erosive powers of airborne pollutants and acid rain. The Acropolis, in Athens, apart from the direct effect of increased acidification in the air, has also been subjected to the rusting of metal clamps inserted in the many columns to stabilize the building earlier this century. The process of oxidization – a direct result of air pollution – has caused the clamps to expand and decay, splitting open the marble columns in the process.

Overwhelming evidence now suggests that acid rain derives from two main sources; burning of fossil fuels and motor vehicle exhausts, the former producing 70 per cent of sulphur dioxide whilst the latter creates 63 per cent of nitrogen oxides. As coal burning emits more sulphur dioxide than oil burning, coal rich nations such as West Germany, France and Great Britain are likely to remain the main environmental culprits until binding legislation is introduced to control emissions. Whether such legislation will transgress major political boundaries into Eastern Europe – where Poland, East Germany and the Soviet Union are major contributors of sulphur dioxide to the atmosphere – remains to be seen (Table 8.2).

Fortunately, pressure is mounting throughout Europe with a number of measures being introduced to reduce and control air pollution at its source. Additionally, maximum permissible pollution standards have been imposed for a range of pollutants, including sulphur. In 1979 the United Nations Commission for Europe held the International Convention on Long-Range Transboundary Pollution in Geneva, which all the EC countries signed as well as the great majority of other European nations. In 1982 an international conference was held in Stockholm on the 'Acidification of the Environment'. Its aims were to strengthen international co-operation and the political will to ensure concrete counter measures to combat acidification.

Steps have also been taken recently to reduce the main pollutants from motor vehicles. Once again technology is available for this purpose and in 1983 the Scandinavian and Benelux countries, West Germany, Switzerland and Austria agreed to remove lead from petrol and to detoxify the exhaust fumes from motor vehicle engines by the start of 1986. Other countries in Western Europe have not been so prompt in taking action, but there is a general consensus to reduce lead levels to negligible amounts by 1991 and it is more than likely that emissions of nitrous oxides and hydrocarbons will also be controlled in the near future.

b) Water pollution

Throughout Western Europe the demand for water has increased annually. Once considered a free and plentiful gift, it is now apparent that water is becoming a scarce resource – especially in Luxemburg and the Netherlands – and demand is starting to outstrip supply. Indeed, the EC has estimated that by the

Table 8.2 Air pollution – who pollutes whom?

'Receivers' \ 'Emitters'	Austria	Belgium	Czechoslovakia	Denmark	East Germany	Finland	France	Greece	Hungary	Ireland	Italy	Luxemburg	Netherlands	Norway	Poland	Portugal	Spain	Sweden	Switzerland	UK	USSR	West Germany	Yugoslavia
Austria	52	3	35	0	24	0	20	0	14	0	56	0	2	0	13	0	2	0	3	8	0	32	35
Belgium	0	67	0	0	2	0	28	0	0	0	0	0	4	0	0	0	0	0	0	18	0	24	0
Czechoslovakia	22	10	483	2	195	0	45	0	68	0	39	0	7	0	95	0	4	0	2	32	14	108	53
Denmark	0	0	3	39	12	0	3	0	0	0	0	0	0	0	4	0	0	0	0	11	0	11	0
East Germany	0	8	56	3	497	0	20	0	4	0	3	0	6	0	23	0	0	0	0	22	3	85	6
Finland	0	2	8	3	19	77	5	0	3	0	2	0	2	2	15	0	0	14	0	14	13	15	3
France	0	33	7	0	20	0	629	0	0	2	41	4	12	0	4	0	66	0	7	99	0	98	6
Greece	0	0	4	0	3	0	3	93	5	0	14	0	0	0	2	0	2	0	0	0	4	3	27
Hungary	9	0	46	0	18	0	7	0	194	0	27	0	0	0	21	0	0	0	0	3	4	13	75
Ireland	0	0	0	0	0	0	53	0	0	18	0	0	0	0	0	0	0	0	0	12	0	0	0
Italy	6	2	12	0	11	0	3	0	11	0	793	0	0	0	6	0	17	0	7	9	2	22	65
Luxemburg	0	0	0	0	0	0	15	0	0	0	0	3	0	0	0	0	0	0	0	0	0	2	0
Netherlands	0	16	2	0	6	0	11	0	0	0	0	0	40	0	0	0	0	0	0	27	0	45	0
Norway	0	4	8	6	22	2	26	0	2	0	0	0	3	20	10	0	0	10	0	40	8	20	2
Poland	7	7	136	8	213	0	0	0	43	0	18	0	7	0	565	0	2	4	0	32	34	80	40
Portugal	0	0	0	0	0	0	0	0	0	0	0	0	0	0	0	20	17	0	0	0	0	0	0
Spain	0	2	0	0	4	0	38	0	0	0	2	0	3	0	0	9	367	0	0	15	0	20	0
Sweden	0	5	18	16	42	10	15	0	5	0	4	0	5	8	31	0	0	83	0	35	24	35	7
Switzerland	0	0	2	0	3	0	23	0	0	0	47	0	0	0	0	0	2	0	14	5	0	13	3
UK	0	7	4	0	11	0	27	0	0	7	2	0	5	0	2	0	4	0	0	675	0	21	0
USSR	14	18	189	25	283	56	37	22	159	2	94	3	23	5	386	0	10	38	3	103	3,610	190	208
West Germany	8	37	48	4	118	0	108	0	6	0	24	5	23	0	18	0	6	0	6	72	3	561	15
Yugoslavia	14	2	38	0	25	0	20	6	72	0	131	0	2	0	21	0	7	0	0	7	8	22	557

(*Source: Science et Vie*, April 1984)

The matrix shows the 'emitters' and 'receivers' of sulphur pollution in Western and Eastern Europe. The figures in the matrix denote the average sulphur deposit in hundreds of tonnes per month. To obtain the corresponding quantity of sulphur dioxide, multiply the figure by 2. (Thus UK exports 9,900 tonnes of sulphur to France = 19,800 tonnes of sulphur dioxide.)

On the matrix the horizontal axis shows European countries and gives the quantities of sulphur that they receive from those countries listed on the vertical axis.

Self assessment exercise

By adding up the rows and columns of this matrix you will then be able to answer a number of questions.

  i) Ignoring their own sulphur output, list the top five countries which receive the most pollution.
 ii) Which countries receive the least sulphur pollution?
iii) Comment about the geographical distribution of the top and bottom five nations.
 iv) List the top five 'exporting countries' i.e. 'the culprits'.
  v) Which countries pollute the least in terms of sulphur?
 vi) Why should this be so?
vii) Select one country from Question i) and one from Question iv) and construct a choropleth map to show the receipt and emission of sulphur to and from other European countries.

beginning of the twenty-first century the demand for water will have doubled since 1970. The problem of future water supply is further exacerbated by its declining quality as a direct result of mistreatment by a variety of industrial and domestic processes. Paradoxically, recent years have witnessed an increasing use of inland water by a multitude of leisure activities such as angling, sailing and other sports, for which *clean* water is of paramount importance. Inevitably, recreational requirements conflict markedly with other water uses – in particular industrial processes which use water for cooling and cleansing purposes as well as a means of effluent disposal, which seriously disrupt the water quality and the immediate river and lakeside environment. Despite the value of water for recreational activities many of Western Europe's rivers and lakes, both great and small, are seriously polluted. Additionally, water pollution, like atmospheric pollution, does not stop at national boundaries and all too frequently one country's waste becomes the pollution of its neighbour. The extent of the problem can be gauged by the fact that 80 per cent of all lakes and rivers in the EC are shared by two or more member states.

Case study: The River Rhine

The River Rhine flows for 1,320 kilometres from its source in the Swiss Alps through the industrial heartland of Western Europe to its delta in the North Sea. As the waters of the Rhine move downstream, the oxygen content declines as sections of the navigable river have been canalised and in its middle and lower reaches the absence of swamps or waterfalls does not allow re-oxygenation. Levels of pollution rise swiftly – within a hundred kilometres of its source Swiss industries deposit quantities of cellulose, dyestuffs and chemical wastes. Flowing across the border and bisecting France and West Germany the river receives a massive input of salt pollution from the potash mines around Mulhouse extending further into the Alsace region towards Colmar. The five mines in this area are responsible not only for 40 per cent of the salt pollution in the entire river but also the contamination of aquifers which extend far to the east of the river towards Baden in West Germany. Through West Germany the river receives a mixture of pollutants from a variety of industries situated on the river or its tribu-

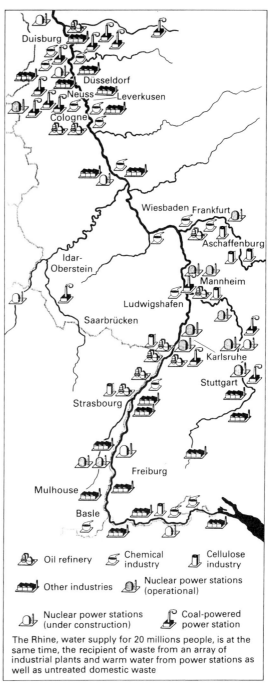

Figure 8.9 Pollution of the River Rhine

taries: chemicals, the wastes from coal mining, iron and steel production, cellulose, metal compounds, more salt and domestic sewage. By the time the Rhine reaches the Dutch border at Bimmen-Lobith, the ingredients for this noxious cocktail have been well and truly mixed with the 12,000 tonnes of oil which are flushed into the river annually and then warmed by the return of water used in the cooling of nuclear (see Fig. 8.9), coal and oil fired power stations which border the river at regular intervals from its source in Switzerland. Despite the disgraceful quality of the water at this stage of its journey to the North Sea it still has another major task to perform – to provide the Dutch with drinking water . . .! Severe ground water deficits in the Western Netherlands mean that the Dutch are largely dependent upon surface water for domestic consumption as well as for the vital irrigation of farmland and market gardening in that part of the country. As a result 65 per cent of water in the Netherlands comes from the Rhine, 7 per cent from the Meuse, 25 per cent from rainfall and 3 per cent from smaller secondary water sources. Not surprisingly, the Netherlands has been one of the prime movers for international legislation controlling the pollution of Western Europe's greatest waterway.

The five states which border the Rhine (Switzerland, France, West Germany, Luxemburg and the Netherlands) have, for a long time, looked long and hard at the problem. The Rhine Commission was formed in 1963 and quickly identified the nature and causes of pollution; the remedies, however, were more difficult to implement. After years of intensive discussion and international co-operation the five nations responsible signed the Bonn Convention in 1976, for the first time agreeing on measures to reduce chemical and salt pollution throughout the course of the river by introducing more effective water treatment plants and new technology designed to remove the most serious pollutants. Ironically, the cost of removing salt pollution, the majority of which comes from France, has been borne largely by the Dutch who have agreed to pay 34 per cent of the cost, whilst France and West Germany will pay 30 per cent each and the Swiss will pay the remainder. In this case the principle that 'the polluter pays' has had to be abandoned if there was to be any improvement in the river quality by the time it reached the Netherlands. Despite this apparent inequity, it is now just possible that the quality of the Rhine might well start to improve – but such improvement will only be sustained if there is constant vigilance on behalf of all interest groups.

c) Marine pollution

Case study: The Mediterranean Sea

Inevitably, the river systems of Western Europe drain into the seas and oceans surrounding the continent depositing their load of pollutants collected en route. Although marine pollution occurs in all European coastal waters, it is in the tideless Mediterranean that the effects are most severe and disturbing. Supported by a northern coastline which contains some of the most heavily industrialised nations in the world and to its south by a series of countries on the brink of industrialisation, the Mediterranean countries contain a resident population of 100 million swelled annually by a further 30 million tourists who choose to spend their holidays in the region (see Fig. 8.10).

Over the years, the increasing intensity of population and industry has exerted considerable pressure upon the marine environment, to the extent that the Mediterranean has been described as a 'gigantic sewage dump' collecting the wastes from 120 coastal cities and thousands of tonnes of industrial and agricultural chemicals washed down by the major rivers which flow into it. Without a strong tidal action – which helps to disperse the pollutants that enter other seas in Western Europe – the Mediterranean has no natural cleansing mechanism. In fact very little seawater actually passes to or from the Atlantic Ocean through the narrow, shallow gap of the Strait of Gibraltar. Unfortunately, the prevailing currents that do exist flow from west to east, compounding the pollutants

ENVIRONMENTAL CONCERN 245

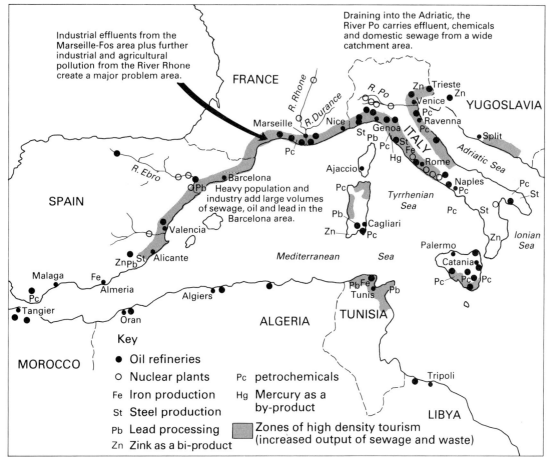

Figure 8.10 Sources of pollution in the western Mediterranean

deeper into the Mediterranean Basin further away from the only corridor which connects it with the open sea. Furthermore, environmental pollution is most severe in the shallower, coastal waters adjacent to man's activities. This littoral zone is the main centre of attraction for the great majority of holidaymakers and those who choose to spend their leisure time in and around the azure sea. Yet, apart from the major conflict of interests which exists between the forces of industrial development and tourism, it is here that the more sinister risks to human health occur.

Four main sources of pollution have been identified in the Mediterranean; domestic wastes, metals, organic pollutants and oil. It is estimated that 430 billion tonnes of domestic wastes are poured into the Mediterranean annually, not only organic sewage, but also detergents and chemicals. Of this, about 90 per cent remains untreated and, as a result, has the capacity to contain dangerous microbes which can cause swimmers stomach disorders such as viral hepatitis and dysentry or, worse, typhoid and cholera. Such microbes are particularly dangerous if they contaminate seafood; a recent survey of shellfish taken from 50 areas in France, Greece, Italy and Yugoslavia showed that some 90 per cent were unsuitable to be eaten raw.

A second major source of Mediterranean pollution is caused by a variety of metals which filter into the sea from a multitude of coastal industries as well as being carried in

Figure 8.11 Beach debris

suspension by one of the many rivers pouring into the basin. In particular, industries involved with the processes of oil refining, chemicals and steel making are the guilty partners, accounting for 5,000 tonnes of zinc, 1,400 tonnes of lead, 950 tonnes of chromium and 100 tonnes of mercury being deposited every year. Additionally, atmospheric pollution probably dumps four times as much mercury in the Mediterranean as the rivers.

Organic pollutants, thirdly, include a vast array of chemicals which are used in a wide range of industrial processes, as well as those used in modern farming practices pursued along the coastal agricultural zones. These, particularly poisonous toxins, enter the sea via the atmosphere and all the normal water courses, imparting a major threat to the whole marine ecosystem.

The final source of pollution emanates from the transport of oil – around 400 million tonnes of oil are transported across the Mediterranean annually (nearly a quarter of the world's oil that is transported by tanker) to any one of a number of major oil refineries such as those at Marseilles, Genoa, Trieste and Athens. Although oil spillage resulting from accidents at sea has yet to occur on the same scale as the *Amoco Cadiz* or *Torrey Canyon* disasters, the possibility of such a disaster remains extremely high in the crowded shipping lanes around the Mediterranean ports. Furthermore, the two major disasters which have so far occurred off the coast of Western Europe had the benefit of tidal action to help break up the oil spill – the Mediterranean will not, of course, be so fortunate! The majority of seaborne oil occurs as a result of the flushing of oily ballast water from the tanks of vessels before they fill up with a fresh cargo.

When considered in totality, these pollutants provide a formidable challenge to the viability of the marine environment throughout the region. Recent years have witnessed the regular closure of beaches and recreation areas in many countries, in particular adjacent to major estuaries such as the Rhone, Po and Ebro and heavy industrial areas around Marseilles and between Venice and Trieste. In areas such as these, coastal lagoons and estuaries of rivers such as the Rhone and the Rhine, the water has become algae-ridden and sulphurous due to conditions which preclude a normal diversity of organisms. Along the Côte D'Azur on the French Riviera, the undersea habitat has become little more than a wasteland as pollution, combined with commercial fishing, private spearfishing, coastal developments and tourism have rapidly destroyed one of Europe's greatest natural assets through carelessness and neglect.

The current situation would have been even worse had not eighteen of the nineteen nations surrounding the Mediterranean decided to make a determined effort to improve the quality of the sea. Albania was the only nation not to sign the Barcelona Convention in 1976. The signatories enlisted the help of the United Nations Environmental Programme to co-ordinate scientific research into, and the monitoring of, marine pollution. In so doing they aimed, firstly, to pinpoint the main sources and types of pollutant and the magnitude of their effects and, secondly, to combat pollution and enhance the marine environment through a planned and co-ordinated policy of prevention and abatement. Not surprisingly, at this time many countries on the northern coast were becoming increasingly worried that their lucrative tourist industries – requiring a high quality environment – were liable to ruin if conditions deteriorated further.

Since 1976 considerable efforts have been made to investigate the sorts of pollutants that are being dumped into the sea and to determine what effect they have upon marine organisms and how the sea itself can cope. Despite the immense difficulties involved a detailed inventory has been assembled providing a clear basis for action. Other developments such as the building of comprehensive sewage treatment plants in Marseilles and Athens, the establishment of a Regional Oil Combatting Centre in Malta and the introduction of tough environmental legislation – especially in France – are notable successes in the fight against environmental pollution in the Mediterranean. However, the willingness of countries to pay for such measures and the tighter controls needed to enforce this policy depends, largely, upon a country's wealth and whether tourism plays a fundamental part in its economy. Whilst towns along the Côte D'Azur, for example, have collectively organised an early warning system to guard against pollution drifting in from northern Italy, other countries such as Algeria and Libya, on the other hand, which have much less interest in international tourism have no plans for spending money on such schemes.

Although the news is heartening there is still a great deal to achieve before the Mediterranean is returned to pristine condition. More recently, steps have been taken by the United Nations to *forbid* the dumping of pollutants such as DDT, oil, mercury, cadmium, persistent plastics and radioactive wastes and to stringently control a wide range of other metallic and chemical compounds. However, despite the good intentions of such policy there are grave difficulties in enforcement. The introduction of legislation is both costly and time-consuming. Yet in the long run it may be money well spent because some of Western Europe's poorest regions skirt the Mediterranean and for them tourism is perhaps their last chance to escape from the grips of poverty and to reassert their new found dynamism.

6 A response to environmental degradation

The problems so far discussed represent the major environmental issues in Western Europe today. In each case the international dimension is of fundamental importance in terms of both cause and effect. Fortunately, there does seem to be a belated response to these issues and positive suggestions have been proposed in order to tackle some of the worst excesses of international environmental pollution.

Although international bodies, such as the United Nations and the European Community, are the prime movers of environmental legislation much of the initial impetus for such improvement has occurred as greater awareness from an increasingly well informed West European public has developed. More and more people have started to question the mounting environmental costs of unbridled economic growth and, as a consequence, have attempted to re-assess non-material values

Eat, drink, swim and be wary

Whoever first exhorted the jaded traveller to see Naples and die almost certainly did not have "raw municipal waste-water effluents" in mind. Although the place had already been notoriously pestilential for centuries, and was to remain so for several more, it required all the linguistic alchemy of modern science to transform the indescribable into words that fall almost benignly, dry as dust, from the lips of academics.

Today's sun-seeker is equally unlikely to set off for a Mediterranean holiday with visions of fecal coliforms and streptococci dancing in his head. This year's tourist population will probably total, as usual, about 100 million, effectively doubling the resident coastal population over the short holiday season. But some of those transients, at least, are bound to carry as unwanted baggage a vague suspicion that the Mediterranean may be hazardous to their health.

They are absolutely right about that. The latest word from the scientists, however, while confirming many of our worst fears about the uncleanliness of Mediterranean waters, is also cautiously optimistic. It is still possible to see Naples, even to swim there, and live.

First the bad news. About 85 per cent of the sewage from 120 coastal cities is discharged into the sea with inadequate treatment, or none at all, and most often close to the shore line. That output, appalling enough to begin with, is of course increased beyond any hope of control by the annual influx of tourists.

Poisonous wastes from factories (primarily heavy metals) and from oil refineries meet and mingle with the sewage; four of the world's largest river systems – the Nile, the Rhône, the Po and the Ebro – carry, with numerous smaller systems, a volume of municipal, agricultural and industrial waste that far exceeds the contribution of the coastal cities and towns. About 120,000 tons each of mineral oils, nitrogen and phosphorous, 60,000 of zinc, 100 of mercury, 3,800 of lead enter the sea each year. As much as a quarter of all the world's marine oil pollution, most of it from merchant shipping, may end up in the Mediterranean.

"Twenty per cent of Mediterranean beaches are unsafe"

The most recent estimates are that 20 per cent of the Mediterranean's beaches are unsafe for bathing, although "unsafe" merits careful definition. And the general view is that most of the region's shellfish, contaminated by bacteria, are unsafe by any definition. Because the sea is virtually landlocked, with the nine-mile-wide Strait of Gibraltar its only inlet and outlet, even if all pollution stopped tomorrow, it would take at least 80 years for the waters to renew themselves. And, at present rates of growth, in the next 40 years the resident population could increase five-fold and the number of tourists double.

The United Nations scientists who study the pollution patterns of the Mediterranean most closely are duty-bound to report their findings as diplomatically as possible, so that no country will take umbrage (and lose valuable tourist income) by being singled out as "dirtier" than any other. The UN Environment Programme, Unep, has mapped 13 obfuscatory "regional divisions" for sampling purposes; armed with metres of data and a computer it is just about possible to work out which areas are more or less salubrious. But the reckoning is perforce vague and, if taken at face value, would mean writing-off every one of the most popular European coastal resorts between the Balearics and Corfu.

The good news is that there is more room for manoeuvre than the bare statistics might suggest. Civilizations have been polluting the Mediterranean, after all, as long as man has lived there; ancient Rome must have generated, via the Tiber, a fair amount of sewage, and the canals and lagoons of Venice have been renowned for their putridity throughout the city's recorded history.

If the native peoples of the Mediterranean have survived and even flourished in all that muck it is not only because they have developed some immunity to the more toxic aspects of the environment, but because they have learned how to cope with the consequences of their mistakes. They follow, almost instinctivly by now, commonsense rules which are simple for the tourist to learn.

Because the absorptive capacity of seawater is almost infinite – the open ocean is hardly the natural habitat for bacteria that have adapted happily to a sheltered life in the inner sanctum of the human gut – pollution by sewage is almost by definition localized and short-lived. As the microbes disperse they succumb rapidly to the combined effects of salinity, sunlight, cold, dissolved minerals and even predators – natural marine micro-organisms that eat them. Some scientists think it may be only a matter of hours before the hostile marine environment "inactivates" any number of invading germs. Unfortunately, the flow of sewage into the Mediterranean is so continuous that the cleansing power of the sea can never quite catch up.

Unep and the United Nations World Health Organization spent five years collecting and studying 12,500 water samples from 700 stations in 14 countries to reach their conclusion that three-quarters or more of the Mediterranean's beaches are relatively safe, microbially speaking.

Applied to the same data, the stricter EEC criteria would reduce the percentage of safe beaches by half, to 37. It all depends on what you mean by "safe".

One danger about which the organizations are unanimous is that of contaminated shellfish. Oysters, clams, mussels and the like are worst offenders; being filter feeders, they extract nourishment from seawater by passing it through their bodies, concentrating bacteria, viruses and chemicals in their

Figure 8.12

succulent flesh. Even those few shellfish that come to market in an unsullied state may be "freshened" on the stall with bucketfuls of almost certainly polluted seawater, Crustacea — shrimps, lobsters, crabs — have different feeding habits and digestions, so are safer.

Less than four per cent of the stations monitored by Unep and WHO were considered to be safe for shell-

"You can catch the same 'diseases' in distilled water"

fish. EEC standards, curiously, would increase the figure 10-fold, to 40 per cent; but the potential consequences of a mistake (typhoid and cholera, for example) are so dire that the percentage might as well be nil.

One man who should know takes a refreshingly insouciant view on the matter of holiday health. Dr Stjepan Keckes, who runs Unep's Regional Seas Unit, likes his seafood and he likes his daily swim. He has seen his home town of Rovinj in Yugoslavia, develop into a popular tourist resort in one of the "dangerous" areas of the region; yet he still swims there, a few metres upcurrent of the local outflow.

He is cynical about some of the horror stories, especially where swimming is involved. The swimmers may have eaten tainted shellfish, he says. Or they may have eaten perfectly clean shellfish — or anything else — to excess. ("I know of one man who ate 40 oysters and then blamed the Mediterranean when he got sick.")

As for diseases spread by swallowing seawater: "Unlikely. Polio, for example, does not survive in the sea; it spreads in swimming pools." Some "diseases", he argues, can be caused by pathogens naturally occurring in or on the body that spread to the wrong places when the body is immersed for too long or at too great a depth. "You can catch the same 'diseases' in distilled water."

Inevitably, the differing EEC and Unep/WHO standards will be brought into line one day. Meanwhile, the really good news concerns mercury levels in the Mediterranean, also the subject of a Unep study involving more than 2,700 fish samples, 700 molluscs, and 600 crustaceans. High mercury levels have always occurred naturally in the region and the scientists found no significant differences between current levels and those in some — very elderly museum specimens.

They concluded that pregnant women, perhaps, should go easy on the tuna but otherwise there is little danger from mercury, however frequent or extended the visits. There is even some talk now of abandoning present restrictions on maximum mercury levels in fish altogether, although direct discharge of mercury by industry would continue to be forbidden.

Unep, not always the most effective of the United Nations agencies, has been campaigning since its formation in 1972 for a cleaner Mediterranean. On paper at least it has achieved remarkable success: every Mediterranean country except Albania has been pressured, wheedled and cajoled into a formidable network of protocols, treaties and conventions committing them in principle to refrain from fouling the sea.

The stumbling block, as always, is money: cleaning up Alexandria's sewerage system alone would cost about £200m. But the bureaucrats can claim with some justification that through their good offices the Mediterranean is, if nothing else, degenerating a bit more slowly into the world's largest cesspit.

Safety rules

● Never, under any circumstances, eat *raw* shellfish. They are perfectly safe if well cooked. Crustaceans, either cooked or raw are probably safe but if you are prone to hypochondria give them a miss too.
● Avoid swimming near the centre of town, in any harbour, in front of your hotel, or anywhere the water looks unusually cloudy or the shore is very littered. If the beach in front of your hotel looks inviting, find out where the wastes are discharged before you take the plunge. In exceptional cases it may have a drainage pipe that goes well out to sea. If not, move upcurrent of the outlet. Cannes, Nice and Monaco, alone in the Mediterranean, are safe for virtually the entire length of their beaches.
● If you uncertain where to swim, find out where the locals go.
● If you are still worried, have polio, typhoid and perhaps cholera shots before the holiday.
● Always wash fruit and salad.

The Sunday Times,
29 July 1984

such as 'the quality of life'. Greater prosperity has allowed people to move freely, away from the constraints of their immediate material needs, and to pursue more non-material aspects of their lives. Such new found material security, reflected by increased personal and residential mobility, has enabled many people throughout the sub-continent to enjoy the fruits of their labours rather than struggle solely for the accumulation of wealth just to provide the necessities of life. Furthermore, environmental issues have become popular issues, attracting nationwide media coverage and the growth of a number of influential pressure groups. Environmental issues, in whatever form, have become emotive issues – people throughout Western Europe have realised that it is not only environmental quality that is at risk, but also human health, safety or even survival. Paradoxically, environmental concern is felt, often most strongly, from those who have benefited from economic prosperity derived from the very industrial society which caused such problems in the first place. Environmental concern and an appreciation of the environment has evolved as a result of the economic growth which has occurred in Western Europe since the Second World War. Nevertheless, whatever its origins, the new environmental awareness – represented by political parties especially in West Germany – has strongly challenged the depletion of resources, the destruction of nature and the loss of amenities, refuting the assumption that 'progress' is reflected automatically by an increase in material prosperity. In so doing the social and environmental costs of economic growth and technological advancement have been seriously questioned by those who see such 'progress' as ultimately harmful to the quality of life, resulting inevitably in its own self destruction. Although environmental protection has often been directed towards some of the most vulnerable and attractive areas of land and seascape in Western Europe, equal concern has also been shown to less beautiful areas which also contain their own set of peculiar environmental problems. In particular, considerable emphasis has been placed upon the improvement of many urban environments throughout the sub-continent. Urbanisation is, undoubtedly, one of the most noticeable characteristics of twentieth century Western Europe and the Eurocore has become one of the most densely populated regions in the world. Despite the competing claims of the more obvious penalties of urban living, the availability and proximity of open space has become a problem of increasing importance. Furthermore, increasing suburbanisation since 1945 has, in many cases, proceeded without due regard for facilities which could have provided a valuable recreational resource for those people faced with the prospect of high density living in so many West European cities. The massive congregation of people and industry has exerted severe pressure upon the landscape surrounding many city regions, and each year more land is acquired for housing and industrial development with the result that the visual intrusion of urban sprawl bites deeply into the countryside. In regions where towns and cities have, historically, grown up together, as in the Ruhr in West Germany, or the Randstad in the Western Netherlands, the problems are intensified as open space is consumed by the process of urban coalescence. Fortunately, in both the examples mentioned above positive steps have been taken to prevent such a 'merger' by the introduction of far-sighted planning policy aimed at maintaining the essential ingredients of open space as well as corporate identity. Exactly how these policies have evolved is worth closer examination.

Environmental improvements in the Ruhr and West Netherlands

From an economy founded 150 years ago upon coal mining and the production of iron and steel, the Ruhr still remains Western Europe's largest industrial conurbation, home for nearly 5.5 million people based around the major cities of Duisburg (population in 1981

ENVIRONMENTAL CONCERN

West Germany

Greening the Ruhr

FROM OUR BONN CORRESPONDENT

The Greens are jubilant, the liberals are glum, the Christian Democrats are worried and the Social Democrats do not know what to think. These were the main parties' reactions to the local election results on September 30th in North Rhine-Westphalia, West Germany's biggest and most industrialised state. As always in local elections, the outcome was complicated, but the overall pattern was clear: there were two losers and one and a half winners. Last Sunday's results may prove to be an important pointer to the state election in North Rhine-Westphalia next spring.

The main winner was the pro-ecology, anti-nuclear Green party. It can now make or break council majorities in various cities, including the state capital of Düsseldorf. The Greens won 8.6% of the votes, a seemingly small increase from their national total of 8.2% in last June's European election. In North Rhine-Westphalia's previous local elections, which provide a better yardstick, the Greens won only 1.5% of the overall vote, although they ran candidates in only a handful of districts.

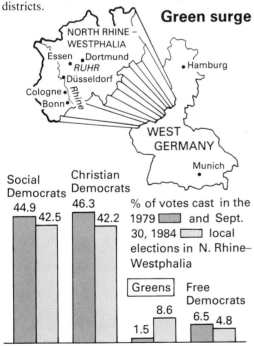

Figure 8.13 *The Economist*, 27 April 1985

of 554,377), Essen (643,360), Bochum (398,492) and Dortmund (605,418). Successive periods of growth and industrialisation have witnessed the expansion of these cities and a series of larger towns, all of which have encroached upon valuable open space between them, creating the appearance of a single urban agglomeration. The very nature of industrial development in the Ruhr, characterised by pitheads, slagheaps, derelict land, blast furnaces, air and water pollution, has coloured the image of the region for much of its history. As elsewhere in Western Europe, recent years have witnessed a personal re-evaluation of living conditions as residents have had time to take stock of their immediate urban environment. This increasing environmental awareness, the fact that there were decreasing amounts of open space in the heart of the region and that levels of air and water pollution posed a direct threat to human health, provoked an inevitable public response. Fortunately, the machinery for action had been established in the 1920s when all local communities in the region banded together to form a central regional planning body. From the outset this unique body made environmental protection and regeneration one of its main priorities. In the 1960s positive planning was introduced to control not only the growth and expansion of the urban area, but also to improve the tainted environment. To control urbanisation and to prevent the joining up of adjacent towns and cities, the planning authority proposed the creation of a number of 'buffer zones' to be established in the intervening countryside which still remained between the expanding centres of population and industry. By designating areas sacrosanct from development the threat of urban coalescence diminished overnight. Today a number of 'green wedges' run parallel in a north-south direction separating the major communities on the one hand whilst, on the other, providing valuable open space for recreation and agriculture over which public access is guaranteed. Additionally, the regional planning authority has taken over the protection and development of all green areas, water bodies and woodland of recreational and ecological significance, to care and maintain the countryside so as to make it more accessible to the public. The authority has also been charged with the responsibility of creating and operating a series of public leisure and recreational facilities established, in many cases, within or in close proximity to the main urban areas, often upon derelict land and water reclaimed from worked out industries (see Fig. 8.14). In a short space of time the results have been impressive and the region can now boast that 58.4 per cent of land is in non-urban use, and this figure is likely to increase in the foreseeable future as the reclamation of derelict land continues apace. In terms of open-space provision, the reclamation of derelict land and the reduction of atmospheric and waste pollution and the containment of urban growth, the region provides a positive example of what can be achieved.

A similar problem of urban sprawl exists in the Western Netherlands (see Chapter 4). Although this region is not an area scarred by heavy industrialisation, the density of population alone exerts severe pressure upon a region desperately short of open space and recreational facilities. In spite of strict planning control, which has existed since the mid 1960s to ensure the viability of the green heart, many of the towns and cities have extended their boundaries. In some areas – such as between Rotterdam, Delft and the Hague – the extent of open space between the three urban zones is little more than 5 kilometres. Elsewhere the situation is little better and without strict enforcement of planning control the cities are likely to coalesce. Apart from the fear of urban coalescence, the population (in excess of 5 million) has to be provided with open space as well as housing and employment opportunities. Traditionally, the area was served by the sand dunes and beaches of the North Sea coast, plus some

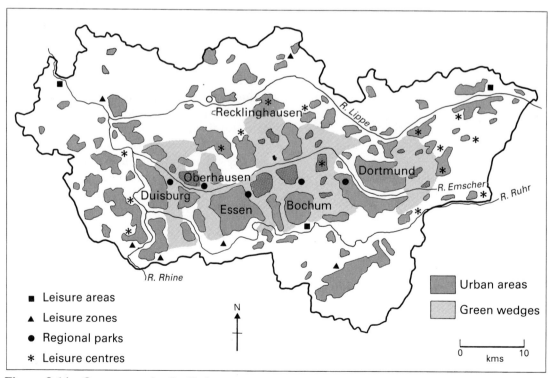

Figure 8.14 Open space in the Ruhr

additional lakes and woodland areas inland. Although many of the cities, too, had made provision at an earlier stage for urban based recreation, such as the Amsterdam Wood and Kralinger Wood in Rotterdam, resources such as these have come under increasing pressure as the population has continued to grow.

To solve both these problems of urban coalescence and the provision of open space drastic action was required. An immediate solution was found, as in the case of the Ruhr, in the creation of a number of 'buffer zones' where urban and industrial development would not be permitted. Originally eighteen buffer zones were designated but only seven have been established to date. In each case they act as defensive barriers to urban development but also serve as important rest and recreational areas for inhabitants of the

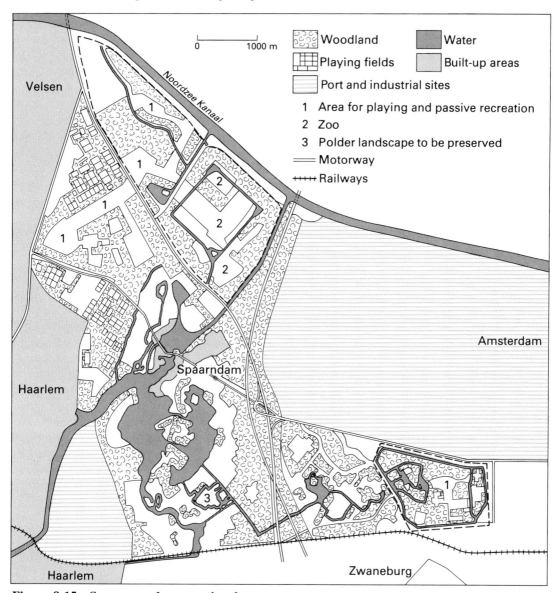

Figure 8.15 Spaarnwoude recreational area

surrounding towns and cities. The success of such a scheme can be appreciated by the example of the recreation area at Spaarnwoude, lying in a triangle bounded by Amsterdam, Velsen and Haarlem (see Fig. 8.15). Here, the possibility of urban and industrial coalescence along the southern bank of the North Sea Canal from Amsterdam to the coast was a very real threat. By creating a buffer zone of some 2,750 hectares to maintain existing agriculture and encourage outdoor recreation, the possibility of such a threat was removed. The site itself has been laid out to cater for a wide range of water and land based recreational activities and is supported financially by central government and the surrounding municipalities. A similar project, although smaller in scale, lies north of Amsterdam – Het Twiske – and an area twice the size of Spaarnwoude is to be established in Midden Delfland on the intervening land so much at risk between Rotterdam, Delft and the Hague. In this case, though, two-thirds of the land will remain in agricultural use.

Both these examples provide proof that positive action can go some way in alleviating the environmental impact of urbanisation in some of the most densely populated regions of Western Europe, providing an urgent breathing space for an industrialised society in which recreational resources are at a premium.

7 The growth of environmental awareness

Greater environmental awareness has also been brought about by the growth of tourism which has occurred since the 1950s. Greater amounts of disposable income, reduction in the number of working hours, paid annual holidays and a decline in the relative cost of transportation have enabled many more Europeans to take regular holidays away from home and escape, albeit temporarily, the drudgery of life in our industrialised society (see Fig. 8.16). Not surprisingly, tourism has become a vital part of the economy of many countries in Western Europe and is vigorously promoted in some of the more depressed, peripheral regions of the sub-continent such as Brittany, southern Ireland, the Greek Islands and southern Italy. In regions such as these, where unemployment has traditionally been high and incomes low, tourism has been viewed as a solution to these problems for not only does the industry provide direct benefits in terms of providing employment – particularly for unskilled labour and women – but it also indirectly supports other sectors of the local economy such as building and construction, catering and food processing.

Tourism is, nevertheless, a paradoxical industry. On the one hand a high quality environment is essential to attract visitors, yet on the other the 'environment' itself is a perishable commodity. Apart from certain intrinsic qualities, such as culture, a region must have an attractive landscape and a pleasant climate including clean air and water as well as a restful atmosphere. However, environmental quality is threatened not only by the number of visitors, but also by developments associated with the expansion of the tourist infrastructure. Impoverished regions of high environmental quality are faced with the dilemma of promoting tourism at the expense of a certain degree of environmental disruption, or protecting the environment and in so doing forfeiting potentially lucrative income from tourism. Inevitably, the lure of tourist expenditure has, more often than not, tipped the balance and tourist development has proceeded regardless of environmental disruption. In Spain, development has often taken place in a series of small, economically and socially under-developed regions. As a result tourist resources have been over-exploited, the natural environment has been damaged and infrastructural developments have not kept pace with the rapid increase in visitors. As a consequence, air, water and noise pollution have occurred, together with the loss of natural landscapes, as agricultural

ENVIRONMENTAL CONCERN

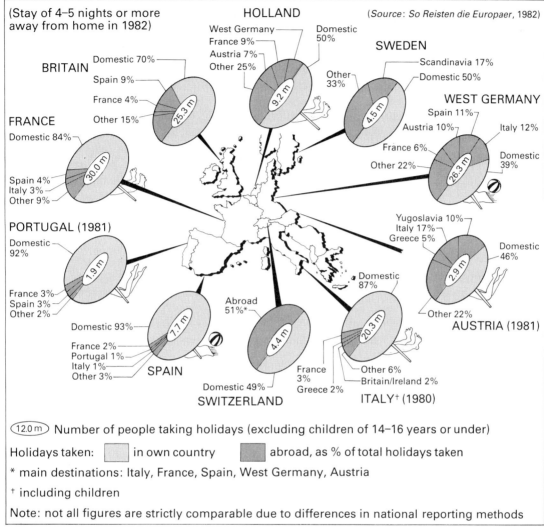

Figure 8.16 Western Europe on holiday, 1982

and pastoral land, forests and beaches have been consumed by the spread of tourist accommodation. Additionally, the destruction of various plant and animal species as a result of irresponsible tourist behaviour (forest fires, for example) and the visual intrusion of tourist amenities, often built without due regard to local building styles, add to the detriment of the situation.

The combination of these factors has led the Spanish government to re-assess land-use policies in the main tourist areas with special emphasis upon environmental protection. In future, drastic measures are to be taken, not only to improve the existing tourist fabric, but also to plan consciously for the next phase of development.

Despite the hard lessons which have been learnt from uncontrolled tourist growth, there is no doubt that vulnerable tracts of land have been retrieved from the clutches of the developer. One such example is the Camargue in southern France, an extensive wetland system lying west of Marseilles and the River Rhône.

This region, formed by the Rhône delta, was once an inter-connecting series of freshwater marshes and salt water lagoons, but modern agricultural practices – in particular land drainage and the cultivation of rough pasture – has threatened the very existence of an area characterised by wild horses, bull fighting, gypsies and flamingo's, containing one of Europe's most important wildfowl reserves.

Wetlands, such as the Camargue, are one of the world's most threatened landscape types accounting for some 6 per cent of the world's land surface, marking the transitional zone between aquatic and terrestrial ecosystems. Being of little economic significance to man, these fragile places have become susceptible to 'improvement' in terms of agricultural modification as well as urban, industrial and recreational development. Not surprisingly extensive wetland systems have diminished to such an extent that their protection has become a matter of international concern, being given top priority by a number of influential conservation organisations.

In the Camargue, although land drainage and cultivation have changed much of the wild character of the region, the reclaimed land is still of poor quality and agriculture, in general, is depressed. The area has, however, always been a tourist attraction and in 1970 in order to promote this industry – as well as to protect the region from wasteful development – the French government designated the Grande Camargue together with the Petite Camargue as regional parks. In 1972, the heart of the park became a national nature reserve owned by the government with financial assistance from the World Wildlife Fund.

Within these protected areas the traditional character of the landscape was to be maintained so that visitors could experience one of Western Europe's few remaining wilderness areas. Today over a million holidaymakers visit the region annually making a vital contribution not only to the local economy, but also to the protection of the landscape.

The increasing interest in nature conservation and the desire to protect some of Western Europe's most vulnerable and attractive landscapes – such as the Camargue – have given rise to a system of national and regional parks, nature reserves and other protected areas throughout the region (see Chapter 9). Despite considerable achievements in this area there is little room for complacency as the ever present demand for development continues unabated. Furthermore, the increasing popularity of these and other designated areas of protected landscape receive more and more visitors annually, many of the more popular parks and reserves are already approaching levels of saturation at peak periods of the year. To safeguard the natural environment of Western Europe requires careful and constant management and administration. In addition, new areas must be set aside to absorb the ever increasing number of visitors, perhaps the only penalty of greater environmental awareness.

8 Western Europe – an environmental future?

Environmental problems and issues are of central concern to all nations in Western Europe. The spread of pollution, the depletion of resources, the destruction of nature and the loss of amenity land have fuelled the passions of a public increasingly concerned about the state of the environment. Environmental awareness has now started to challenge traditional assumptions about progress and advancement which relate general well-being with material prosperity. Such opinions question not only the costs of economic growth and technological development but also the whole issue of the quality of the environment which may be hindered or even destroyed by a policy of sustained growth.

Despite this overall perception of the main environmental problem areas, many governments throughout Western Europe have been almost reluctant to act upon the blatant indicators charting the main threats. Reaction has often been muted unless the issues at hand provide a direct assault upon the fundamental ideology of economic growth. Lessons which have been learnt have often been done so in the hardest possible way. The oil crisis of the early 1970s serves as an example of this as the pessimism theoretically expounded by the authors of 'The Limits to Growth' became an overnight reality. Rising oil prices forced all governments of Western Europe to treat energy conservation as a foremost matter of urgency. The crisis, serious though it was, did give rise to what we might call a 'conservation ethic' – the need to save and use energy in the wisest possible way was the order of the decade, not to do so would drain the life-blood from the whole economy. As a result governments throughout the sub-continent hurriedly reassessed their energy budgets and implemented a series of extremely sound conservation measures in order to offset the increasing price of oil. More cynically, one can question the motives behind such policy – after all such an impetus was little more than would be expected from a group of nations whose continual economic growth was, and still is, dependent upon a cheap, efficient and endless supply of fuel. The same devotion to duty has not been applied to other major environmental threats – the cleaning up of water courses, the control of atmospheric pollution, the conservation of wildlife and the maintenance of attractive landscape – and attempts to remedy these equally serious problems pale into insignificance by comparison to the efforts taken to resolve the energy crisis.

Although the growth of environmental consciousness has fostered a number of conferences concerned with a wide range of national and international issues in Western Europe, international agreement has been absent or hindered by a general lack of agreement. The main environmental problems are, after all, global problems and it is only with international co-operation that solutions will be achieved. In spite of the good intentions of the United Nations Environmental Programme and the EC Environmental Policy, the villains of trans-frontier air and water pollution are still prepared to allow neighbouring countries to meet the costs of their pollution. The effects of acid rain, in particular, have been recognised and well documented for nearly a quarter of a century, yet only now has the problem started to generate any serious attention from the majority of West European governments (even though Great Britain – one of the main culprits in the emission of sulphur dioxide – is still reluctant to face up to its environmental responsibilities). Similarly, the harmful effects of leaded petrol, supported by volumes of conclusive scientific evidence collected over many years, has still not persuaded the majority of West European governments to take the necessary action and ban its use. Unlike the United States of America, which reacted swiftly in banning the toxic additive, nations on this side of the Atlantic have been painfully slow in implementing similar measures. In fact there is little hope of a complete ban occurring in Western Europe until well into the next decade.

Whilst governments across Western Europe have dithered in their approach to environmental problems and issues, more forceful demands for improvement have come from other sources. For the first time in Western Europe's lengthy political history new political parties have started to form, championing the cause of environmental concern. In West Germany, France, Denmark and the Netherlands these 'green parties' have attracted considerable support from a public increasingly aware of their environment and, at the same time, increasingly frustrated by traditional governmental inactivity. At best the impact of these and similar parties might herald a new era of environmental consideration in which deeply rooted, materialistic values are replaced by a more harmonious outlook upon the environment. At worst, existing governments and political parties might be frightened into re-examining their basic philosophies and in so doing pave the way to greater environmental respect and the removal of the main threats of degradation.

9 The Resource Issue

What have the following in common?

- A Danish Member of the European Parliament arrested for fishing in British waters in the North Sea.
- American anger at Western European countries buying natural gas from Russia via the Siberian pipeline.
- International concern for the conservation of Venice which is threatened by periodic flooding and was gradually sinking beneath the sea due to underground water abstraction.
- The rise of the 'Greens', anti-nuclear groups and other Ecology parties in protest in part against the development of nuclear power.

The thread that links them all together is that they are the end-product of disputes concerning the exploitation of *natural resources* (see Box 9.1). The seemingly-straightforward process of extracting sufficient raw materials for our various needs is in fact fraught with difficulties producing many conflicts between interested parties.

No longer can it be assumed that society can dip into large, seemingly-infinite reserves of oil, coal, forests, iron ore, water and so on to build and heat its homes, fuel its cars and industry, and provide all the other essentials of modern life. There is now a growing realisation that such resources will not last forever or will be so badly damaged by other actions and side-effects that they will be rendered useless. The issue of resource exploitation is not simply a problem of how to extract materials from a sometimes hostile environment, nor is it just a question of the economics of supply and demand; increasingly it revolves around complex environmental and political debates. This chapter will be primarily concerned with an analysis of such issues in Western Europe but, firstly, it is necessary to concentrate on the global dimensions of the debate which will illustrate many of the fundamental issues that are applicable in the European context.

1 The depletion of global resources: optimistic versus pessimistic scenarios

Widespread interest in the issue of resource exploitation was sparked in the early 1970s with the publication of the 'Limits to Growth' report. This was an attempt to forecast the development of the entire world system over a 200 year period from 1900 to 2100 by using a sophisticated computer model. It tried to predict changes in five basic elements – population, agriculture, industry, resource use and levels of pollution. The first three of these were positively related, in that an increase in one brought forth growth in another, whereas resource use and pollution had a negative effect to slow down or even reverse growth. For example, acid rain in destroying forests and fisheries cuts timber and food production. The predictions were overwhelmingly gloomy; even when very optimistic forecasts were built into the model (such as unlimited nuclear energy) it presented a picture of dwindling resources, widespread pollution, with dramatic reductions in population levels and agricultural and industrial production, and hints at the social crisis that would probably follow.

> **Box 9.1 Natural resources**
>
> *Natural resources* are that part of the total stock of the Earth's materials that are of use to society under existing social, economic and technological conditions, a part of which will constitute the *reserves* that are actually available for use at any one time given the state of technology and the economy. For example, there are considerable deposits of potential resources, notably metallic ores in nodules, on the sea-bed of the deepest oceans but as yet these are not part of reserves as neither the technology needed to develop them nor the economic demand for them has been forthcoming. Perception of what is a resource changes through time: materials which were once considered of no use to society can rapidly become valuable resources as the technology to exploit them develops and the associated economic demand grows. For example, uranium was of no apparent use for centuries until scientific advances enabled the production of energy from it and the insatiable demand for power could not be met from conventional sources such as coal, oil and gas. Thus, the term resource does not relate to the material itself as such but to the value placed upon it because of the function it can perform. In other words, resources have to be subject to a human or cultural appraisal. This can change either to identify new resources or to make others redundant. For example, the wind mills that powered many of the activities of the Middle Ages became obsolete with the advent of water, steam and electrical power from the Industrial Revolution onwards, but the 'energy crisis' of the late twentieth century may see a revival of interest in a modern form of wind power.

Although the methods and findings of the 'Limits to Growth' team have been widely criticised because they based their work on debatable assumptions (notably the idea of exponential growth, to be discussed later) and failed to take account of economic reactions to resource crises (shortages can lead to price increases and the use of substitutes), the report played an invaluable role in stimulating interest and debate into the issue of resource depletion and dependency. It highlighted the long-term problems that modern society faces in trying to maintain its present way of life and showed that seemingly beneficial developments can have potentially harmful long-term effects. For example, nuclear power may overcome an energy crisis but unless safe means of disposal are found for its highly toxic waste it may do irreparable damage. Generally the response has been to divide those involved into two strongly opposed schools of thought, the pessimists who foresee a collapse of society based on long-term predictions, and the optimists who put their faith in technological advance and the workings of economy.

The pessimists base their arguments on the concept of exponential growth. This envisages the world's population growing at an ever-increasing rate, for example, current estimates are that the world's population doubles every 35 years so that the estimated figure of 4,000 million in 1975 will reach 8,000 million in 2010 but will be 16,000 million in 2045 and 32,000 million in 2080 and so on. (You can see this more graphically if you plot a graph with estimated population on the vertical axis against time on the horizontal axis to obtain the exponential growth curve.) Moreover, it is argued that this situation is compounded by the fact that standards of living are rising and that we are demanding and consuming ever-greater quantities of resources per head of population. So the exponential growth curve in the use of resources is even steeper than that for population. On the basis of such arguments the pessimists predict that we will be running out of certain key resources in the very near future and that further industrialisation will lead to untenable levels of population and environmental destruction. They conclude that fundamental changes in lifestyle will have to be made to prevent the breakdown of modern industrial society.

The optimists counter these views by arguing that exponential growth takes no account of how people react to a given situation. Population does not tend to double in a fixed manner, experience has shown that its rate of growth tends to fall markedly once a high level of material wealth has been achieved. More specifically, they argue that the economics of supply and demand will intervene so that as resources become more scarce their price will increase. This in turn will prompt a variety of reactions. Firstly, it will encourage the careful use of dwindling resources, exploitation of lower grade deposits, and recycling of waste products. Secondly, it will prompt the use of substitutes. Both of these are short-term reactions whilst the final, long-term solution may be found in the technological advances which will have been encouraged to provide new discoveries and alternative resources.

What then is the relevance of the global debate to the issue of resource exploitation in Western Europe? As a major industrial power, second only to North America, the countries of Western Europe are massive consumers of resources. In fact the whole basis of modern urban-industrial society is underpinned by the continuing availability of cheap and abundant supplies of raw materials. During the early phases of industrialisation these were mainly provided by the European nations themselves (Ruhr coal, Lorraine iron ore, Swedish timber, and so on), although the colonial empires that the major powers built up also played an invaluable role in providing cheap and guaranteed supplies of resources. Indeed it has been argued that this was one of the reasons why the British, Dutch, French and Germans sought control of far-flung Third World countries. Today the situation has gone one step further. The European nations tend not to be major producers of raw materials, instead they rely increasingly on supplies from elsewhere. These are then transformed in Western industrial complexes for internal consumption or for sale back to the resource-producing countries. One result of this transformation is a tremendous degree of interdependency between the raw material producers and industrial consumers as the recent fluctuations from oil scarcity in the West following the Arab-Israeli War in 1973 to the recent oil glut have shown.

Within Western Europe the results of the transformation have been twofold. Firstly, the urban-industrial areas have become more specialised concentrating on manufacturing and service activities rather than the extraction of raw materials. Thus, for example, iron and steel producers tend no longer to rely on expensive, locally mined ore and coal but on cheaper supplies imported from elsewhere. Hence, as Chapter 3 has shown, the indigenous industry in Lorraine has suffered as a consequence of the development of the French iron and steel industry favouring coastal locations such as Dunkirk using imported raw materials. Similarly, Chapter 2 illustrated the run-down of the extractive industries in the Ruhr. Clearly, resource producers within Western Europe find it increasingly difficult to compete with overseas competitors and this has resulted in economic and social problems as pit closures and high unemployment have become rife in once prosperous, urban-industrial areas.

A second product of the transformation has occurred in those regions which have traditionally specialised in resource production rather than manufacturing. These tend to be peripheral to the urban-industrial Eurocore and have proved just as vulnerable to outside competition. Here there is an absence of alternative employment opportunities once the extractive industries have begun to suffer. On the other hand, the situation can be reversed and new areas of specialist resource production can spring up, as in the case of North Sea oil and gas, giving rise to a whole range of employment opportunities. Once again, the economic and social fate of these peripheral regions is inextricably linked to what is happening in the Eurocore.

Should the pessimists be correct, it is probable that the first indications of their predictions will be seen in Western Europe. How then does the situation here match up to their forecasts? In terms of population growth it is clear that it is not growing exponentially.

However, although the rate of population growth has fallen dramatically, the standard of living of most Europeans has markedly improved so that we are consuming more and more resources per head. All the manifestations of modern society such as cars, televisions, freezers and refrigerators were not available to the vast majority of previous generations and, of course, it takes ever-increasing quantities of resources to produce them. Moreover, new innovations are constantly pressed upon us which further increase the strain on resources. As the countries with some of the most prosperous communities in the world are situated in Western Europe, it is from here that a considerable proportion of the spiralling demand for raw materials originates.

One point from the global debate is of particular relevance to Western Europe, that of the production and impact of pollution. As one of the major industrial areas of the world it is also one of the major pollution producing areas. Chapter 8 has shown the strenuous efforts made by governments to combat the menace of pollution but considerable environmental damage is still taking place. No longer is this a local phenomenon and it is affecting increasingly larger areas, as the debate concerning the export of acid rain from Britain and West Germany to Scandinavian forests and lakes has illustrated. Furthermore, pollutants are tending to become intrinsically more toxic and hazardous as the dilemmas of nuclear waste and the dioxin disaster at Seveso have demonstrated. Pessimists argue that this is pointing the way to the dangers yet to come, whereby the environment will no longer have the capacity to absorb this waste and will become so badly damaged that it will be incapable of producing the raw materials needed in the first place.

2 Resources and resource depletion

To examine these issues in more detail a series of case studies will be analysed but prior to this it is necessary to take a more refined look at the concept of resources, for the various types of resource present problems of a rather different nature. Basically there are three types of resource: *stock resources* (e.g. oil) are non-renewable and once extracted cannot be replaced; *flow resources* (e.g. forests) as the name suggests can provide a steady supply but are susceptible to human interference to increase or decrease it; and *continuous resources* (e.g. solar power) derive from sources generally unaffected by human actions and can be harnessed in a variety of ways (see Box 9.2).

Different problems and issues emerge according to the type of resource involved. The dilemma of non-renewable stock resources has two faces; either the depletion or exhaustion of traditional sources which leaves areas prone to serious economic and social problems, or the sometimes frantic rush to develop new deposits can disrupt both physical and human environments. Flow resources are basically concerned with questions of sensible management but a growing issue is the impact of pollution on supplies. The major problem concerning continuous resources is how they can be safely and economically harnessed but also human modification of amenity landscapes is emerging as an issue. The following sections take case studies from Western Europe to examine the issues involved in the exploitation of each resource type in detail.

Case study: Stock resources

THE DEMISE OF THE SOUTH LIMBURG COALFIELD

It would be naive to think that the exploitation of a particular stock resource only stops when its supply in that location has been exhausted. In modern industrial society economic factors tend to halt the extraction process should the cost of mining or quarrying be considered too high in relation to similar supplies from elsewhere or alternative materials; consequently, valuable but 'expensive'

> **Box 9.2 Typology of natural resources**
>
> *Stock resources* consist of those materials that have been created so slowly that to all intents and purposes they can be considered non-renewable. They comprise the fossil fuels (coal, oil and natural gas), metal ores (e.g. iron ore, copper, aluminium, gold, etc.) and other useful minerals in the earth's crust such as salt, potash, china clay, and even sand and gravel. In essence, they represent the core of the resource problem as once extracted from the earth they cannot be replaced except over geological time. The fact that such natural resources that took over 100 million years to form by geological processes will be totally consumed in only 100 to 200 years of industrialisation sums up the dilemma. Although a small proportion of metal ores can be reused following recycling, this does not change the fundamental nature of the problem of stock resources. Fossil fuels in particular are totally dissipated in use and these are the basis of modern industrial society.
>
> *Flow resources* are those materials which are continually replenished but whose supply can be maintained, increased or decreased by human action, such as forests, soil, or animal life. For example, fish constitute an important food source but their supply can be depleted by commercial over-fishing or pollution of their habitat, whereas a ban on fishing in breeding-grounds can actually increase fish stocks. Certain flow resources need human interference to increase supplies. Careful husbandry of plants such as rubber trees involves the transfer of production from the natural forest habitat to plantation agriculture and increases yields many times over. As such, flow resources do not constitute a problem as long as they are carefully managed and not subject to destruction by over-cropping or pollutants, but experience shows that this has not always been the case.
>
> *Continuous resources* are always available irrespective of human actions and can be harvested or used by society without disrupting their supply. The sun's energy and tidal movements are prime examples, they can be harnessed in various ways but the level of supply cannot be interfered with. Also included in this category are amenity landscapes (areas used primarily for recreation) but in this case they can be modified by human actions. For example, an area within a National Park can be changed by quarrying, afforestation or over-use for recreation – the amenity still exists but in a rather different form. The problem concerning continuous resources is how to harness them effectively so as provide a long-term alternative to renewable stock resources and depleted flow resources.

reserves are left in the ground. Moreover, as British readers will only too painfully be aware, the closure of 'uneconomic' pits is not just a question of the economics of supply and demand in the coal industry but depends on the resolution of powerful forces in society, notably major employers (both private and state) and trade unions as well as governments. This is primarily due to the economic and social havoc such closures can bring about in communities heavily dependent on the extractive industries and the attempts made by workers to prevent this by preserving their jobs. Thus the issue of the exploitation (or non-exploitation) of stock resources should not only be seen in the narrow context of a particular location and its economic viability but also in the wider economic and social debate surrounding the consequences of such actions. The case study taken here is of the complete shut-down of a small but thriving coalfield in the South Limburg region of the Netherlands during the late 1960s and early 1970s. It shows how the fate of the industry and the area heavily dependent on it was in large part determined by outside forces over which it had little or no influence and how the future prospects of the region were greatly improved by outside intervention to create replacement employment.

Limburg is the most southerly part of the Netherlands, a narrow strip of land sandwiched between Belgium and West Germany remote from the main centres of activity in the

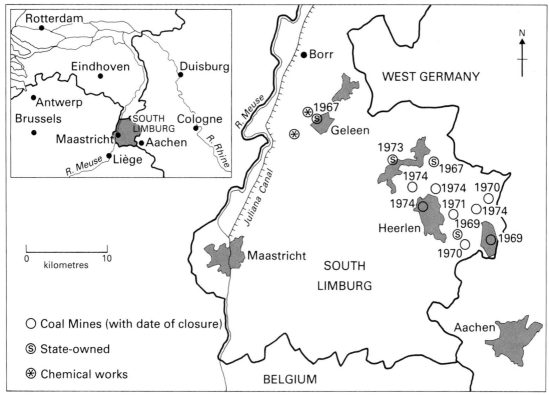

Figure 9.1 The South Limburg coalfield

Randstad. It contains an extension of the disjointed coalfield that runs from north-east France through Belgium, and large scale mining began here relatively late during the second half of the nineteenth century. Prior to this date Limburg was considered a remote and inaccessible region badly served by poor communications to the bulk of Dutch industry, which was concentrated in the ports in the west of the country where it was cheaper to import British and Belgian coal. An influx of foreign capital prompted exploitation of its coal reserves and eight private mines were established in the second half of the nineteenth century. These were followed by four state-owned mines in the first part of this century and a series of transport improvements, notably the completion of the rail network linking to the main Dutch system, improvements to the navigation of the Meuse, and the construction of the Juliana Canal (see Fig. 9.1). As a result coal mining in South Limburg expanded rapidly and employed nearly 60,000 workers at the height of its prosperity in mines concentrated principally around Heerlen in the west and also at Geleen, where a large processing plant was set up by the state-owned mines to manufacture nylon, fertilisers and other chemicals from locally mined coal.

By the mid-1960s the South Limburg coalfield continued to prosper, employment in the twelve mines had been slimmed down to 45,000 but they were still producing a total of over 10 million tonnes of coal each year. Yet within the space of only eight years all of these pits had been closed down despite the fact that there were another 100 years of coal reserves left to be mined and that they were among the newest, safest and most modern mines in Europe. This apparent paradox can only be explained by events outside the coalfield which rendered it uneconomic. Firstly, there was the impact of cheap coal imported in to Europe from the United States and elsewhere. Despite the fact that Limburg is at the heart of the industrial complex of the EC, it found it extremely difficult to compete with coal mined by inexpensive open-cast methods and transported cheaply in vast quantities by bulk carriers. Secondly, fuels other than coal were becoming available in increasing quantities and at lower cost. The vast refining capacity built in the Europoort complex at Rotterdam and the newly discovered gas fields in the north of the country around Groningen had a major impact on energy use, so that the percentage of energy consumption in the Netherlands provided by coal fell from 78 per cent in 1950, to 30 per cent in 1965, and only 6 per cent in 1980, whereas the corresponding figures for oil and natural gas together were 22 per cent, 70 per cent and 90 per cent respectively. Lastly, the coalfield was not exclusively controlled nationally as the Netherlands was a member of the European Coal and Steel Community, a forerunner of the EC set up in 1951 to create a common market for the coal industry and to coordinate its expansion initially, then its rationalisation. Hence, South Limburg was competing in a much larger market and decisions concerning its future were being taken in the context of this larger area and not simply with regard to the Dutch situation.

Immense economic problems resulted from the extremely rapid run-down of the South Limburg coalfield, especially in the eastern mining district around Heerlen where half of the economically active population were employed in the mines. Moreover, there was great social disruption in a community long dominated by a single industry around which their whole lives had revolved. This prompted rapid and vigorous action from the Dutch government which in concert with the ECSC identified South Limburg as a priority area for economic development. Generous incentives were offered to industry to locate there and retraining programmes were set up. These attempts at regional development have proved successful; the chemical plant at Geleen has expanded output but now uses oil and gas instead of coal, a new car factory (originally DAF, now owned by Volvo) employing 5,500 was set up at Born in the late 1960s, and the government decentralised some of its activities from the Randstad (for example, offices of the government pension fund and the Central Bureau of Statistics were established in Heerlen, the latter on the reclaimed site of a former mine). Overall, this process of industrial restructuring aimed at the absorption of 45,000 workers appears to have been largely successful aided by natural wastage and migration but the situation in South Limburg still gives cause for concern as unemployment rates stubbornly remain at around double the national average.

South Limburg, then, can perhaps be seen as a successful example of how the run-down of natural resource extraction can be managed. However, the region was fortunate in two respects; firstly, its period of decline coincided with a time of economic prosperity when new jobs and investment were available and, secondly, it is situated in the heart of the Eurocore in close proximity to major urban markets – in fact Maastricht is closer to Brussels, Cologne, and Düsseldorf than it is to Amsterdam and Rotterdam. However, the prospects for an equally successful run-down in extraction from other European coalfields during the recession of the 1980s are very slim.

There are a number of points of a general nature that can be drawn from this study. In natural resource terms exploitation ends not at the point of exhaustion but at the time that it ceases to be of economic value to society. This value is being determined in progressively larger markets so that the future of resource-producing areas is increasingly dependent on distant places where the same material or alternative resources can be produced at lower cost and imported cheaply. Governments have found it increasingly difficult to intervene to prop up struggling regions by subsidising raw material production and are being forced into closure plans and attempts to restructure the economy of ailing regions. It is here that the basically economic debate becomes widened into a social and political issue as to the best means of helping the affected locations. This has led to bitter political disputes with wider ramifications. For example, the announced closure of one of the few remaining Belgian coalmines in September 1984 was greeted with widespread anger and riots not only because of its economic impact but also because this was the last remaining pit in the French-speaking Walloon southern part of the country, and its closure further fuelled the sectarian divisions with the Dutch-speaking Flemish north. It is not so much the dilemma of leaving potential resources in the ground so that they are virtually unrecoverable for future generations that is at the core of the issue of stock resources, rather it is the human reactions to the social and economic consequences of such actions that are central to the debate. Essentially, it is between whether economic criteria alone should determine the fate of resource-producing regions or whether state intervention is necessary to reduce its harsh consequences.

THE DEVELOPMENT OF THE NORWEGIAN OIL AND GAS INDUSTRY

Phrases like 'North Sea oil bonanza' give the misguided impression that the discovery and exploitation of a new stock resource is overwhelmingly beneficial and non-problematic. But experience shows that this is not necessarily the case and that even the development of new, badly needed reserves frequently involves both positive and negative aspects which are subject to fierce debate and controversy. It is possible to draw up a balance sheet of the primarily economic benefits as against the more varied costs of any resource development, as shown in Table 9.1. Frequently it is impossible to make an overall judgement as to whether an actual or potential development is beneficial or damaging because, whilst it is

Table 9.1 Balance sheet of the advantages and disadvantages of stock resource developments

Benefits	Costs
Primarily economic	*Environmental damage*
• reduction in unemployment as new jobs are created not only in the extractive industry but also in processing, transportation, and services	• visual intrusion of large scale structures, quarries, etc. plus derelict land and spoil heaps
• a multiplier effect deriving from above can stimulate the general economy	• pollution leading to possible destruction of other natural resources
• exported output brings in foreign currency and the government benefits from taxation and royalty payments	*Economic side-effects*
• wealth generated can be used to increase individual earnings and finance state services, especially welfare services	• these depend on the degree of domination of the overall economy but it can: 1 push up wages levels making other firms uncompetitive 2 inflate the value of the currency to make imports cheaper and make exporting more difficult 3 leave a country heavily dependent on foreign multi-nationals
• carefully sited development can foster growth in locations where need is greatest. This can have the social advantage of stemming depopulation of depressed areas	*Social impact*
	• disruption of long-established communities by the sudden influx of migrant labour
	• despoliation of recreational areas

Table 9.2 Output of major European oil and gas producers (in million tonnes oil equivalent)

		1972	1974	1976	1978	1980	1982	1982 % of World output
United Kingdom	Oil	0.1	0.1	11.8	53.3	80.5	103.3	3.8
	Gas	23.3	30.6	33.7	33.7	32.0	32.4	2.4
Netherlands	Oil	—	—	—	—	—	—	—
	Gas	44.2	63.3	73.6	67.1	68.9	54.6	4.0
Norway	Oil	1.6	1.7	13.8	17.2	25.8	24.1	0.9
	Gas	—	0.1	0.3	13.4	23.5	20.9	1.5

Other minor producers include West Germany (4.3 mtoe oil and 14.6 mtoe gas in 1982), Italy (1.6 and 12.3 mtoe) and France (1.6 and 7.8 mtoe).

possible to put a price on the economic attributes, it is not possible to estimate the cost of many of the social or environmental losses (for example, how can you put a price on the loss of a beautiful landscape or the disruption of a close-knit, long-established community?). Hence, the issue of stock resource development tends not to be judged in strictly economic terms but includes a more wide-ranging debate covering less tangible matters.

Norway has been successfully exploiting its natural resources such as fish, timber and hydro-electric power for many years and the addition of an oil industry would seemingly add to this list of successes, but its development became an issue both within Norway itself and in the wider economic community also. Production of oil began on a fairly small scale in the early 1970s but it has expanded steadily and has been supplemented by the rapid exploitation of natural gas reserves in the late 1970s, so that in 1984 Norway produced nearly 50 mtoe (million tonnes of oil equivalent) of oil and natural gas. This makes Norway the third largest European producer with more or less the same output as the Netherlands but far behind British output, but in world terms the country is very much a minor producer (see Table 9.2).

Initially the major production area was the Ekofisk field in the south-western corner of the Norwegian sector close to the median lines which divide up the North Sea between the various countries, as shown in Fig. 9.2. A string of smaller fields were then opened up to the north mainly adjacent to the median line between the British and Norwegian sectors. The next major field to come on stream was Statfjord in the early 1980s and Nowegian output has been supplemented by gas finds, notably the Anglo-Norwegian Frigg field. Oil and gas are mainly exported to North America (22 per cent of Norwegian output in 1982), Britain (16 per cent), the Netherlands (15 per cent), Sweden (11 per cent), France (10 per cent) and West Germany (6 per cent) with only a small proportion (18 per cent) being consumed internally. The pipelines to St Fergus (north of Aberdeen), Teesside and Emden (to supply the Ruhr and parts of the Netherlands, France and Belgium) are evidence of this market demand but also of the physical difficulties in bringing production ashore in Norway – all the major fields are separated from the mainland by a very deep marine trench. A considerable on-shore industry has also developed to build exploration and production rigs and platforms and to provide services for the off-shore facilities with Stavanger emerging as the main Norwegian centre.

During the early 1970s a diversity of public opinion was ranged for and against the oil developments. The economic advantages of such off-shore developments to a country with a large part of its population strung out along its narrow coastal strip and with a strong maritime tradition spoke for themselves. However, perhaps predictably, environmentalists were gravely concerned about the possible destruction of Norway's natural environment and, together with the fishing industry, feared the pollution of the coastline. The social and cultural impact of oil-related

THE RESOURCE ISSUE 267

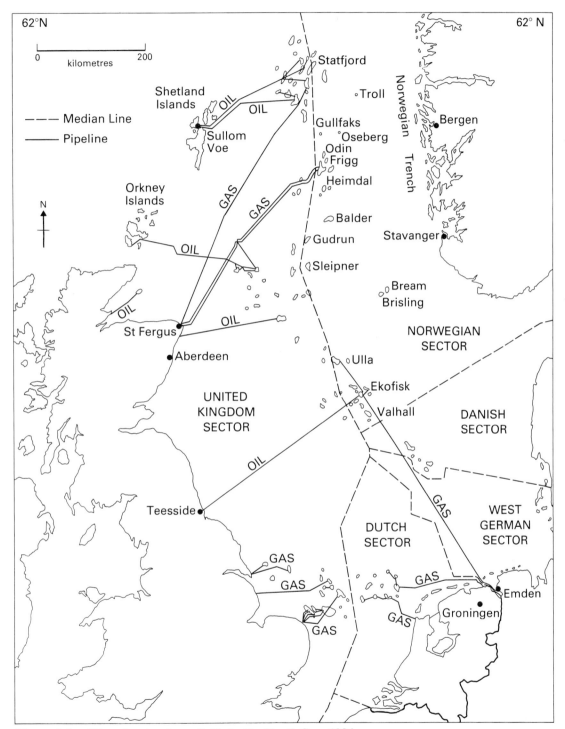

Figure 9.2 Oil and natural gas fields in the North Sea, 1984

Figure 9.3 A Norwegian oil rig in the Ekofisk field

developments on the mainland (for example, the construction of rigs involving the import of migrant labour out-numbering long-established, closed communities) was anticipated with much concern. Doubts were also voiced by industrialists in other sectors of the economy that wages would be forced upwards and imperil their businesses. In fact, this amalgam of forces had some success in persuading the political parties to restrict the amount of oil and gas that could be produced each year, but, in practice, maximum permissable output was set so high (90 mtoe per annum) that there is little prospect of it ever being attained. However, in recent years resistance to the growth of the oil industry has evaporated as, firstly, the anticipated environmental, economic and social damage has not taken place on anything like the scale envisaged and, secondly, Norway, like every

Western country, has begun to experience the worst effects of the recession. Although its unemployment rate is unquestionably low by European standards, the off-shore oil and gas industry together with its on-shore rig and platform construction, refining, servicing and other oil-related developments are seen as vital job creating agencies. Hopes are particularly focused on the more remote Arctic regions of the Norwegian coastline where unemployment and associated depopulation are highest due to the decline of traditional activities such as fishing.

Norwegians are now looking to the economic benefits of North Sea oil production, seeking to open up new fields and to explore progressively more remote and hostile areas. Considerable amounts of oil and gas still remain in the existing fields to the south of the 62nd parallel, only 1,100 mtoe out of an estimated 5,000 mtoe of recoverable reserves are within fields already being developed, and it is thought that as much again will be found to the north. The mid-1980s saw the opening up of three new fields – Ulla, Heimdal and Gullfaks – with Oseberg to follow shortly. Two further fields represent the greatest future potential, the Sleipner gas field and the giant Troll field (believed to be even larger than the massive Groningen gas field in the Netherlands), but the development of the latter involves considerable technical difficulties in tapping gas from great depths. A further promising find has been made in the far north at Askeladden but further exploration is now taking place to try to find more oil to turn it into a commercially viable proposition capable of financing a pipeline south through Sweden to the Baltic or down the Norwegian coast (Fig. 9.4). Together with the newly completed road from Narvik to Kiruna in Sweden, these oil-related developments are seen as key factors in attempts to revitalise the economy of the Arctic region. Despite the commercial and technological difficulties that will have to be faced, it seems that the future prospects of the Norwegian oil and gas industry are rosy but such a view is too

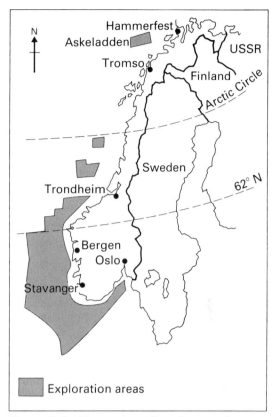

Figure 9.4 Norwegian oil and natural gas exploration areas

simplistic as it faces both internal and external pressures that make the issue far from clear-cut.

The internal economic pressure to expand production in order to create jobs is clear but the emerging problem is the impact of the growth of the oil and gas industry on the rest of the Norwegian economy. To all intents and purposes it dominates the remainder of Norwegian industry accounting for one-third of all exports and one-fifth of total investment. As the industry only employs 2.7 per cent of all workers (7,000 off-shore and 45,000 in oil-related on-shore activities), it can be seen as a capital-intensive industry with little job creation potential and considerable detrimental side effects. In practice, the oil revenues inflate the value of the Norwegian currency making imports cheaper and

Norwegian exports more expensive. Thus, Norway's non-oil sector is not only losing its export markets but is also suffering increasingly from competition from imports, so without the oil revenues there would be a large balance of payments deficit. Oil money has also helped the Norwegian government to finance a considerable expansion of its welfare state provision to levels comparable with the highest in the world but this has recently been cut back in an attempt to revive the non-oil sector of the economy. It seems clear that the very success of the oil and gas industry can contribute to excessive problems elsewhere in the economy, and present politicians and public alike with conflicting pressures and issues that are extremely difficult to resolve.

At the same time Norway faces a series of external pressures, both commercial and political, concerning its oil and gas industry. The commercial issue revolves around the Norwegian policy of trying to extract the greatest possible national benefit from their resources. By demanding a progressively larger share in the ownership of each field at the time when licences for exploration and development are issued, Norway has built up a steadily expanding proportion for its state-owned company, Statoil. Together with the imposition of stiff taxation and royalty payments, it is considered by some experts that Norway is employing too nationalistic an approach which will scare off the large oil companies and leave it with oil and gas resources that it will be incapable of developing itself. The external political pressure derives from the anxiety of certain Western European and the American governments over the extent of Western European dependency on natural gas supplies from the Soviet Union. They see the strategic importance of an expansion of Norwegian production as an alternative to that purchased from Russia, but to do so would probably mean Norway having to accept a lower price for its gas and less state involvement in its exploitation.

Overall, the Norwegian experience illustrates the problems faced in exploiting rich, newly-discovered stock resources. Superficially the riches underneath the North Sea can only appear to bring benefits but when probed a little deeper a series of dilemmas begin to emerge. How much social and environmental damage will be caused? How many new jobs will be created? Will the new jobs be in the right places? Will the remainder of the economy suffer unduly? It is really this last question which represents the fundamental issue – are the short term gains worth the possibility of serious long term damage which will transform the country into an extremely vulnerable, virtually single-industry nation? In other words, the question still remains what happens when the oil runs out?

Case study: Flow resources

THE CONTINUING SAGA OF FISHING DISPUTES IN EUROPEAN WATERS

It has already been noted that flow resources constitute potentially renewable sources of raw materials but only so long as they are carefully harvested and managed to prevent overexploitation and possible exhaustion. In addition, flow resources can prove vulnerable to pollution from outside sources which can disrupt and diminish supplies. This latter aspect has been dealt with previously in Chapter 8 where the impact of acid rain on forests and pollution of waterways has been examined at length. Consequently, the question of flow resource management will be highlighted here together with a detailed analysis of the disputes that such policies give rise to at both a national and international scale.

In recent decades, stocks in fishing grounds in close proximity to densely populated urban areas have come under increasing strain, not least those in the north-east Atlantic and North Sea. Previously they had been generally regarded as a common resource which was not under any sort of ownership and upon which quite distant countries could draw. This attitude has been questioned recently as countries have sought to bring 'their' sector of the

Continental Shelf under national control together with the fish stocks to be found there, but in the past it led to a situation where fisheries were open to all-comers and virtually unmanaged. The introduction of modern fishing techniques with large vessels assisted by sonar to locate fish and trawling techniques which swept up fish irrespective of size or type, together with the development of industrial catches (for fertilisers, animal feed and fish oil) resulted in severe over-fishing, destruction of breeding grounds, and drastic falls in the number of certain fish types. For example, in the North Sea the crucial spawning stocks of herring and sole which provide future generations of fish were dramatically reduced to only one-tenth of their level in the 1930s. The impact of this can be seen in Fig. 9.5 which illustrates the effects of over-fishing and the closure of distant fishing grounds in the North Atlantic during the 1970s on herring and cod catches, as opposed to the continued growth in industrial fishing mainly for other species. In addition, new fishing fleets emerged on the scene during the 1960s notably from the USSR and Eastern Europe. Today the USSR takes a major catch from the north-east Atlantic, second only to Norway. The other principal fishing nations are the traditional maritime countries of Denmark, Iceland and the United Kingdom (Fig. 9.6).

This geographical distribution of catches was only arrived at after a long and hard process of bargaining and conflict which at times even involved the use of rival navies and gun-boat diplomacy. The issue of who fishes for what and where has involved complex ecological, economic and political debates as rival nations sought to get their hands on what they believed to be their fair share of what appeared to be a dwindling flow resource – a situation brought about by their own short-sightedness and lack of management in the first place. Disputes have been manifested on two geographical scales; individual nations have sought to keep their fishing grounds primarily for their own fleet with the total or partial exclusion of vessels from other countries (as during the Cod Wars between Great Britain and Iceland 1958–1976) and, secondly, supra-national agencies have tried to bring nations together to try to harmonise their fishing so as to protect stocks and at the same time allocate them equitably, as the Common Fisheries Policy attempts to do.

The Common Fisheries Policy (CFP) has been a bone of contention ever since the mid-1960s, through the negotiations for the enlargement of the EC beginning in 1970, until the 'final' agreement introduced in January 1983. It is an attempt by a supra-national body to determine how fishing resources should be allocated and conserved in the interests of the Community as a whole. As such it has given rise to intense conflicts as member states have sought to push their own national interests. The original CFP embodied the principle that all Community fishermen should have equal access to all fishing grounds under their jurisdiction. This meant that although individual nations still controlled their territorial waters they had to apply any regulations equally to all Community fishermen. After some years of wrangling the original six members of the EC agreed to this policy, but the timing of this decision was most

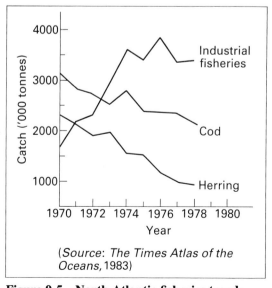

(Source: The Times Atlas of the Oceans, 1983)

Figure 9.5 North Atlantic fisheries trends

272 GEOGRAPHICAL ISSUES IN WESTERN EUROPE

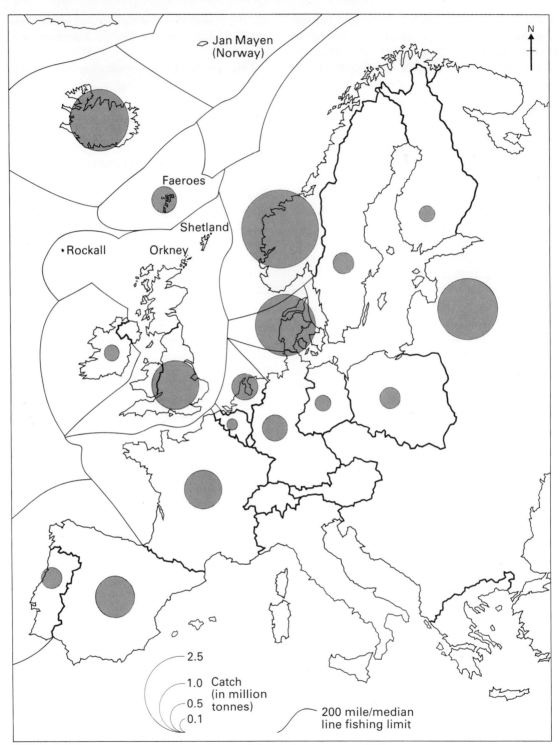

Figure 9.6 Fish catches in the north-east Atlantic by country, 1978

unfortunate in that it was just before the maritime nations of Britain, Denmark, Ireland and Norway began negotiations to enter the Community. This raised suspicions that it was a device to gain greater access to the increasingly restricted waters of the four applicants who, not unnaturally, saw it as an attack on their resources. The British government was caught between continuing to support open access (as it had done against Iceland) and wishing to keep in-shore fisheries for its own fleet. The Danes were generally in favour but wanted to keep foreign vessels out of the fisheries around its dependencies, the Faroes and Greenland. Ireland was strongly opposed as was Norway, who fiercely tried to protect its fishing industry. The outcome was that a transitional agreement to last for 10 years was introduced under which member states would maintain their own six-mile limit (extendable to 12 miles in some in-shore fishing areas) for those who operated from ports in the geographical area plus those who had traditionally fished in those waters and from which foreign vessels would be excluded. This satisfied three of the applicants but Norway chose not to enter the EC largely on the basis of opposition to this fishing policy.

Events overtook this agreement in the mid-1970s when many countries including those in the EC declared 200-mile or median-line Exclusive Economic Zones which included fishing rights. Once again, the issue of who should be allowed to fish in European waters was raised with Britain and Ireland fighting a rear-guard action to restrict access to their fisheries by their Community partners. On one side, those in favour of equal access pointed out that membership of the EC with its common market involved the scrapping of national systems that discriminated against fellow members, stressing that any costs incurred would be offset by benefits in other spheres, and suggested that effective conservation measures required Community rather than national controls. On the other hand, Britain and Ireland argued that they did not get access to the resources of other EC countries (for example, gas, coal, iron ore) so why should fish be treated differently? They also argued that in having the largest sections of the continental shelf their catches should reflect the size of this contribution and that conservation policies were most effectively policed by individual states. After long and acrimonious bargaining an agreement was finally reached and introduced in January 1983 under which the EC sets annual total catches for the main fish types and allocates them in each fishing ground according to quotas for each country reflecting traditional fishing patterns and the needs of regions heavily dependent on fishing. The six-mile limit was also maintained.

These examples show how difficult it is to manage a flow resource so that its continued existence is not endangered. Iceland had to fight off the claims of other nations in order to secure what it considered to be adequate protection and the EC has faced extremely bitter and protracted discussions before a compromise agreement could be hammered out. One special difficulty in this case is the very idea of 'ownership' and the difficulties countries face in asserting that certain fish types are 'their' resources when they may only be in their waters for a few months each year before returning to spawning and feeding grounds elsewhere. In this sense true resource management can only be achieved at a supranational level, otherwise it can all too readily recede into international conflicts. The case study shows that national interests tend to be thrust forward in any dispute, and the fact that flow resources are potentially renewable may make the issue of how they are exploited even more important to those involved.

Case study: Continuous resources

THE SEARCH FOR ALTERNATIVES TO FOSSIL FUELS

Fossil fuels are the basis of modern urban-industrial society but, as stock resources, their lifespan is limited and ultimately they will be exhausted. Consequently, the search for

Table 9.3 Primary energy consumption in Western Europe (in million tonnes oil equivalent)

	1972	1974	1976	1978	1980	1982
Oil	701.8	699.3	710.3	715.1	680.1	601.1
Natural gas	113.5	147.2	163.6	174.6	184.4	174.3
Coal	245.7	247.8	248.3	246.5	265.5	264.3
Water power	92.0	96.4	91.7	104.3	103.1	107.1
Nuclear energy	15.9	18.6	29.2	36.1	46.3	70.5

renewable alternatives to coal, oil and natural gas has assumed paramount importance in Western society. Ever since the Industrial Revolution the sophisticated economies of Western European countries and the standard of living of their inhabitants have been dependent on fossil fuels. The average European uses about four to five times as much energy as people in Third World countries. Moreover, the location of such power supplies has proved a powerful influence on the geographical distribution of industry and population. Price increases and fuel shortages have illustrated the vulnerability of society to disruptions in supply and have driven governments to seek other sources with great urgency. The level of dependency remains high though; in 1982 oil still provided 49.4 per cent of primary energy consumption in Western Europe, gas 14.3 per cent, and coal 21.7 per cent, whereas water and nuclear power only produced 8.8 and 5.8 per cent respectively. Only the development of nuclear power in recent years has made an impact on the domination of fossil fuels, as Table 9.3 demonstrates.

For the greater part of recent history Western Europe depended almost exclusively on coal as a source of all its energy requirements and the bulk of this was mined within the sub-continent itself. However, from the 1950s onwards oil began to take over and a vast proportion of this was imported cheaply from abroad. This left certain countries who were without their own coal or oil resources with massive 'energy gaps' with consumption far outstripping domestic production. The full implications of this precarious situation became apparent after 1973 when the era of cheap oil came to an end as the oil-exporting countries (predominantly in the Middle East) disrupted supplies and tripled the price of oil as a response to the Arab-Israeli conflict. This short-term crisis highlighted the economic and strategic vulnerability of many Western nations in the wake of the subsequent recession, rampant inflation and balance of payments crises. It also triggered off speculation about the long-term future of fossil fuels and spurred on the search for alternatives. Today, very few European countries actually produce more of a given fossil fuel than they consume – only the Netherlands (gas), Norway (oil and gas), the United Kingdom (oil and coal) and West Germany (coal) are in this fortunate position – see Fig. 9.7. As this map shows, oil dependency still leaves many nations, notably France, Italy, and West Germany, with yawning energy gaps and some countries – Denmark, Finland, Ireland, Portugal, Sweden and Switzerland – have no fossil fuel resources at all. The vast discrepancies between energy production and consumption shown here underpin recent moves to find suitable alternatives. One response, of course, has been the anxious rush to exploit newly-discovered oil and gas reserves within Western Europe itself but this as we know is fraught with difficulties. However, the replacement of one fossil fuel with another can only be considered a stop-gap measure, in the long-term renewable energy resources will have to be found. What are the possibilities and how much of a contribution can they make? It is to these two questions that we turn next.

HYDRO-ELECTRIC POWER
Hydro-electric power has long provided an alternative source of energy in some parts of

THE RESOURCE ISSUE 275

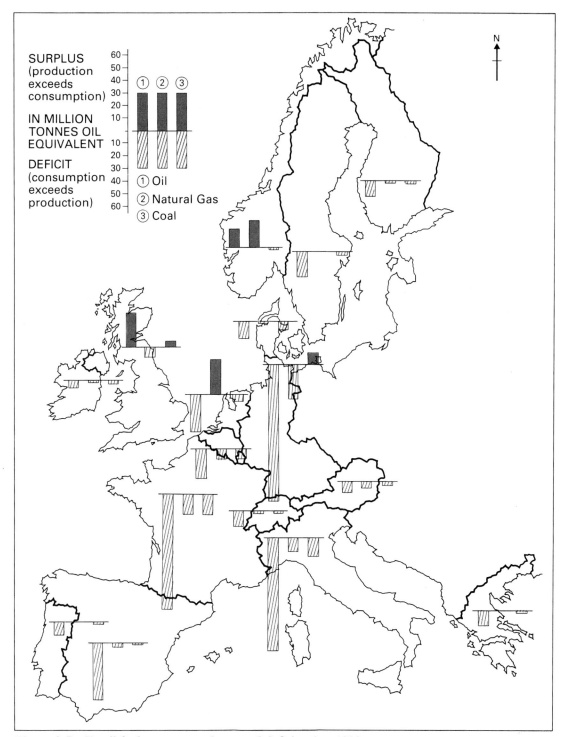

Figure 9.7 Fossil fuel energy surpluses and deficiencies, 1982

Western Europe. It is restricted to regions where suitable environmental conditions are found – high relief, high and regular rainfall and suitable valleys for flooding. Consequently, the Alpine and Scandinavian countries are the only major producers, as illustrated by Fig. 9.9, but only the leading producer Norway provides over one-half of its energy requirements in this way (74.5 per cent of primary energy consumption in 1982) with a much smaller proportion in the other major producing countries – France (8.6 per cent), Italy (7.8 per cent), Switzerland (37.4 per cent) and Sweden (25.2 per cent). There is little potential for future expansion of HEP generation in Western Europe, many of the suitable sites in France, for example, have already been fully harnessed. Moreover, it has always faced the problem that by its very nature generating stations tend to be in remote locations distant from the major centres of demand in lowland Europe and electricity is difficult and costly to transport.

SOLAR POWER

Solar power appears to be an attractive alternative. In theory the energy of the sun's rays is a continuous, virtually limitless resource if only it can be successfully harnessed but, although it can generate power in overcast conditions, in the more northerly temperate latitudes of Western Europe with relatively few days with clear skies it is hardly feasible on a commercial scale. For example, it has been calculated that in order to generate the output of a small conventional power station about 50 square kilometres of land would be required. Experiments have taken place in the sunnier parts of the continent, and France built Europe's first experimental solar furnace at Odeillo in the Pyrenees-Orientales in 1970. However, such prototypes have shown that production is not yet possible on an industrial scale. At present the use of solar power is restricted to providing background heat captured by solar panels to save on consumption of conventional fuels primarily for domestic purposes.

WIND POWER

Wind power is a transformation of solar energy but it too suffers from the same problems of unreliability of climatic conditions and restricted geographical potential. Coastal areas are favoured sites and in parts of Western Europe both experiments and practical applications have recently been developed. The Dutch traditionally utilised wind mills to drain low-lying reclaimed land and today in parts of the Netherlands their modern day equivalents are being employed to do the same job but again not as a complete alternative to conventional fuels but rather as an energy saving device.

TIDAL POWER

Tidal power also offers some hope of continuous energy production to coastal regions of Western Europe. Experimentation is pressing ahead in some countries and France has had a tidal power station operating in the estuary of the River Rance in northern Brittany near St Malo since 1966. Further development has not taken place as the projects remain expensive in comparison to conventional fuels but in the not too distant future tidal power appears likely to be important in those locations which possess suitable geographical conditions.

GEOTHERMAL ENERGY

Geothermal energy deriving from the heat at the earth's core can also be used as a substitute for fossil fuels. Where the heat is trapped in natural pockets near to the earth's surface underground hot springs can be tapped by drilling. This has been used as a source of heat for many years in Italy at Lardarello but it is not restricted to regions of volcanic instability and now other Western European countries are attempting small-scale local schemes.

The last four continuous resources discussed above share similar characteristics in that at present they are essentially theoretical alternatives which have yet to be proved on a large scale. They remain hopes for the

Figure 9.8 Nuclear power station, Switzerland

future whilst providing energy-saving devices on a small scale in the short term. Together with HEP, they share the property of being restricted to certain localities where conditions are suitable. It is difficult to see them as substitutes for fossil fuels in the near future and furthermore a great deal of research and investment will have to be devoted to them in the long term before they can become feasible alternatives. What then can Western Europe turn to, to try to reduce its energy gap? At present the source which offers the greatest potential both in the short and long term is nuclear power. It is a proven generator of electricity on a large scale and free from many of the geographical restrictions but, of course, its development is a very contentious issue.

THE DEVELOPMENT OF NUCLEAR POWER IN WESTERN EUROPE

Heat can be generated from the processes of atomic fission (splitting the atom) or atomic fusion (combining atoms) and this can be used to produce steam which then generates electricity in a conventional manner. At present only the nuclear fission process has been mastered sufficiently but research is proceeding urgently on the safer fusion process. Although reactors initially depended on a relatively rare stock resource (uranium 235), this handicap has been overcome by the development of breeder reactors which can enrich more plentiful materials so that they are suitable for use. In this way the major shortcoming of stock resources has been overcome. However, the development of nuclear power has become a very controversial issue in Western society. Its proponents argue that it will provide a clean, cheap and virtually infinite source of energy which will be vital if the quality of life and economic growth are to be sustained before fossil fuels become scarce. This

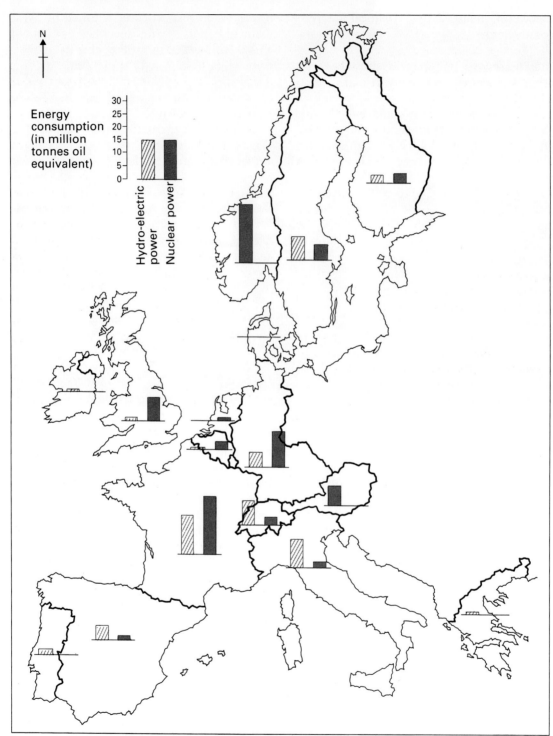

Figure 9.9 Hydro-electric and nuclear power generation, 1982

alliance of industrial and political interests sees the growth of nuclear power as essential for the continuance of modern urban-industrial society. Its opponents stress the threat to humans and the environment that it poses either from accidental leakages during the production process which could emit highly-toxic, radioactive material into the environment (as exemplified by the incident at Three Mile Island near Harrisburg, Pennsylvania in 1979 and at Chernobyl, USSR, in 1986) or, more importantly, the long-term problem of the safe disposal of dangerous nuclear waste that needs to be totally isolated for many years. Moreover, the safe transportation of nuclear material poses a further problem as the sinking of the French ship *Mont Louis* off the coast of Belgium highlighted in 1984. Opposition to nuclear energy can also be viewed as part of the wider movement opposed to nuclear weapons. The rise of ecology parties, such as the Greens in West Germany reflects also a questioning of the basis of modern industrial society.

This formidable clash of interests has produced intense political debates throughout Europe and occasional direct action (for example, mass demonstrations at the sites of nuclear power stations in France and West Germany) and consequently development of nuclear power has varied markedly from country to country. In 1982 France, West Germany and Britain were the major producers following their early commitment to nuclear power. The geographical pattern of production is shown in Fig. 9.9 and the highest *relative* contribution is in Finland and Sweden where nuclear power provided 19.1 per cent and 17.1 per cent of primary energy consumption in 1982 with slightly lower proportions in Switzerland (14.4 per cent) and France (12.7 per cent). However, the rate of installation has varied markedly in recent years. Nuclear expansion is taking place relatively slowly in West Germany and Britain in the 1980s whereas France, in particular, and Italy and Spain are all pursuing vigorous nuclear programmes. By 1990 France will be by far the largest producer followed by West Germany, Italy, Spain and then the United Kingdom. Sweden too will double its number of reactors from 6 in 1979 to 12 in 1986 but, following a referendum in 1980 which voted to scrap any further developments, future increases are unlikely.

Of all the Western European nations it is France which has embarked upon the most ambitious programme. In 1962 nuclear power provided only 1 per cent of French energy needs but ten years later this had increased twelve fold and the rate of growth was accelerating. France is now at the forefront of nuclear developments. By 1984 it had 38 nuclear plants in operation and another 24 in the process of completion scattered throughout the country (see Fig. 9.10). In 1981 nearly 30 per cent of electricity was nuclear generated, by 1983 this had risen to 48 per cent and in 1984 it exceeded 60 per cent. This gave France the highest proportion of nuclear-generated domestic electricity in the world (as compared with 17 per cent in Britain and 13 per cent in the USA in 1983) and in terms of installed capacity the French industry (25.6 thos. MW in 1983) exceeded the USSR (19.3 thos. MW) and was second only to the USA (63.1 thos. MW). It is envisaged that France will produce three-quarters of its electricity from nuclear power stations by 1990.

French governments have pressed ahead at such a rate primarily because the oil crisis of 1973 exposed the political and economic vulnerability of a nation which imported three-quarters of its energy needs. To many French people the issue was clear-cut, nuclear power was seen as an important element in the modernisation of the country's economy which will help to guarantee its economic and political independence. The degree of consensus is such that all the main political parties and two out of three of the public in opinion polls support the nuclear programme. This political will has been backed up by a system of strong central government planning which has pushed through proposals quickly without long drawn-out public inquiries. French

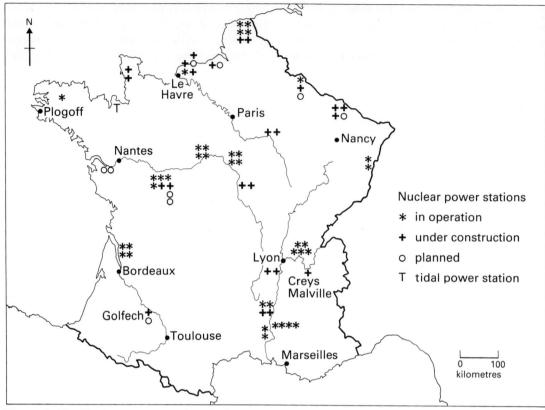

Figure 9.10 Nuclear power stations in France, 1984

nuclear expansion plans have rarely been held up by political opposition from environmental pressure groups worried about safety, as has happened in some other Western European countries. Sporadic anti-nuclear protests have taken place, notably at Golfech in the Midi-Pyrenees and at Creys-Malville east of Lyon (the site of the new Superphenix fast breeder reactor) where violent clashes have resulted, but such incidents have been few and far between and local opposition has generally evaporated as attractive economic advantages (such as 10 per cent cuts in electricity prices in a 10 kilometre radius) have been offered. Only in one case, at Plogoff on the coast of Cap Sizun in western Brittany, has protest prevented the development of a nuclear site. Here a large power station was planned but large-scale demonstrations in 1979–80 by a combination of Breton nationalists and ecologists persuaded the incoming President Mitterand to abandon the scheme, but it must be stressed that the anti-nuclear issue in this case was clouded very much by the nationalists who seized on it as a means of fighting Parisian dominance.

France, then, has been able to take a large step towards energy independence in a relatively short period and it seems that other countries are beginning to follow its example pushing aside environmental doubts in favour of nuclear expansion. Strong pressure from ecological groups in the late 1970s had been successful not only in Sweden but also in Austria, where it prevented the opening of a nuclear plant at Zwentendorf; in Switzerland, where it almost produced a majority against further nuclear development in a referendum; and in West Germany where a three-year moratorium on building new nuclear power stations was imposed. But the tide appeared then to turn. In Switzerland, for example, in 1979 a proposal to give local communities a veto over the building of nuclear power stations received 49 per cent of votes in a referendum and effectively forestalled further growth, but in a similar exercise in September 1984 a proposal to build no more nuclear power plants was rejected by 55 per cent of those voting. Even in West Germany, perhaps the most environmentally-conscious nation in Western Europe, a strong ecological and anti-nuclear movement centred around the 'Greens' had little overall impact on the growth of nuclear power. In effect, they were successful in delaying construction work at a number of reactor sites by invoking legal actions using laws that only permit construction if safe disposal of nuclear waste can be shown, but the underlying nuclear expansion remained. By 1984, fifteen nuclear power stations were in operation, a further eight were in the process of construction, and another three were planned, (their distribution is shown in Fig. 9.11). This programme will install a large nuclear capacity, second only to France, which will go a long way towards satisfying West Germany's future energy needs. Paradoxically this expansion is paralleled by the rise of the 'Greens' from an obscure pressure group advancing mainly anti-nuclear and environmental protests into a major political force which gained 28 seats in the *Bundestag* (West German parliament) in the federal elections of March 1983, followed by representation in the European Parliament and for a short period in 1984 entered into a coalition to run the state government of Hesse. The disaster at Chernobyl in the USSR in 1986 reinvigorated the issue of nuclear development. A cloud of nuclear waste heading west across Europe, brought the reality of the dangers of radio-active material to millions of Europeans. Will this nuclear accident have the effect of changing expansion plans?

In conclusion, the fundamental dilemma remains; dependency on rapidly disappearing fossil fuels continues to be high, posing the

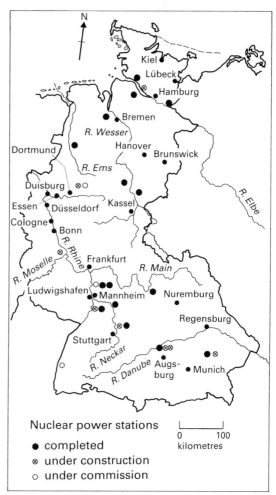

Figure 9.11 Nuclear power stations in West Germany

question of how modern society will maintain present day levels of affluence without such power sources, but the only feasible large-scale alternative (nuclear power) brings with it such frightening possibilities of human and environmental catastrophe that many see it not as a saviour but as a threat to existence itself. Unless and until scientists can find means to ensure the safety of reactors and solve the problem of the disposal of nuclear waste, public concern will continue. The countervailing views of the economic necessity of providing cheap and continuous supplies of energy in the future are not as emotive and do not have the impact of the anti-nuclear stance which is also bound up with the wider issue of the proliferation of nuclear weapons together with fears of nuclear war and a 'nuclear winter'. Even if the scientific advances in reactor safety and waste disposal can be made, the issue will still remain. It is not simply a matter of technology but it is also about winning over the minds of people over a whole range of economic, social, political and environmental issues.

THE MANAGEMENT OF RECREATIONAL AREAS

The increasing disruption of environmentally-sensitive areas due to industrial pollution and the encroachment of urban areas has already been discussed but a further threat derives from the attractive nature of such locations themselves. In recent years with increased prosperity and leisure time the more remote scenically-attractive regions of Western Europe have become popular recreational areas for millions of urban dwellers. As such they can be conceived as valuable resources, not in the tangible sense of objects that can be consumed but as continuous resources that can be places of enjoyment to be revisited time after time. Whether coastal regions, like the Camargue or mountainous areas such as the Alps, such locations are often termed 'amenity landscapes' but their continued existence can be threatened by their very success in attracting visitors. The qualities that attract people in the first place (beautiful scenery, remoteness, peace and quiet) come under pressure which can change the nature of the landscape as a resource. One way of overcoming such pressures is careful management of these areas to try to prevent disruption by attempting to balance the need to provide access with the aim of environmental preservation. Here, a brief case study of the Eastern Alpine region is taken to illustrate the issues involved in greater detail.

Case study: Managing amenity landscapes – the Eastern Alps

The attractions of the Alpine regions for both active and passive recreation are many and varied. In summer the mountains, forests and lakes provide beautiful settings for walkers and climbers and in the winter the ski-slopes take over. Alpine tourism began as far back as the early nineteenth century when the aristocracy and upper-middle classes sought out places for hunting, mountaineering and spa cures, and fashionable resorts such as St Moritz and Kitzbühel developed. But recent years have seen the advent of mass tourism as accessibility has improved with the construction of the European motorway network and most Europeans now have more opportunity through greater leisure time and affluence. In particular, the Eastern Alps are in close proximity to three cities with populations exceeding one million – Vienna, Munich and Milan. Austria has in fact now overtaken Switzerland in terms of the number of tourist visits.

The impact of the mass influx of tourists has not been distributed evenly throughout the

Figure 9.12 Tourism in the Alps

Alpine region. Tourist traffic is heavily concentrated in the most popular resorts such as St Moritz and Davos in Switzerland; Innsbruck, Kitzbühel, Salzburg, Gastein and Bregenz in Austria; Garmisch Partenkirchen in West Germany, and the Cortina d'Ampezzo area in the Italian Dolomites (see Fig. 9.13), whilst away from these main tourist centres much of the Alpine region remains virtually untouched by tourism. This geographical pattern has created strains in the areas of concentration. Within the popular resorts there is pressure on land for modern hotels, shops and second homes and on community services so much so that it is threatening the scenic beauty of such areas as urban development invades picturesque valleys. There is also pressure on the surrounding recreational areas as the demand increases for access roads, ski-lifts and associated features. Both of these intensive physical developments imperil the quality of the landscape but there are other conflicts hidden beneath the surface. To many of the long established residents of the Alpine region tourism is seen as a life-line in an area of marginal farming which has long suffered from depopulation. It provides valuable employment to some and supplementary income to others, although at the same time it takes labour from an already hard-pressed farming sector. These economic benefits can be balanced against the need for conservation of the Alpine environment.

One means of resolving these conflicts is by the careful management of the area so that the landscape, flora and fauna are conserved by providing appropriate facilities for the public whilst at the same time rigorously protecting the regions under threat. In this way an attractive landscape can be maintained for both tourists and inhabitants which will, on the one hand, continue to provide a valuable resource for tourism by protecting it from the worst excesses and, on the other, will preserve the living and working environment which gave the region its attractive characteristics in the first place. Considering its landscape quality it is perhaps surprising that national governments have appeared slow to take such initiatives in the Eastern Alps where

Figure 9.13 The Eastern Alps

the long-established Swiss National Park in the south-east of the country (established 1914) and the adjoining Stelvio National Park in Italy (established 1935) have only recently been supplemented by the Berchtesgarden National Park in West Germany (1978) and the Austrian Hohe Tauern National Park (1981) (see Fig. 9.13). However, for many years private initiatives and local authorities have built up an impressive array of small-scale nature reserves, protected areas and nature parks scattered throughout the region to protect the environment whilst at the same time providing recreational opportunities. It is hoped to provide for the long term survival of the vast porportion of the Alpine landscape by retaining its essential character by protection measures and by channelling tourists to recreational zones with the capacity to cope with their demands.

Throughout Western Europe similar amenity landscapes have been under pressure. Their plight has been highlighted by environmental pressure groups, and governments have begun to respond by setting aside areas for the public whilst at the same time protecting the natural heritage. Foremost has been the creation of national parks, a movement that began slowly in Sweden (1909), Switzerland (1914) and Spain (1918) but which has gathered pace during the last 40 years so that in 1983 some 133 national parks existed in Western Europe. They are mainly concentrated in the northern and southern periphery of the sub-continent but they serve increasingly as vital recreation areas for the urban regions of the Eurocore. As such they constitute a valuable continuous resource which if carefully managed will provide enjoyment for millions of people for many years to come.

3 Resource issues in perspective

What lessons can be drawn from this discussion? We hope that the case studies will have challenged your perception of the conventional wisdom that our environment has plenty of natural resources available for use and that the only problem is how to exploit them. The issues raised in this chapter have shown that we can no longer assume the cheap and plentiful supplies of raw materials which have fuelled economic growth in the past, nor that there are any easy alternative sources. Moreover, natural resource exploitation cannot be viewed in the narrow context of the laws of supply and demand but brings with it such economic and social or environmental side-effects that governments feel obliged to intervene. Many of the conflicts in Western Europe today revolve around resource issues and this perhaps is a hopeful sign as it means that some of the fundamental dilemmas affecting society both now and in the future are being debated. In the final analysis the bulk of Western Europe is an artificial urban society divorced from its physical base but ultimately its quality of life is dependent upon these resources, if they are despoiled or exhausted the flow of resources for future generations is likely to dry up.

Conclusions

Twenty years ago any future projection of European geography would have been based on the expansionary trends of the time – population growth, rising demand for labour and seemingly limitless growth in output. It could not have foreseen such events as the oil crises of the early 1970s and subsequent worldwide recession, because these arose largely from actions taken outside the West European economic and political systems. Yet their effect inside Europe was virtually to bring the whole growth process to a standstill from which it has never properly recovered. Hence the dangers of futurology.

At the same time, this book has raised several key themes which are bound to play an important part in public debate about which way the future *ought* to go. One of the most vital of these is the question of *access*, which has cropped up repeatedly. Here is one of the great paradoxes of modern Western society. In a literal, geographical sense, the dominant trend is one of increasing accessibility, with developments in transport and communications and the growth of supranational institutions guaranteeing ever more contact between people in distant places. Geographical isolation is on the wane. But on the social plane the trend is in the opposite direction. In some senses, people are becoming more isolated, both from the source of power in society and from one another.

Economic decision-making

An outstanding feature of the modern economy is that fewer and fewer people have any real control over their means of livelihood. The rise of big business has meant that the number of self-employed workers – peasants, small farmers, craftsmen, independent business proprietors – has dwindled to a small minority, many of whom struggle to survive. More and more they are employed by giant enterprises, both public and private, where they usually do not participate in major decisions affecting their jobs and prospects. For workers in transnational companies, control may come from hundreds, even thousands of kilometres away across international frontiers. Much of this book has been concerned with the human consequences of this distant impersonal relationship: rural workers displaced by technology and farm rationalisation (Chapter 5); industrial regions undermined by foreign competition and changing consumer demand (Chapters 1–3); industrial regions devastated by technological progress and written off by company decision-makers in some distant boardroom. More often than not the victims of such 'progress' are powerless to defend themselves, even when, as in south Belgium (Chapter 1), and Lorraine (Chapter 3), they are members of large trade unions. In the old industrial regions much union energy is devoted to resisting job losses but in the long run there seems little they can do to stem the twin tides of: a) automation; and b) the flight of investment to new cheap labour regions in southern Europe and the Third World.

Already various Western European governments have been obliged to take an expanding role in support of the deprived. Unemployment and welfare payments take up an increasing proportion of the growing state budget. In some cases the state welfare system has serious defects and reports of growing homelessness in Paris have identified many of the victims as long term unemployed, too poor to afford a roof over their heads. Even in countries like West Germany or Belgium, where state support is more generous, the plight of the unemployed is still a matter of the gravest concern. Apart from the obvious point that it is a pitiable waste of human resources, it is also an affront to the very principles which the democratic state claims to uphold and a grave economic and social handicap to those afflicted.

At various points of the book we have discussed many of the responses to this worsening problem.

- State support for the growing army of unemployed and deprived, without which certain regions like the Italian Mezzogiorno would face complete economic collapse.
- The trend towards earlier retirement and shorter working hours (Chapter 1), which somewhat eases the demand for jobs. Demand will be further reduced by the decline in numbers of job seekers, brought about by current demographic changes (Chapter 6).
- Undoubtedly the most hopeful response has occurred where ordinary people have, themselves, attempted to take the initiative in job and enterprise creation. The rise (or rebirth) of the small-scale economy is hailed by some as the ideal antidote to the large-scale economy, an alternative source of jobs and a means by which ordinary individuals can participate in ownership and decision-making (Chapter 2). Furthermore, it has the potential for beneficial geographical change – deconcentration, decentralisation, a revival of rural areas (Chapter 5) and an easing of urban congestion and environmental pressures (Chapter 8).

Political decisions

These attempts to move decision-making back to the 'grassroots' are paralleled in the political sphere in moves towards decentralisation. Two aspects of this have been touched here.

- *Regionalism/minority nationalism*: Discontent on the part of peripheral dwellers at the concentration of national power in capital cities: this reaches its height in regions like Catalonia or Brittany where there is consciousness of a separate nationhood (Chapter 1). During the 1970s several national governments – Spain, Italy, France – were obliged to respond to this feeling by devolving certain political responsibilities to the regional level. There is a widespread feeling that the growing influence of supra-national institutions such as the EC (highly beneficial in itself as a means of greater international co-operation and contact) need to be counteracted by moves to give regions and local communities a greater say in managing their own affairs
- *Urban protest* (Chapter 4): Discontent at the way in which many cities have been replanned to give priority to offices, commercial interests and 'up-market' housing. Here we see the rise of local community-based resistance, action groups demanding a greater say in their own local environment. In many cases, as with squatters' groups, action transgresses the bounds of legality but in many cases community organisations have forced authority to recognise their participation as legitimate.

Thus a multitude of decisions are being compounded to create sweeping changes in the geography of Western Europe, which in turn is having a significant impact on many of the decisions being made. As we stated at the outset a geographical perspective can make a valuable contribution to the understanding of the major issues in Western Europe today and we hope that we have demonstrated this to you.

Index

Abruzzi-Molise 137–8, 149, 152
accessibility 195, 197, 202, 207, 219, 221, 285
acid rain 229–231, 237–41, 257
agglomeration economies 49, 50
agribusiness 133–4, 232
agriculture
  agricultural change 133–6
  agricultural overdependence 21–2
  agricultural regions 136–41
  farm fragmentation and reform 141–6
    in France 144–5
    in Spain 142–4
    in West Germany 145–6
  underdeveloped agricultural regions 22–7, 31–2
air transport 192, 200, 211–6
airports 213–6, 232
Alentejo 140–1
alpine regions
  land fragmentation 141
  recreation and tourism 149, 282–4
Alsace 181, 243
Amoco Cadiz 226–9
Amsterdam 98, 101, 104, 117, 118, 119, 121, 123, 124–6, 181, 209, 214, 215, 216, 218, 221, 231, 253, 254
Andalucia 16, 142, 144
Antwerp 104, 209, 218, 238
Athens 16, 126, 234–5, 241, 246, 247

baby boom 167
Baden-Wurttemburg 41, 81, 145
Barcelona Convention 247
Basle 190
Basque provinces 16
Bavaria 41, 89, 145
bidonvilles *see* shantytowns
birth rates 24, 162, 163, 164, 166, 167, 168, 169–71, 173–6
black economy 27
Black Forest 239
bocage 144
Bochum 35, 110, 251
Bologna 129–30
Bonn 208, 209
Bonn Convention 244
branch plant economy 67
Bremen 39, 42, 43
Brittany 23, 31–3, 52, 85, 86, 88, 144–5, 149, 226–8
Brussels 28, 85, 98, 103, 120, 209, 221
buffer zones 251–4

car ownership 199, 211–2
Casa per il Mezzogiorno 90, 91, 92
Catholic Church 175, 176
centrality 47
Channel tunnel 202
chemical industry 196
CIBA-Geigy 55
citizenship rights 185
clean air legislation 237
coal industry 28–30, 61, 250–2, 261–5
Cologne 42, 110, 186, 201, 209, 221, 241
COMECON 197, 198
command region 55
Common Agricultural Policy 75, 81, 93, 154–9, 232
Common Fisheries Policy 76, 271–3
Common Regional Policy *see* European Regional Development Fund
Common Transport Policy 221–3
containerisation 202, 216–9
Continuous resources 273–84
  alternatives to fossil fuels 273–82
  definition 262
  management of recreational areas 282–4
Copenhagen 98, 101, 208
core-periphery 36–7, 44, 55, 59–60, 174, 186, 220, 260
Cote d'Azur 246, 247
counterurbanisation 114–7, 150–2

Danube 236
death rates 163, 168, 171, 172–3
deindustrialisation 60
demographic change 162, 166
demographic transition 168–72, 180
deurbanisation 182–3
diffusion 173, 175
Dortmund 110, 196, 201, 208, 251
Duisburg 110, 196, 250
Dunkirk 82, 83
Dusseldorf 39, 41, 42, 47, 110, 121, 188, 251

Ebro 246, 248
economic potential 47
emigration 163, 177, 181
employment 16–20
  activity rate 27
  changing structure 19
  hierarch 18–9
  marginalisation 23–4
  occupational position 16–8
  status 18

energy 227, 229, 257, 273–82
  energy crisis 229
  energy gap 274
  production and consumption 274–6
  from non-fossil fuels 274–6 *see also* nuclear power
environmental awareness 250, 254–6
environmental concern 5, 224–57
environmental costs 226–8, 235–47
environmental disasters 226–9, 233
environmental policy 234, 244–7, 250–4
Essen 110, 208, 251
European Coal and Steel Community 79, 83, 84, 264
European Community 53, 54, 75–81, 83, 84, 132–3, 162, 198, 220, 221–3, 229, 241, 243, 247, 248–9
European Community Environmental Policy 257
European Conference of Ministers of Transport 220
European Free Trade Association 53–4, 74
European Investment Bank 78, 92
European Regional Development Fund 75–8

family planning 163, 165, 166, 173–4, 175, 176
female emancipation 176
fertility *see* birth rates
fishing industry 270–3
Florence 231, 241
flow resources 270–3
  definition 262
  fisheries 270–3
foreign workers *see* immigration
forestry 239–41
Fos 82, 197
fossil fuels 234, 237, 241, 273–4
Frankfurt 41, 42, 43, 98, 121, 207, 214, 215, 220

Galicia 23, 142, 144
gastarbeiters *see* immigration
gender 34
Geneva 190, 202
Genoa 128, 200, 218, 246
geothermal energy 276
Golden Triangle 53
Gothenburg 73, 216
Grenoble 64–6, 94, 220
growth poles 215
guestworkers *see* immigration

Hamburg 39, 77, 104, 177, 181, 182–3, 216, 218
handicraft industry 180
Helsinki 73, 74
housing 99–100, 225, 250
hydro-electric power 274–6

immigration 33–4, 69, 121–3, 166, 177, 184–91
industrial decision-making 45, 55–9
industrial linkage 48
industrial restructuring 85
industrialisation 10–11, 27–30, 32–3, 35, 50–2, 192, 200, 224–5, 229–31, 245
infant mortality 163, 174
inner city problems 120–3, 231
iron and steel 29–30, 61–2, 82–5, 196, 250

labour reserves 185
Lapland 16, 85
latifundia 139, 140, 142
Le Havre 218
leading and lagging regions 136–9
life expectancy 163, 172–3
Limits to Growth 229, 257, 258
Lisbon 16, 101, 104, 128
literacy 176
Lorraine 10, 35, 81, 82–5, 86, 96, 181, 202, 213
Luxemburg 77, 81, 221
Lyon 197, 201, 202, 209, 220

Madrid 104, 128, 178, 214
Marseilles 113, 197, 200, 202, 216, 218, 246, 247
Massif Central 88, 114, 144, 148, 149
Mediterranean Sea 244–7
megalopolis 110
mercantile cities 104
metropoles d'equilibres 85, 88
metropolitan regions 20, 42–3, 104–17
Mezzogiorno 34, 78, 90–4, 128, 149, 174–6, 221
migration 7, 24, 27, 35, 69, 121–3, 176–91
Milan 113, 128, 207, 209, 233
minifundia 142
minority nations 7–8
mortality *see* death rates
motorways 200, 202, 205, 206–8, 221, 223, 232
Munich 42, 43, 208

Naples 90, 92, 101, 104, 207, 248
national parks 283–4
natural resources *see also*
 continuous, flow and stock definitions 259
 exponential growth 259
 resource depletion 226, 233, 258–61
 types 261–2

Nord 35, 86, 181, 224, 231
Nordic Council 71–4
Nordic Investment bank 72
nuclear power 277–82
 in Austria and Switzerland 280–1
 in France 279–80
 in West Germany 281
Nuremburg 197, 209

OECD 235
oil and gas industry 196, 218–9, 246, 265–70
Oslo 44, 73

Paris 20, 46, 84, 85, 86, 88, 96, 100, 101, 103–4, 106–10, 114–7, 120, 128, 181, 201, 202, 208, 209, 214, 215, 221
 Paris Basin 81, 137, 144–5
 Paris region 16, 44–7, 77, 114
part-time farmers 150
pipelines 192, 195, 196, 197, 200
Po 246, 248
pollution 226–57
 air pollution 211, 225, 231, 234–41, 246
 effluents and waste 229, 231, 232, 243–4
 freshwater pollution 238–44
 lead pollution 241, 257
 marine pollution 244–7
 noise 231
 transfrontier pollution 237–47, 257
polycentric connurbations 110
Population *see also* demography
population crisis 162, 166–8
population explosion 163
Portuguese regions
 agricultural underdevelopment and reform 139–41
 rural depopulation 147–8
poverty 13–6
public transport 205, 208–9, 235
pull-push factors in migration 180–1

quality of life 34–48, 249 *see also*
 standard of living
 geographical variations 34–5
 index of social well-being 36
 indicators 34–5
 in West Germany 37–48

racism 188
rail transport 192, 195, 197–205, 212, 221
Randstad 100, 106, 110, 117–20, 123–6, 250, 252–4
rationalisation 61–2
regional accounting 96
regional division of labour 16, 59
regional nationalism 30–33
regional planning 85–95, 202, 223, 264
 effectiveness 94–5
 in France 87–8, 89–90

in the Mezzogiorno 90–4
in South Limburg 264
in West Germany 88–90
inter-regional planning 85–6
intra-regional planning 95
reproduction of labour power 95
Rhine 193–7, 198, 243–4
 Rhine Commission 244
Rhine-Ruhr 100, 106, 110
Rhone 246, 248, 256
road accidents 209–11
road transport 192, 195, 198, 199, 200, 205–11, 221
Rome 90, 214
Rotterdam 117, 118, 119, 123, 181, 196, 200, 218–9, 238, 252, 253, 254
Ruhr 10, 27, 29, 35, 41, 42, 60–4, 81, 82, 89, 196, 221, 224, 225, 231, 232, 235, 238, 250–2
rural depopulation 146–50, 178, 225
rural deprivation 152–4

Saarland 10, 41, 42, 77, 81, 82, 89
Sambre-Meuse 10, 28–30, 35, 77, 82, 231, 238
Sardinia 23, 90, 175
seaports 216–9, 247
seasonal workers 188–9
second homes 151–2
self-sustaining growth 47
Seveso 233
shantytowns 187
Sicily 23–7, 47, 85, 90, 92, 175
solar power 276
South Limburg 231, 261–5
standard of living 9–10, 12–16, 261
 see also quality of life
stock resources 261–70
 definition 262
 Norwegian oil and gas 265–70
 South Limburg coalfield 261–5
Stockholm 44, 73, 85, 214
Strasbourg 197, 221
Stuttgart 207

TGV 201–2, 204
The Hague 117, 118, 120, 209, 252, 254
Ticino 16, 190
Torrey Canyon 226
tourism 33, 149–50, 212, 226–8, 244–6, 248–9, 250, 254–6
traffic congestion 207–11, 225, 231, 234
Trans Europ Express 220
transfer payments 96
transport 192–223
 competition 195, 196, 197, 198, 199, 212, 214
 costs 196, 198–9, 207, 212
 environmental impact 207–11
 integration 193, 206, 209, 219–223
 mobility 205, 206, 207, 220
 regional growth 197